Mixed Reality – Merging Real and Virtual Worlds

Figure 1.3

Figure 1.4

(c)

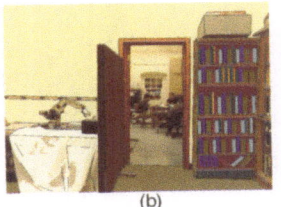

(b)

(d)

(a)

(e)

Figure 1.6

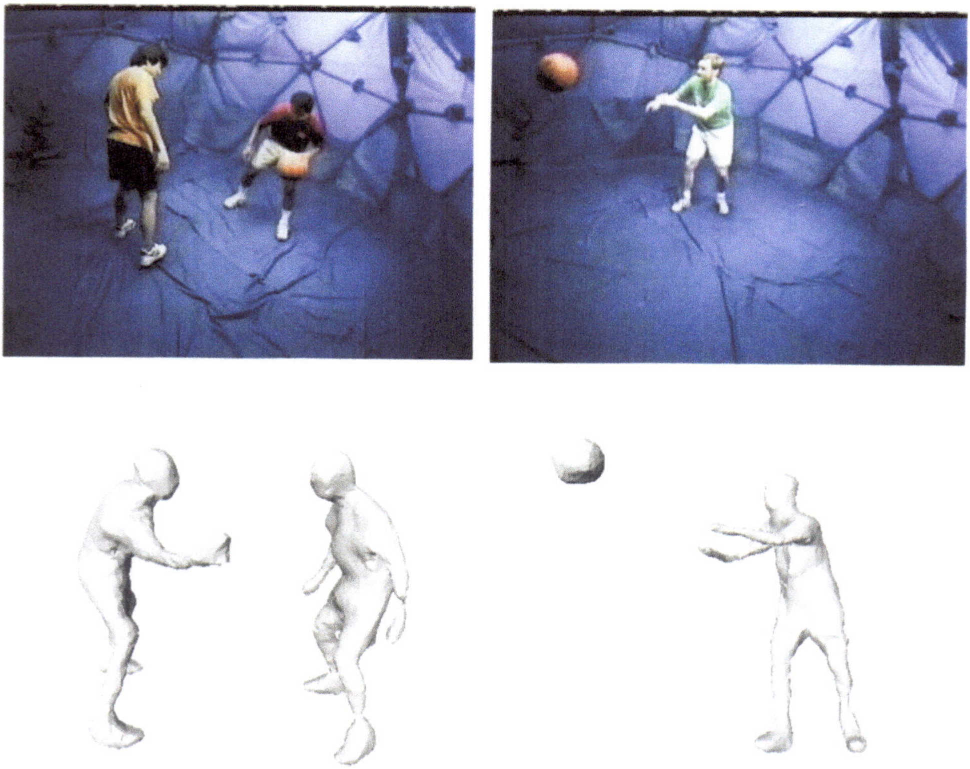

Figure 3.7

Figure 3.10

Figure 4.2

Figure 4.12

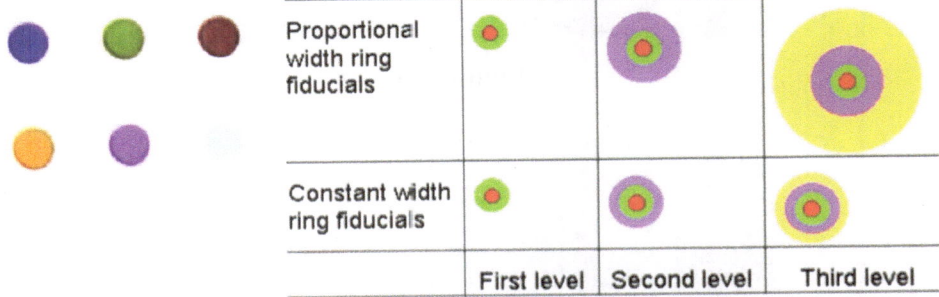

Figure 6.5

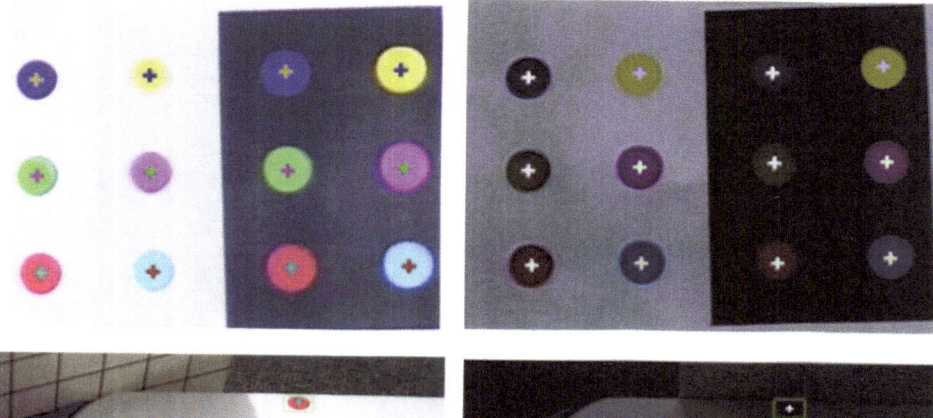

Figure 6.4

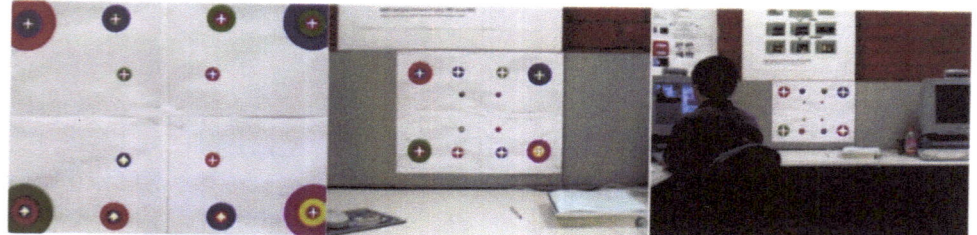

Figure 6.8

Figure 6.13

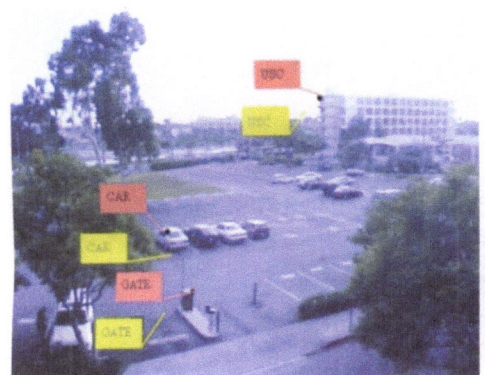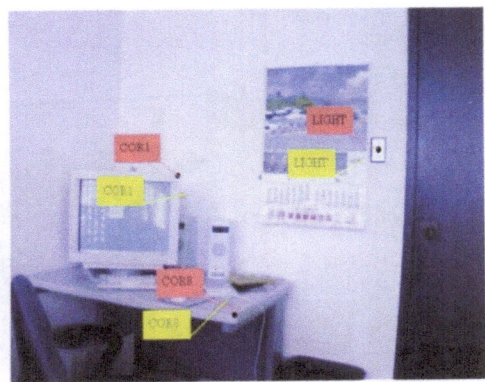

Figure 6.17

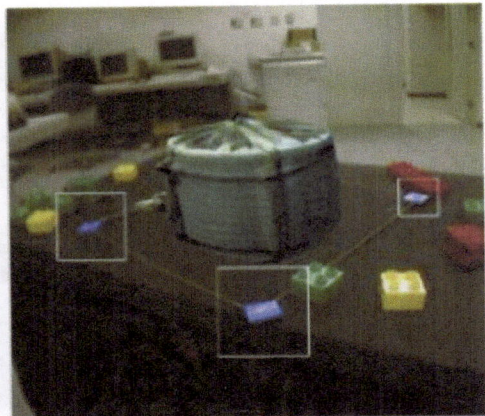

Figure 7.2

Figure 9.13

image 2
image 3
interpolated image
image 1
forward
left
right
center
turn left
turn right
back

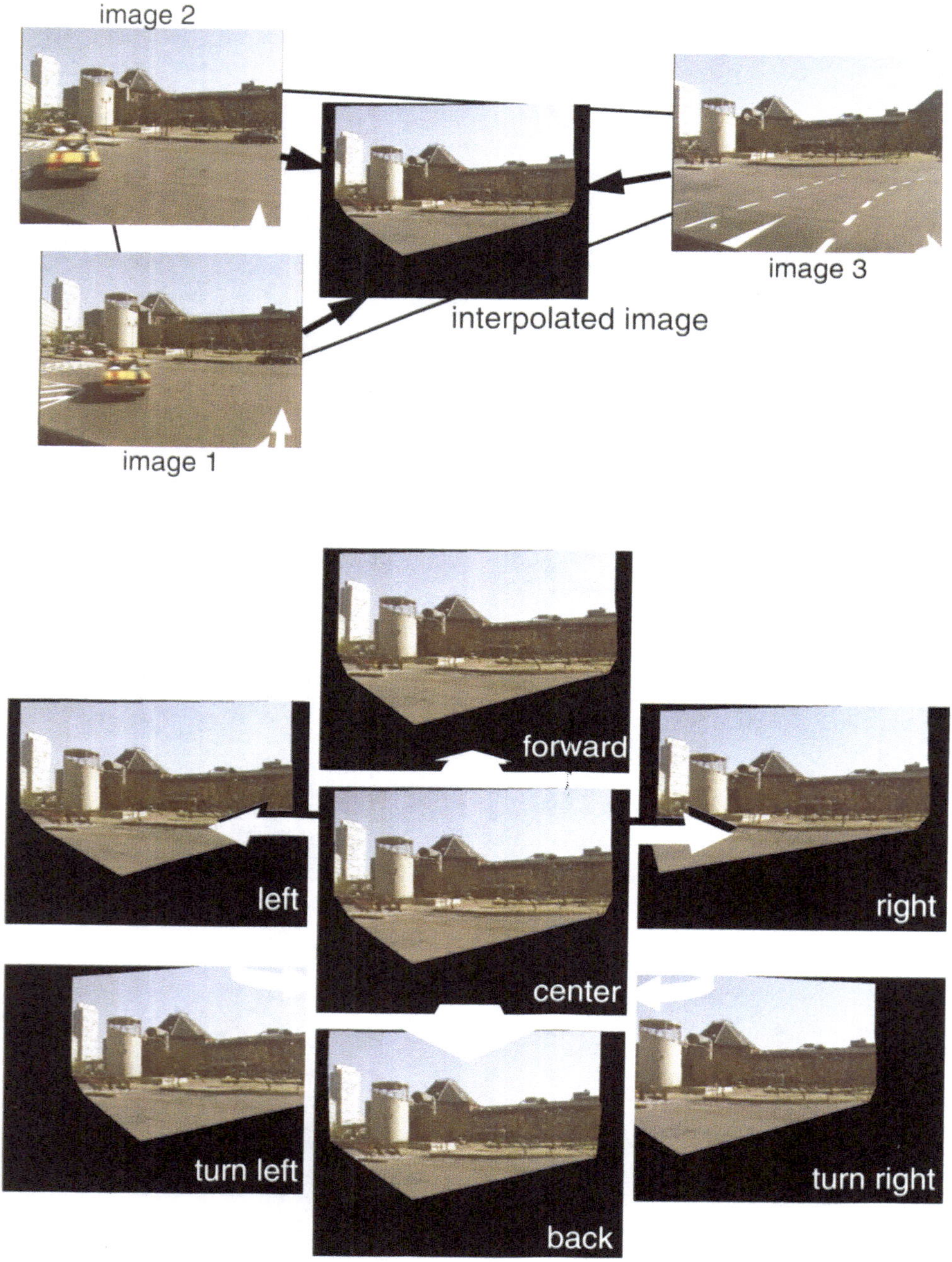

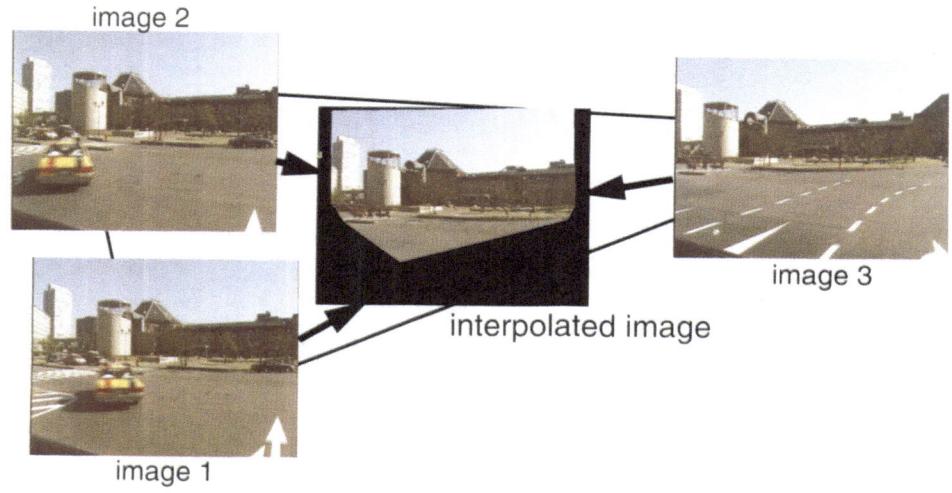

Figure 10.8

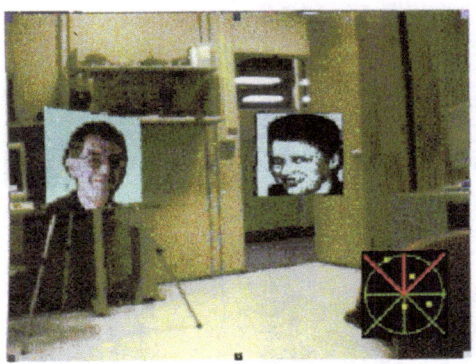

Figure 15.6

Figure 18.2

Figure 18.3

Figure 18.4

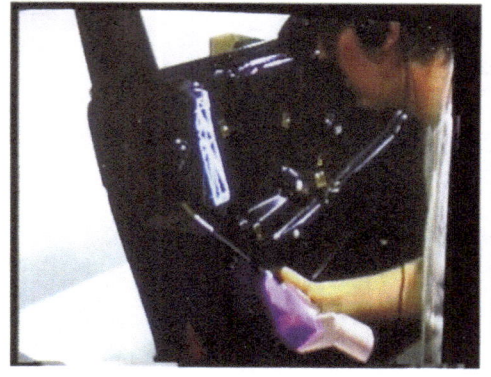

Figure 18.5

Figure 18.6

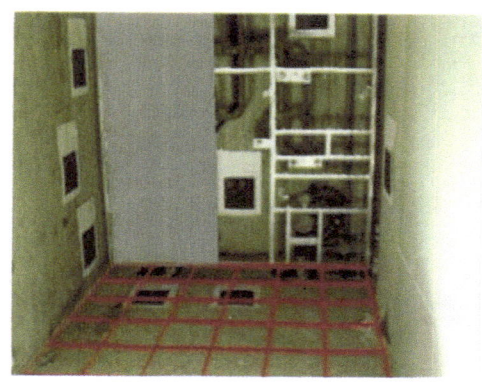

Figure 18.7

Figure 18.9

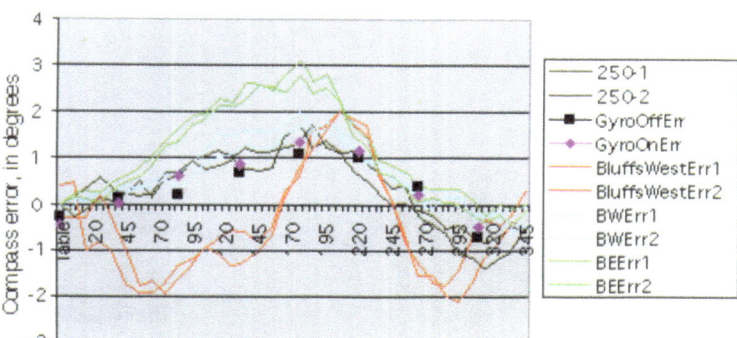

Figure 21.2

Mixed Reality
Merging Real and Virtual Worlds

Edited by
YUICHI OHTA
HIDEYUKI TAMURA

Springer-Verlag Berlin
Heidelberg GmbH

Mixed Reality — Merging Real and Virtual Worlds
Edited by: Yuichi OHTA
Hideyuki TAMURA

Copyright © 1999 by Springer-Verlag Berlin Heidelberg
Originally published by Ohmsha, Ltd. 1999
Softcover reprint of the hardcover 1st edition 1999
3-1 Kanda Nishiki-cho, Chiyoda-ku Tokyo 101-8460, Japan

ISBN 978-3-642-87514-4   ISBN 978-3-642-87512-0 (eBook)
DOI 10.1007/978-3-642-87512-0

Library of Congress Cataloging-in-Publication Data applied for

Preface

The past ten years have seen the popularization of virtual reality (VR) systems, which enable participants to interact with virtual environments synthesized by a computer. Most of the VR systems we have experienced in this decade, however, have conveyed a poor sense of reality, primarily because the environments are synthesized entirely within a computer. Due to this limitation, people started to incorporate the rich information available in the real world into their VR systems. It therefore became essential to identify a technology that deals concurrently with the virtual synthesized world, as well as with the real physical world. Mixed reality is just such a technology, in that it realizes environments that seamlessly integrate both real and virtual worlds.

This is the first book that includes "Mixed Reality" in its title. It will also serve as the proceedings of the First International Symposium on Mixed Reality (ISMR '99), scheduled in March 1999 in Yokohama, Japan. Let us explain the background and circumstances that compelled us to edit and publish this book.

The Key-Technology Research Project to investigate the next generation of VR technology was conceived in Japan in 1996. We, the members asked to participate in this project, decided to adopt the name "Mixed Reality" (MR), which is broader than the term "augmented reality", as the project's main theme. The MR project was started in early 1997 by obtaining financial support from the Ministry of International Trade and Industry, and the Mixed Reality Systems Laboratory Inc. (MR Lab) was established. Soon after that, the Special Interest Group on Mixed Reality was formed within the Virtual Reality Society of Japan. Various Japanese academic societies have also made MR the theme of a number of special sessions at their annual conventions and periodic meetings.

People came to realize the importance of position sensors and computer vision technology in this field. On the other hand, wearable computers have more recently begun to acquire popularity and special attention has begun to be paid to MR systems in which the observer wears a see-through head-mounted display. At the current time, representatives of a large variety of organizations are visiting the MR Lab in Yokohama on a continuing basis including broadcasting stations, video production companies, electrical power and gas companies construction companies, city planning agencies, and museums Among these video game companies and medical instrument manufacturers are at the forefront seeking the most direct route towards practical applications of this technology.

As an extension of our original proposal, we, the editor of this book and promoters of the MR project, decided to combine the demonstration of our intermediate

iii

research results with an international symposium on the topic of MR. Our objective was not that the workshop would necessarily be a place only for presenting original papers, but also a forum in which people from various fields having an interest in this technology could get together and talk freely about its essential elements and the potential for future advances. With this in mind, we have endeavored to identify a variety of participants renowned for their farsighted opinions and outstanding research results in this field and invite them to talk about the MR theme. During our selection of these invited speakers, we realized that the contents of the invited talks would likely be too valuable merely to be a conference proceedings presented only to the symposium participants. We therefore decided to collect these papers and arrange them as a book, to be published widely through an international publisher.

We believe that the authors are well selected and that the book will stand on its own merits. A significant amount of important material is presented throughout the book, which we hope will become a valuable resource for people who want to acquaint themselves with the current orientation of emerging MR technology.

We also draw the reader's attention to the fact that various events have been planned to be held at ISMR '99 to demonstrate the breadth of MR technology and applications, such as a media art gallery, a panel discussion on future entertainment, and technical demonstrations of late-breaking results. We regret that we can not report on all of those events in this book.

Finally, the editors want to express our gratitude to many people for their devotion and cooperation in the creating and publishing of this book. We especially offer our thanks to Prof. Heitou Zen and Dr. Takeshi Naemura, the ISMR '99 Publication Chairs, for their good instructions and suggestions, to Ms. Yuko Wakatsuki, Mr. Mahoro Anabuki, Mr. Tomotaka Kanamori, Mr. Daisuke Kotake, Mr. Noriyuki Sugai, Mr. Masahiro Suzuki, and Ms. Rika Tanaka of MR Lab for their patience in the face of annoying tasks and complicated challenges, and to Dr. Haruo Takemura, who is currently on leave at the University of Toronto, for his voluntary cooperation from abroad. The editors also express their special thanks to Prof. Paul Milgram and Prof. Steven Feiner for their careful proofreading of the editorial introductions. Last but not least, we thank all the authors for their invaluable contributions.

January 1999

Yuichi Ohta
Hideyuki Tamura

Contents

Part I   Overview and Perspective   **1**

1 A Taxonomy of Real and Virtual World Display Integration   5
  1.1 Definition of Mixed Reality   5
  1.2 Centricity and Control Issues Associated With Mixed Reality 13
  1.3 Global Taxonomy of Mixed Reality Display Integration 23
  References 27

2 Displays for Augmented Reality: Historical Remarks and Future
  Prospects   **31**
  2.1 Introduction 31
  2.2 Summary of Augmented Reality Display Technologies and Examples
     of Their Use 32
  2.3 Conclusion 39
  References 40

3 Virtualized Reality: Digitizing a 3D Time-Varying Event As Is and
  in Real Time   **41**
  3.1 Introduction 41
  3.2 Modeling Real Events into Virtual Reality 42
  3.3 Related Work 43
  3.4 Virtualized Reality Studio: From Analog "3D Dome" to Digital "3D
     Room" 44
  3.5 Creation of Three-Dimensional Model 45
  3.6 Combining Multiple Events 50
  3.7 Examples 51
  3.8 Conclusions 54
  References 55

4 Steps Toward Seamless Mixed Reality   **59**
  4.1 Introduction 59
  4.2 Outline of the MR Project 61
  4.3 Approaches to Seamless Augmented Reality 63
  4.4 Approaches to Seamless Augmented Virtuality 69
  4.5 3D Display Technologies 72
  4.6 Concluding Remarks and Future Studies 76

References 77

Part II   Registration and Rendering   81

5 Vision-Based Geometric Registration of Virtual and Real Worlds 85
  5.1 Introduction 85
  5.2 Depth of Real World by Occlusion Detectable Stereo 87
  5.3 Synthesis of Novel Views of a Virtual Object from Several Images . . 90
  5.4 A Linear Method for Euclidean Motion/Structure in Real Time . . . 94
  5.5 Conclusion 98
  References 99

6 Augmented Reality Tracking in Natural Environments   101
  6.1 Introduction 101
  6.2 Indoor Tracking 102
  6.3 Outdoor Tracking 112
  6.4 Summary and Conclusions 125
  References 126

7 Stereo Vision Based Video See-through Mixed Reality   131
  7.1 Introduction 131
  7.2 Algorithms 134
  7.3 Experiments and Discussion 140
  7.4 Conclusion 143
  References 143

8 Photometric Modeling for Mixed Reality   147
  8.1 Introduction 147
  8.2 Creating Models from Observation 148
  8.3 Integrating Virtual Objects with a Real Scene 157
  8.4 Conclusions 160
  References 161

9 The Ray-Based Approach to Augmented Spatial Communication
**   and Mixed Reality   165**
  9.1 Introduction 165
  9.2 Integrated 3-D Visual Communication System 167
  9.3 Ray-Based Representation of Visual Cues 168
  9.4 Applications of Ray-Based Approach 174
  9.5 Conclusions 180
  References 180

10 Building a Virtual World from the Real World   **183**
    10.1 Introduction 183
    10.2 Data Capturing System 184
    10.3 Image Reproduction Systems 187
    10.4 Image-Based Walk-Through System 193
    10.5 Conclusion 195
    References 196

Part III   Multi-Sensory Augmentation   **199**

11 Auditory Distance Perception in Real, Virtual, and Mixed Environments   **201**
    11.1 Introduction 201
    11.2 Information for Directional Localization 202
    11.3 Information for Distance Localization 203
    11.4 Measurement of Perceived Distance Using Perceptually Directed Action 204
    11.5 Externalization of Earphone-based Virtual Sound 206
    11.6 Virtual Sound in Mixed Environments 208
    11.7 An Application of Mixed Auditory Environments: The Personal Guidance System 208
    References 210

12 Feel-through: Augmented Reality with Force Feedback   **215**
    12.1 Introduction 215
    12.2 Basic Idea of Feel-through 216
    12.3 Force Display for Feel-through 216
    12.4 System Design 220
    12.5 Evaluation of Prototypes 221
    12.6 Wearable Force Display and Mobile Augmented Reality 225
    12.7 Conclusions 225
    References 226

13 Tangible Bits: Coupling Physicality and Virtuality Through Tangible User Interfaces   **229**
    13.1 Bits and Atoms: GUI, VR, AR, MR, UbiComp 229
    13.2 Tangible Bits: Key Concepts 233
    13.3 Tangible Interface Designs 234
    13.4 Illuminating Light 234
    13.5 InTouch 238
    13.6 Ambient Media: Water Lamp and Pinwheels 241
    13.7 Conclusions 244
    References 245

Part IV   Communication and Collaboration   249

14 Augmented Telexistence   **251**
  14.1 Telexistence 251
  14.2 Augmented Reality in Telexistence 253
  14.3 R-Cubed & HRP 254
  14.4 Augmented Reality in HRP 256
  14.5 Conclusion 259
  References 260

15 Collaborative Mixed Reality   **261**
  15.1 Introduction 261
  15.2 Motivation: Why Collaborative Mixed Reality 262
  15.3 Collaborative Interfaces for Three Dimensional CSCW 265
  15.4 Our Work 267
  15.5 Computer Vision Methods for Collaborative Mixed Reality 274
  15.6 Conclusions 280
  References 280

16 Virtual Reality Technologies for Multimedia Communications **285**
  16.1 Introduction 285
  16.2 Virtual Metamorphosis System 286
  16.3 Novel View Generation 289
  16.4 Fatigueless Head Mount 3D Display 291
  16.5 User Interface in Virtual Environments 295
  16.6 Conclusion 298
  References 298

Part V   Systems: Design Considerations and Future Trends   301

17 Operator Localization of Virtual Objects   **305**
  17.1 Introduction 305
  17.2 Experiment 1 307
  17.3 Experiment 2 311
  17.4 Experiment 3 315
  17.5 Experiment 4 319
  17.6 Design Considerations 321
  References 322

18 Augmented Reality: A Balancing Act Between High Quality and Real-Time Constraints   **325**
  18.1 Introduction 325
  18.2 Applications / Demonstrations 327
  18.3 System Architecture 333

18.4 Live Optical Tracking of User Motions 336
18.5 Off-line Calibration of Video Sequences 341
18.6 Presentation of Virtual Information 342
18.7 Discussion 343
References 344

19 MR Aided Engineering: Inspection Support Systems Integrating Virtual Instruments and Process Control   **347**
19.1 Introduction 347
19.2 MR Technology Applicable for Manufacturing 348
19.3 Paper Manual to MR-Integrated Instruction 349
19.4 MR-Integrated Instruments 350
19.5 A Desktop Environment for PCB by MR 352
19.6 A Backpacking Environment for Power Parts by MR with HMD . . 356
19.7 Conclusions 360
References 360

20 Wearing It Out: First Steps Toward Mobile Augmented Reality Systems   **363**
20.1 Introduction 363
20.2 Related Work 364
20.3 ARC: Augmented Reality for Construction 365
20.4 A Touring Machine 368
20.5 System Design 371
20.6 Conclusions and Future Work 374
References 375

21 The Challenge of Making Augmented Reality Work Outdoors   **379**
21.1 Background in Augmented Reality 379
21.2 Motivation for Outdoor Augmented Reality 381
21.3 Analysis of Problem Areas 382
21.4 Approaches and Conclusions 387
References 389

22 An Outdoor Augmented Reality System for GIS Applications   **391**
22.1 Introduction 391
22.2 The Related Works 393
22.3 Outdoor AR System for GIS Application 393
22.4 Experiment 395
22.5 Conclusions and Future Works 397
References 398

Index   **401**

Part I

Overview and Perspective

In Part I of this book, we present a collection of four papers that deal with the general concept of mixed reality. The term *mixed reality* used in the title of this book was first used in 1994, by Paul Milgram [1]. At that time, the term *augmented reality* (AR) was already in use, referring to augmenting the real or physical world with electronically synthesized data. As a counterpart to the term AR, Milgram suggested a new term, *augmented virtuality* (AV), which referred to augmenting or enhancing the virtual world produced by a computer with raw data from the real world. Rather than regard the two concepts as simple opposites, he instead proposed that they be thought of as the poles of a continuum of technologies covering a broad variety of mixtures of real and virtual data. Finally, to encompass this resulting so-called "Reality-virtuality (RV) continuum", he introduced the new term "mixed reality" (MR).

In [1], Milgram and his colleague cited six classes of hybrid displays to cover the range of MR interfaces. That classification scheme has now evolved, based on extensive practical considerations, to a broader taxonomy of MR proposed in the present book. Chapter 1 deals with the issue of classifying not only real and virtual world mixtures, but also the relationship between RV mixtures and visual viewpoints of the observer, or centricity, within MR environments. The authors also discuss the importance of the degree of congruence between display viewpoint and control means, that is, the user's means of influencing the observed world.

The innovative research of the Computer Science Department of the University of North Carolina at Chapel Hill has traditionally played an historical role in the development of computer graphics and virtual environment technology. Henry Fuchs, one of the key members of that department, has with his colleagues made several seminal contributions to VR through their development of head-mounted displays (HMDs), position sensors and related application systems. They also are pioneers in the field of AR, especially as it relates to applications for assisting visualization of surgical procedures. In Chapter 2, Fuchs and his colleague provide a history of AR displays, transiting from the initial optical see-through HMD to the video see-through HMD, arriving finally at the topic of immersive telecollaboration using projection displays, the current principal theme of his research group.

Let us now move to the other end of the continuum, to AV. It is interesting to note that traditional texture mapping is a kind of AV method, in that it involves the mapping of image data from natural scenes on to the surface of computer-generated solid objects. Modern AV technology includes both image-based rendering methods without geometric data and methods that map raw moving video images as texture onto geometric models of objects. Takeo Kanade of Carnegie Mellon University, one of the most prominent researchers in the field of computer vision, has taken upon himself a much grander challenge. He and his colleagues are attempting to digitize a set of 3D time-varying events as 4D representations and then place the whole set into a computer database such that one can perform real-time playback of the 3D events dynamically, from any viewpoint and at any time. They even consider editing the events in a physical sense, such as adding 3D objects and changing the speed of actions. This can truly be thought as the ultimate AV! Kanade himself refers to his group's research as "Virtualized Reality."

Chapter 3 explains how to realize the exceptional concept of reproducing a moving 3D object in a computer by extracting the shape of the actual moving object using the multi-baseline stereo method with a large number of cameras. For their first prototype, Kanade's group placed as many as 51 cameras on a wall of a room facing the object. With their method, as long as one can reproduce a moving object in a computer, it becomes possible to perform real-time playback of a 3D virtualized event from any viewpoint. Unfortunately, it is impossible at this moment to perform the extraction and reconstruction in real-time; therefore, one can play back only pre-recorded events. By combining this capability with their existing real-time stereo imaging technology, however, it is expected that Kanade's virtualized reality will eventually become the definitive AV, if they can obtain enough computing power.

Chapter 4 introduces the activities of the Key-Technology Research Project on Mixed Reality Systems in Japan. This project has the term of MR in its name because the key members of the project believe that the RV continuum proposed by Milgram will be the next target of the emerging technology. Note that there is a slight difference between Milgram and them in their way of classifying the continuum from AR to AV into classes and steps. In Chapter 4, Hideyuki Tamura and his colleagues discuss their separate research in both AR and AV, as well as the steps they propose to achieve "seamless MR." They cover the development at the MR Lab of a new see-through HMD and a stereoscopic display whose special optics eliminate the need of viewing spectacles. The authors also discuss how these two displays can be utilized and enhanced in the future. The MR Lab's impact can be measured in part by the comments and citations of other author s in later chapters of this book. At present an increasing number of researchers are contributing to this project directly, while others continue to undertake complementary activities.

References

[1] P. Milgram and F. Kishino: "A taxonomy of mixed reality visual display," *IEICE Trans. Inf. & Sys.*, vol. E77-D, no.12, pp.1321–1329, 1994.

Chapter 1

A Taxonomy of Real and Virtual World Display Integration

Paul Milgram
Herman Colquhoun Jr.

University of Toronto, Canada

1.1 Definition of Mixed Reality

Our primary objective in this paper is to present a number of fundamental display integration and orientation issues related to the nascent field of Mixed Reality. Our approach is motivated first by the need for a more encompassing term to supplement the existing definition of Augmented Reality (AR), which leads us to propose definitions of the associated concepts of Augmented Virtuality (AV) and then Mixed Reality (MR). Following our discussion of the breadth of Mixed Reality displays in Section 1.1, we discuss the associated issues of viewpoint centricity and control-display mapping in Section 1.2. Finally, in Section 1.3, we present a taxonomy which we hope will be useful for differentiating between several of the issues raised with regard to the different classes of Mixed Reality display systems.

1.1.1 Augmented Reality

An examination of current literature in which the term Augmented Reality (AR) appears will reveal two classes of definition, distinguished from each other in terms of breadth. Most common appear to be those for which AR refers narrowly to the

5

class of display systems comprising some kind of *head-mounted display (HMD)* or *head-up display (HUD)*. In the case of HMD's, the viewer observes a direct *"see-through" view* of the real world, either optically or via video coupling, upon which is superimposed computer generated graphics. A number of reviews which focus on this class of displays have been written, including that by Azuma [1] and that by Fuchs and Ackerman [2] in the present volume. Some of the prominent examples of such displays include systems for assisting in manufacturing [3]–[5] and in medicine [6]. It is important to note that head-up displays (HUD's), which have existed in primarily military aviation environments for several years, clearly fall within the realm of see-through AR as well, in the sense that graphic information is superimposed upon the pilot's direct view of the outside real world [1] [7]. More recently, the same HMD concept has been proposed for use also by combat soldiers on the ground [8] [9].

The second, broader class of definitions in the literature relaxes the constraint of needing the equivalent of a HMD and covers "any case in which an otherwise real environment is 'augmented' by means of virtual (computer graphic) objects" [10] [11], thereby encompassing both large screen and monitor based displays as well. Examples conforming to this broader definition of AR include applications in robotics [12] and medicine [13] [14].

In a sense a third, even broader class of AR displays has been proposed by some in the literature, encompassing those cases involving *any* mixture of real and virtual environments. Consistent with this interpretation, Azuma, in his earlier survey of Augmented Reality, referred to AR as "a variation on Virtual Environments that combines virtual and real" [15]. He later refined this, however, to comprise *any* system that "1) combines real and virtual, 2) is interactive in real time, and 3) is registered in three dimensions" [1]. As we discuss in the following section, extending the realm of AR in this direction brings to light an important issue – whether it is in fact *reality* or *virtuality* which is being enhanced – and, together with this, the need for a broader, more comprehensive set of definitions. It is our contention that any useful definition of AR should definitely encompass the first two classes of displays mentioned here, but that a different term is needed to account for the third class.

1.1.2   Reality vs. Virtuality

Before proceeding, it is imperative that we clarify what we mean by the key terms "real" and "virtual" environments. Following the approach proposed in [10] [11], we first contend that, although both purely real environments (RE's) and virtual environments (VE's) certainly do exist as separate entities, they are not to be considered simply as alternatives to each other, but rather as poles lying at opposite ends of a *Reality-Virtuality (RV) continuum*, as shown in Figure 1.1. The location of any environment, or "world", along this continuum coincides with its location along a parallel *Extent of World Knowledge (EWK) continuum*. In using the latter term, we are referring to the extent of knowledge present within the computer about the world being presented. Figure 1.1 illustrates the parallel nature of the two concepts.

As indicated in the figure, at the right end of the RV continuum are virtual environments, which must necessarily be *completely modelled* in order to be rendered. At the opposite extreme we are taking real environments to be representations of a world, or region, which are *completely unmodelled*. In using the term "model", we are limiting ourselves to quantitative computer models. Thus, in relation to real environments, which have not been modelled, we refer to situations in which the computer does not possess, or does not attribute meaning to, any information about the content of an image.[1] RE's therefore encompass any kind of sampled image data, and include as the primary example video, but also photographic images (visible or infrared), radar, X-ray and ultrasound, as well as laser scanned data (both range and light intensity data). Note that we are not necessarily limiting ourselves in our definition to two dimensional (2D) data, but we include, especially with respect to the latter example of laser range data, also three dimensional (3D) sampled images.

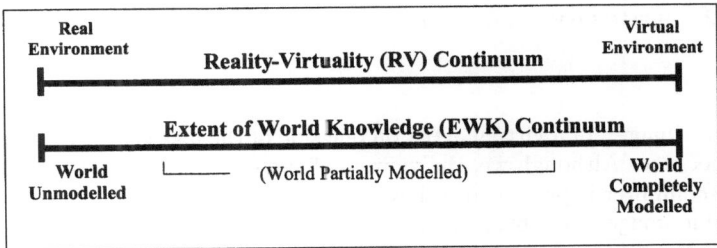

Figure 1.1   Reality-Virtuality (RV) Continuum, in parallel with Extent of World Knowledge (EWK) Continuum.

Returning to the issue of mixing, or combining, real and graphic images, as is the case with AR, this necessarily brings us somewhere towards the middle of the RV spectrum, and concurrently to some region in the middle of the EWK continuum, in which the world is *partially modelled*. This means, for example, that we might know the location of some objects, but nothing about the objects themselves, or we might have elaborate geometric models of some objects, but not know anything about where they are, that is, how they relate to surrounding regions of the image.

As we venture away from the poles of the RV continuum towards the centre, we also eventually begin to encounter the problem of deciding whether in fact what we are doing is augmenting a real world with virtual graphic objects, or whether we are modifying a virtual environment by augmenting it with real data. This issue is discussed further in Section 1.1.4. To the extent that these two cases can be distinguished from each other, in the meantime, it has been proposed that they be labelled Augmented Reality (AR) and Augmented Virtuality (AV) respectively [10] [11].

These concepts are illustrated in Figure 1.2. On the left hand side we have an example of Augmented Reality: a scene comprising a photograph – i.e. a *real image* – of a mountain lake, upon which have been superimposed two computer generated –

---

[1]It is interesting to point out that Durlach and colleagues have made a similar observation in their recent book on Virtual Reality, a portion of which we cite here for convenience [36], pp.59–60): "When a virtual environment application requires a replica of a real environment, it is generally considered preferable to map the real environment rather than build a model of it. Active mapping techniques, such as scanning laser range finders and light stripes, are used to make three-dimensional measurements directly." We must point out that those authors' use of the term "virtual environment" (VE) does not conform with ours, however, since such a collection of sampled data would continue to be termed a "real environment" (RE) by our definition.

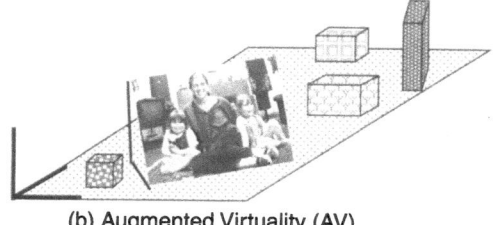

(a) Augmented Reality (AR)   (b) Augmented Virtuality (AV)

Figure 1.2   Illustrations of (a) Augmented Reality (AR) and (b) Augmented Virtuality (AV).

i.e. *virtual* – images, of a virtual artist on one side of the lake sketching a virtual tree on the other side. Although we, the viewers of the picture, comprehend the content easily, the computer is presumed to have no model whatsoever of the content of the photographic image, and thus, to the computer, the location of the virtual images relative to the real image is meaningless.

The converse case, an example of Augmented Virtuality, is illustrated schematically in Figure 1.2b. Here we see a completely modelled (3D) world, comprising a series of virtual 3D blocks located on a virtual plane. In order to draw these objects, the computer must have a model of all of their dimensions and locations. In addition, a photograph of a group of people, comprising real data, has been added, at a specific location. Although the computer must have knowledge about *where* the real image (photograph) has been placed, we can not assume that any knowledge is held about the *content* of that image.

Figure 1.2 is obviously contrived, to facilitate comprehension of the distinction between the concept of an underlying real world and an underlying virtual world. To show how these concepts manifest themselves in relation to actual practical applications, we present two more examples. The first example, of AR, is shown in Figure 1.3, which is from our own laboratory and illustrates ARTEMIS, our Augmented Reality TEleManipulation Interface System [16] [17]. In the figure we see a real robot situated within a real environment. Although the real environment is completely unmodelled, we do possess a model of the real robot, registered to real-world coordinates. This permits us to superimpose a modelled stereoscopically presented virtual robot on top of the real robot, as depicted in the figure. As described elsewhere [16] [17], this set-up enables an operator to pick up and deposit the real objects depicted in the image – even though they are not modelled – simply by aligning the virtual end effector in the 3D work space with the objects to be manipulated and then transmitting the robot joint coordinates to the remote site at the appropriate moment.

As a converse example, of Augmented Virtuality this time, we present Figure 1.4, which is a screen dump of an image produced with Cyberworld® software (www.cyberworldcorp.com), a commercial product which is used to create realistic "3D web pages". To generate the Christmas scene shown here, a 3D virtual world has been created, comprising a large public square. A miniature plan view of the square is shown at the bottom right corner. However, the buildings, the Christmas tree, Santa Claus, the carollers and all of the other objects in the picture are superimposed 2D photographic images, but with known locations in the 3D virtual world.

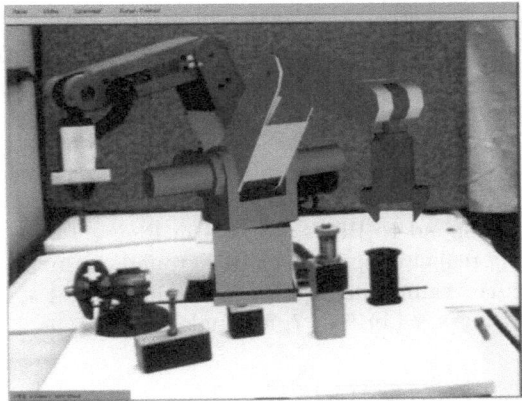

Figure 1.3 ARTEMIS Augmented Reality system: Virtual robot (in this case polygon filled) is overlaid onto a modelled real robot, within an otherwise completely unmodelled world (see color pages).

Figure 1.4 Example of Augmented Virtuality (AV): Superposition of real images and texture mapping onto a virtual 3D world (see color pages). (Composed using Cyberworld® software.)

1.1.3 Exploring the Reality-Virtuality Continuum

Thus far, we have presented only illustrations which purposely emphasise the major distinctions between fundamentally opposing RV mixtures. However, although the two terms Augmented Reality and Augmented Virtuality support these distinctions, in the ensuing discussion we show that it is not always as simple as in the preceding examples to distinguish between AR and AV. We shall thus argue that the more enveloping term, Mixed Reality, becomes necessary, to encompass in a less constrained way all mixtures between the poles of the RV continuum.

To this end, we present Figure 1.5, which illustrates schematically a selection of image composites that could be encountered when one traverses the RV Continuum. On a global level, Figure 1.5 corresponds to the same left-to-right RV Continuum shown along the top part of Figure 1.1. The difference here, however, is that Figure 1.5 highlights the variety of ways in which the real components (R) and the virtual components (V) of an image may be mixed. In terms of our earlier examples, for instance, Figure 1.3 and the left hand side AR example in Figure 1.2a could be considered to correspond to Block #8 or 9 in Figure 1.5, in the sense that we have a predominantly real environment, or background, with "a few" virtual objects superimposed. The AV example in Figure 1.2b and in Figure 1.4, furthermore, would correspond here to Blocks #4 or 6 or 7, for analogous reasons.

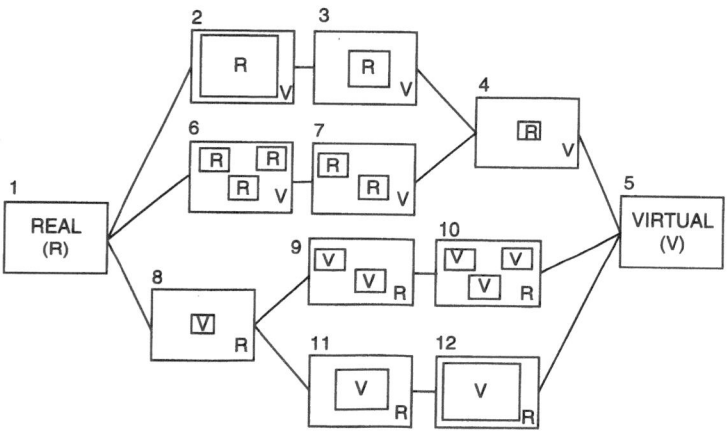

Figure 1.5   Mixed Reality combination space.

In presenting Figure 1.5, we were prompted by a number of cases in the literature for which it is not always obvious whether the primary environment, or "substratum", is real or virtual. One good example of this is the Peloton sports simulator described by Carraro and colleagues [18]. Peloton is a bicycle simulator which simulates a virtual road course for walking, running and bicycle riding. Users stand on a treadmill and locomote through an environment which comprises a computer generated (virtual) surround, in the middle of which is placed a video window, texture mapped onto a large rectangle, or "video screen". In terms of Figure 1.5, the Peloton system therefore corresponds to Block #3. In the SIGGRAPH reference cited [18], the authors describe their simulation of a bicycle path through New York's Central Park, with emphasis placed on their method of blending the internal (real) video window with the surrounding virtual graphics window.

In the following we present an analogous case, but extend the concept somewhat. What we shall do is illustrate what a journey along the RV continuum might look

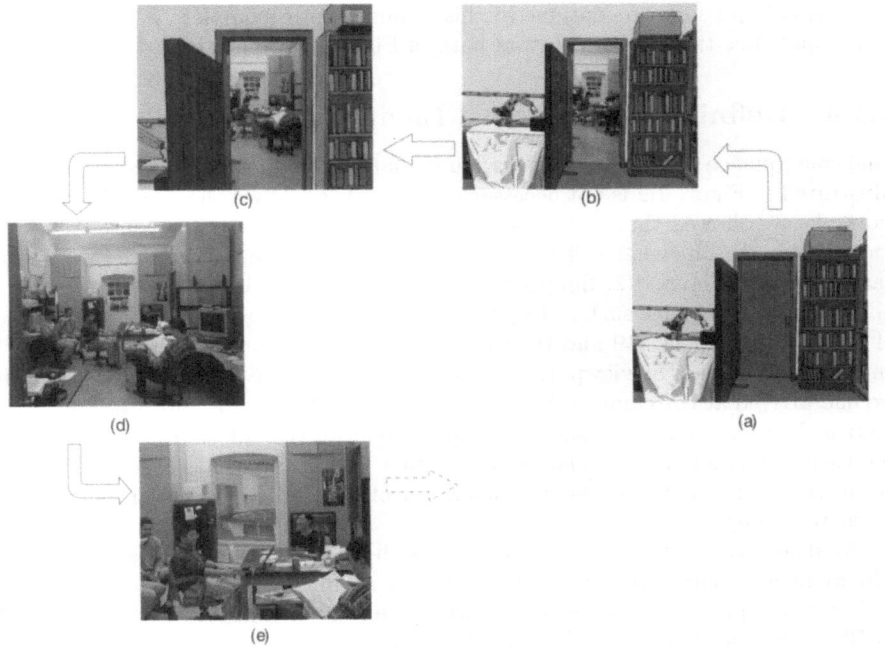

Figure 1.6   University of Toronto ETC Lab, as experienced through a journey along the RV continuum (see color pages).

like, as we travel along the trajectory 5-4-2-1-8 in Figure 1.5. Our point of departure is illustrated in Figure 1.6a, which shows a virtual model of part of our laboratory at the University of Toronto ETC Lab. Because this is a virtual model, the image in Figure 1.6a corresponds to Block #5 above.

To continue the journey, we open the door in Figure 1.6a and project a real image of the adjoining room onto the doorway, as shown in Figure 1.6b. This picture therefore corresponds to Block #4 of Figure 1.5. As we approach the real portal framed by the doorway, in Figure 1.6c, the composite image becomes proportionately more real and less virtual, corresponding to Block #2 in Figure 1.5. Finally, as we advance all the way through the real portal, in Figure 1.6d, we enter into a completely real environment, corresponding to the Real extremum in Block #1 at the left side of Figure 1.5.

Another important aspect of the schematic representation of Figure 1.5 is the essentially circular nature of the RV continuum. That is, in the description above, we have traversed the continuum, from right to left, corresponding to a transition from a completely virtual environment to a completely real one. In Figure 1.6 we also illustrate, however, that it is not necessary to travel along the same trajectory in reverse to return to the virtual side. Rather, as illustrated in Figure 1.6e, we have turned the next doorway into a virtual portal, depicting an adjoining virtual conference room. Figure 1.6e therefore corresponds to Block #8 in Figure 1.5.

Clearly, it is possible to continue in this manner, enter completely into the virtual room, and thus traverse the bottom path of Figure 1.5 to get back to Block #5.

1.1.4   Defining the Principal Environment

One consequence of the above discussion is that the distinction between AR and AV illustrated in Figure 1.2 is not necessarily as simple as shown there. In the preceding example, we showed that virtual and real environments can "flow" into each other recursively. A somewhat different illustration of this occurs if we re-examine the cases shown in Figure 1.2. Suppose that we continue to add more and more virtual objects to an AR image such as Figure 1.2a, thereby moving effectively from the case of Block #8 to Block #9 and then to Block #10 in Figure 1.5. Eventually, if the entire visible image, or viewport, consisted of virtual objects, one could argue that we had arrived at the completely virtual case of Block #5 in Figure 1.5. This would be true, however, only if we possess complete quantitative information about how all the various virtual objects relate to each other within the (3D) space of the image. Otherwise, this would not fit our definition of a completely virtual environment, presented above.

Analogously, if we were to commence with an AV image such as Figure 1.2b, add more and more sampled data images to it, until the scene appeared totally real, it would give the impression that we had migrated from Block #4 to Block #7 to Block #6 and finally to Block #1 in Figure 1.5. Figure 1.4 is a good example of what such an image might look like. However, as long as we retain quantitative information about the spatial relationship among the various real image components relative to each other and/or to the underlying virtual world, we could not consider the final image to conform to Block #1, which requires that the world be completely *un*modelled.

The conclusion to be drawn from these examples is that, according to the operational definition proposed in Figure 1.1, it is not necessarily true that an environment is *completely virtual* if all of the component visible objects in it are computer generated, and it is also not necessarily true that a world is *completely real* if all of the visible objects in it are sample data images. Furthermore, determining whether an image should be considered Augmented Reality or Augmented Virtuality is also not necessarily a matter of simply summating the respective areas of real and virtual images in order to determine a "majority" portion of real or virtual. A practical example of an extreme case of AR is that in which an underlying scene is created from 3D range image data, but then has computer generated polygons mapped onto it to create the appearance of a continuous modelled surface rather than discretely sampled points [21] [37]. Analogously, we have seen that an image may be considered AV, for instance, even if essentially all visible elements in it are derived from sampled real data, but have been texture mapped onto a completely modelled underlying virtual world.

1.1.5   Definition of Mixed Reality

We conclude this section by presenting Figure 1.7, which essentially repeats the RV spectrum of Figure 1.1, but with the generic cases of Augmented Reality (AR),

Augmented Virtuality (AV) and Mixed Reality (MR) indicated explicitly now. The portions of the illustration corresponding to the terms indicated will thus serve as our definitions of AR, AV and MR. Note that the AR segment of the continuum covers a portion of the RV continuum adjacent to, but excluding, the real environment extreme and, similarly, the AV segment lies adjacent to, but excluding, the virtual environment end. Encompassing both AR and AV, the MR portion of the RV continuum covers essentially the entire breadth of the spectrum, but also excludes the end points. In closing, we reiterate that it is our hope that the terms discussed here will serve a useful purpose in distinguishing the various contexts within which diverse research in the field of Mixed Reality is currently being carried out, even though in practice the distinctions are often not always easily recognised.

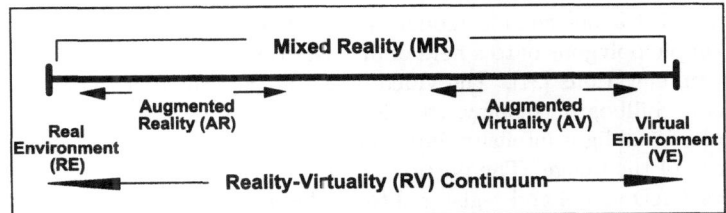

Figure 1.7   Definition of Mixed Reality, within the Context of the RV Continuum.

1.2   Centricity and Control Issues Associated With Mixed Reality

In this section we review some of the issues which may arise when working with complex Mixed Reality worlds, from the point of view of defining an appropriate viewpoint for the observer relative to the objects of interest. One of the fundamental problems which can occur, for example, relates to the fact that virtual environments can generally be presented from any desired viewpoint, whereas the perspective of real data can not ordinarily be changed.[2] This discussion will lead us in a subsequent section to the parallel problems of transitions between virtual and real worlds along the RV continuum and of maintaining suitable control-display relationships when doing so.

1.2.1   Case Study: Remote Mixed Reality Excavation

To illustrate the problem, we shall use as a case study a remotely controlled MR excavator, elements of which have been reported elsewhere, under the acronym VERO (Virtual Environments for Remote Operations) [19] [20]. The original VERO system is currently being followed up by another project dubbed IIRO (Intelligent Interactive Remote Operations), in which we are designing an interface to allow both

---

[2]Note that this problem pertains not only to 2D images, but also to sampled 3D data images which are ordinarily limited in the amount of viewpoint alteration that is feasible, due to problems of missing data and the occlusion of some objects by others.

teleoperation of the remote excavator, as well as a space robotic arm, and the building of scripts for supervised control of either one of these teleoperators over long time delay communication channels.

A sample display screen related to the IIRO project is given in Figure 1.8, where the image shown comprises a virtual model of an actual excavator situated at a remote work site. The joint angles of the real excavator have been transmitted to the local human interface, thereby permitting repositioning of the model with the proper pose, as would be seen from the remotely located camera external to the excavator. That camera is capable of generating both light intensity images (i.e. video) and 3D laser range images. Because communication with the remote site is poor, we assume that the camera image can not be updated frequently, hence the need for interpolation and enhancement using a virtual model.

In Figure 1.8 a deformable model of terrain has been created by mapping a "skin" of simple polygons onto a field of 3D laser range data obtained from scanning actual terrain elevations [21]. The video image in the centre of the figure has been inserted as a "billboard" display, that is, as a virtual 2D screen upon which have been projected real light intensity data from the laser range camera. Note the barrel in the centre of the image. The portion outside the video window has been created from a prior CAD model and registered to the laser range based image. The portion inside the video window represents a recent update of the remote world. Because the external camera is stationary, the two parts of the same barrel match each other seamlessly.

Four important advantages of this mixed reality image are illustrated in Figure 1.8:

- The light intensity image inside the video window shows much more detail than would be possible with the 3D range image data.

- Because the time taken to update a 3D range image scan of a scene is much longer than a light image (video) scan, the portion inside the window is likely to represent much more recent information.

- The virtual portions of the scene provide the means to simulate intervention operations and interactions, in spite of the large time delay.

- Consistent with the points above, the portion inside the window allows one easily to detect departures from any modelled portions of the scene. In particular, Figure 1.8 shows a second barrel (possibly containing toxic waste, for example) within the video window, which had not been expected and thus had not previously been registered to the image, as was the other barrel.

In terms of the definitions presented in Section 1.1, it is interesting to consider whether Figure 1.8 is an example of AR or of AV. Because the terrain has been recreated from *sampled 3D range data*, it is *unmodelled*, such that the foundation of the image, or substratum, is therefore *real*. By virtue of the addition of the modelled excavator, as well as the surface mesh, or "skin", this figure can therefore be classified as a case of Augmented Reality. Note that such a classification might appear counter-intuitive to many, due to the prominence of the virtual excavator and the "virtual-looking" terrain, such that one might be tempted to classify Figure

Figure 1.8   Mixed Reality remote excavation example.

1.8 as an AV image, especially if the video window were not present. Conversely in fact, from the point of view of the modelled excavator, one could alternatively contend that the virtual excavator model has been enhanced through addition of the complementary real world data, making the example arguably a case of Augmented Virtuality. As discussed in Section 1.1, therefore, the exact classification could conceivably be presented either way, depending on ones viewpoint, thus providing further justification for the more encompassing term, Mixed Reality.

1.2.2   MR Design Issues

In this section we briefly outline three problems associated with the MR excavator example described above, all related to the issue of defining the human operator's viewpoint relative to the remotely controlled equipment and all of which generalise in some way to a broader class of MR systems.[3] A much more thorough treatment of the considerations outlined here can be found in the various publications of Wickens and his colleagues [22]–[26].

The first problem associated with the IIRO scenario is the "keyhole effect", an excessive narrowing of an observer's field of view, somewhat akin to peeking through a keyhole [27]. Although the view which is generally best for *local guidance and control* of the excavator is the view from the cab, that viewpoint is unfortunately not usually conducive to maintaining global situational awareness. That is, due to the

---

[3]Note that other technological problems not discussed in detail here must also be overcome, an obvious one being system update rate. Due to the demands of the various levels of detail demanded by the Mixed Reality environment, which includes updating and drawing both modelled objects and real data objects, one of the expected consequences is a slow frame rate, which can clearly have a potentially significant impact on the effectiveness of remote operations.

keyhole effect, the operator is not able to look around the excavator, assess the situation, and plan future operations [23] [24]. This is one of the main reasons for presenting the side view shown in Figure 1.8, rather than a perhaps more conventional through-the-window view.

The second problem is that of distortion. The 2D video image shown inside the window in Figure 1.8 is congruent with the 3D virtual world elements only when the collective viewpoint is located at a station point which corresponds to the location of the remote camera capturing the image. In other words, although the perspective view offered of the virtual screen onto which the video window is mapped is generally correct relative to the surrounding virtual environment, the sampled data content within the video window does not necessarily match that perspective. A view from any angle other than that of the original camera attitude will therefore result in a discontinuity between the virtual environment and the video image contents, resulting in some degree of distortion. It is interesting to note that one of the principal aims of the Peloton system mentioned earlier is to overcome this type of problem [18].

The third problem is that of control reversals, a situation in which perceptual confusion causes an operator to elicit a control action which is opposite in direction to the appropriate action at a particular moment [28]. When presenting the kind of scene shown in Figure 1.8, for example, the potential exists for a mismatch between the outside-in display depicted and conventional inside-out excavator control displays. As discussed further, in Section 1.2.3, this is most likely to have an adverse impact whenever the offset angle between the control action and the observed display is very large.[4]

1.2.3   Effect of Control-Display Compatibility in MR Environments

The principal conclusion to be drawn from the preceding discussion is that, in conjunction with the various advantages associated with mixing real and virtual images in MR, such as improved visualisation of real-world images using RE viewports and added viewpoint flexibility using VE viewports, significant operational problems may result if careful consideration is not given to defining the user's frame of reference relative to the MR display and to the mapping between the user's control actions and the responses of the MR display.

Display centricity

Before delving into these design issues, we must first clarify the meaning of the term "*centricity*", which we use here to refer to the extent to which a human observer's viewpoint is removed from the "*ownship*", that is, from the *nominal viewpoint* with respect to the viewer's own avatar, or own vehicle, or own manipulator within the task space. Referring to the examples presented thus far in the paper, the nominal viewpoint in Figure 1.8 would be within the cab looking out, for local control, and

---

[4]According to research by Ellis and colleagues, control errors are most likely to increase significantly for angular mismatches between control and display axes in the vicinity of 120 to 180 degrees [29].

from outside looking at the excavator, for global navigation and planning. With respect to Figure 1.4 and Figure 1.6, the nominal viewpoint would be through the eyes of an observer's avatar moving through the various scenes presented. With regard to Figure 1.3, the nominal viewpoint will depend on whether one is controlling all joints of the manipulator independently, in which case the nominal viewpoint might be from behind, consistent with ones view of ones own arm, or whether control is resolved directly through the end-effector, in which case the nominal viewpoint could just as well be from in front of the manipulator, as shown in the figure.

The concept of centricity is generalised schematically in Figure 1.9, which shows yet another continuum, this time a *centricity continuum*, together with illustrations corresponding to different combinations of viewing perspectives of an arbitrary excavator. The point at the left of the figure corresponds to an *egocentric* viewpoint, which occurs when the display viewpoint is *ego-referenced*. On the assumption that the nominal viewpoint of the excavator system in this case is at the driver's seat inside the cab of the excavator and looking out, the egocentric case corresponds to the view which would be seen by that operator. This is depicted in the figure by the camera being mounted within the cab and looking out the window of the cab. The *exocentric* case, in contrast, corresponds to a *world referenced* framework, as shown at the right side of Figure 1.9, where the cameras are fixed with respect to the external world. The prefix "exo' thus refers to the state of being outside of and looking at the "ownship", or "own vehicle", that is, the state of looking at the nominal viewing position.[5]

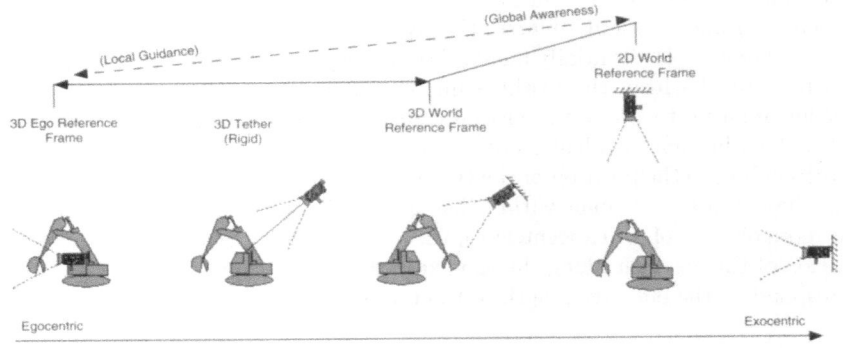

Figure 1.9 Centricity continuum: Illustration of transition from ego- to exocentric viewpoints. (Adapted from [23] [24])

Figure 1.9 conveys the very different effects which would result in the egocentric versus the exocentric cases when the excavator or parts of excavator move. In the egocentric case, movement of the excavator would cause the visual display to change in the same manner as if the operator were inside looking out through the window of

---

[5]An interesting effort to combine elements of both exocentric and egocentric viewing concurrently, within a single viewport, for large virtual environments, is the work of Kitamura and colleagues [38].

the cab; that is, the *world moves* within the viewport whenever the excavator moves. This view is thus often referred to as an *"inside-out"* view. In the exocentric case, in contrast, we have an *"outside-in"* view and, because the cameras are fixed, we are able to see movements of the "ownship" , as if we were looking at the excavator from above, or from the side, or from some other external viewpoint. In that case, in other words, the *excavator moves* while the rest of the world remains fixed within the viewport.

As with the other continua discussed earlier, the centricity continuum also encompasses a variety of interesting intermediate cases. Wickens and his colleagues have treated this topic very thoroughly, together with the implications of centricity considerations with respect to interface design, most prominently in relation to aviation displays [23] [24] [26], but also as it relates to scientific visualisation [30]. One of the important metaphors introduced by Wickens and colleagues in this context is that of a *tether* joining the virtual camera with the nominal viewpoint, to cover the range of cases along the centricity continuum between the ego-referenced and world-referenced extrema. In Figure 1.9 this is depicted as a rigid tether. The reason for using the tether metaphor to represent intermediate cases between pure egocentric and pure exocentric, as indicated in Figure 1.9, is that movement of the excavator will cause corresponding movement of the world as the camera is dragged along by the tether, as with an egocentric (inside-out) display; however, the observer will also have a more encompassing view of his/her own excavator (or aeroplane or automobile, as the case may be) and its surroundings, as with an exocentric (outside-in) display.

It is interesting to note that, although the concept of exocentricity and a world-referenced frame are highly related, they are not equivalent concepts. This is because world referencing is essentially an absolute concept, relating to *whether* the (virtual) camera is fixed within the world, while exocentricity can be considered a relative concept, relating to *where* the virtual camera is fixed relative to the nominal viewpoint. This becomes evident as one extends the length of the tether. As the view corresponding to the position of the (virtual) camera moves farther and farther away from the nominal viewpoint within the "ownship", one gets the sense of being more and more outside of, or exocentric to, the vehicle. Using this analogy, therefore, the location of the observer along the ego-exo centricity continuum can be considered to correspond to the effective length of the tether.

Control-display mapping

In conjunction with establishing the user's viewpoint, it is critical to take into account the mapping between that viewpoint and the user's ability to manipulate objects when designing a MR system. A practical discussion of these considerations is provided in the following section, based in large part on earlier survey literature [31] [32], primarily by Zhai [33]. In the present section we define some of the basic concepts.

As illustrated in Figure 1.10, the *congruence* of mapping a user's input actions to responses in the display space can also be regarded as a continuum. The basic idea is that, depending on the means provided and the circumstances, a user can effect changes in the observed scene either congruently with or, to varying degrees, incongruently with respect to the form, position and orientation of the device(s) provided, as shown across the top of the figure. Ordinarily, a highly congruent control-display relationship will correspond with a natural, or *intuitive*, control scheme, whereas an incongruent relationship will compel the user to perform a number of *mental transformations* in order to use it. The degree of congruence depends on a number of factors, depicted by the individual arrows beneath the all-encompassing congruency continuum shown in the figure.

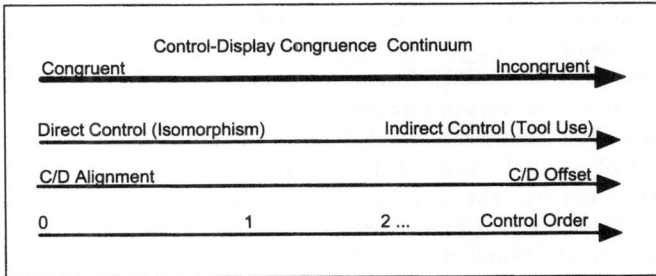

Figure 1.10   Control-Display (C/D) Congruence Continuum.

The most encompassing factor, *directness*, relates to whether the user's control actions map directly (or isomorphically) onto the display space or whether some real or metaphorical device lies between the user and the environment. The former case is naturally quite relevant to the field of Mixed Reality, which includes the broad category of see-through AR display environments, in which the user can interact with the environment with maximal directness, by using her own hands or feet. As one departs from isomorphism, the metaphor is that one is using some kind of a "tool" to manipulate the environment. Physically, this may comprise a variety of manipulanda, such as mice, joysticks, steering wheels, gloves, wands, etc., with the degree of directness of those tools affected by the other factors shown in the figure.

The next factor refers to the *alignment*, or relative location and/or orientation of the control device relative to the display space. A control/display (C/D) offset refers to a displacement between the location of the control device and the corresponding controlled object, and/or to a difference between the orientation of these. A completely aligned mapping therefore corresponds to direct control, in the line above. Conversely, as mentioned earlier, research has shown that performance degrades significantly as the size of the C/D offset increases [29] [39]. As we shall see in the following section, this factor is especially important when dealing with exocentric displays.

The bottom line in Figure 1.10 refers to the transformation between input commands to the control device and the resulting responses of the system being controlled. At a basic level this corresponds to the *control order*, that is, whether the controller has a zero, first, second or higher order transfer function. The zero order control case, at the left or congruent side of the continuum, corresponds to position control, in which case there is a simple gain factor relating control and display response. First order, or rate, control is less direct in the sense that all inputs to the control device are first integrated before control is effected. Second order control involves passing all inputs through two stages of integration, and so forth.

Example of control-display compatibility effects

In Section 1.2.2 we mentioned some of the practical issues involved in designing a particular MR system, the IIRO remotely controlled excavator, where the display involves a mixture of real data and modelled virtual objects. Thus far in Section 1.2.3 we have defined some of the basic concepts which contribute to those design issues. In the present section we bring these concepts together and discuss how they affect each other within a unified framework and how familiarity with these factors can be used to determine an appropriate viewpoint in MR systems. To simplify the discussion, as well as emphasise the generality of these considerations beyond the narrow context of remotely controlled excavators, we present the discussion within the framework of remote vehicle control, or a generic vehicle simulator.

In Figure 1.11 we present four cases of how a simple vehicle simulator might be implemented, from the point of view of real and virtual images, display centricity and control display mapping. In the two quadrants on the left (1 and 3) we have the case of an egocentric (out-the-window) display of a roadway, showing real data. In the two quadrants on the right (2 and 4) we have the same vehicle being controlled, but this time using an exocentric, or world referenced, map display. In both of the displays on the right, the "own vehicle" to be controlled is depicted by an arrow superimposed on the map. Although it is not necessary to determine in this case precisely to what extent the entire map being displayed has been created from modelled or unmodelled data, it is important to note that both the scale of the map and its orientation are modelled, in that the map is presented in a canonically conventional *north-up* fashion.

Turning to the rows, the control device for the top two quadrants is a standard steering wheel, in addition to an accelerator and brake pedals. Because the metaphor with a steering wheel is that one is situated within ones own vehicle and steering it, the nominal control mapping with the steering wheel is considered egocentric, or ego referenced. Note that this is defined to be the case also for the map display in quadrant 2, even though one is looking down on the vehicle (the small arrow shown pointing southwest in the figure). The control device shown in the bottom row (quadrants 3 and 4) is a simple computer mouse, which is exocentrically world-referenced, in a superordinate north-up manner, as indicated in the figure. This means that, by moving the mouse to the right, for example, the vehicle being controlled would be forced in an eastward direction relative to the real world, regardless of its current heading.[6]

In spite of the real vs virtual and the two ego-exo-centric distinctions shown in Figure 1.11, classifying the four cases shown in relation to the various dimensions of the MR taxonomy presented thus far is not straightforward. Although it is evident that neither of the controls are completely isomorphic (refering to Figure 1.10), since both involve the intervention of some kind of a device between the operator and the

---

[6]Note that the effect of using other control devices, such as a fixed joystick, either isotonic or isometric, would be very similar to that of the mouse presented here.

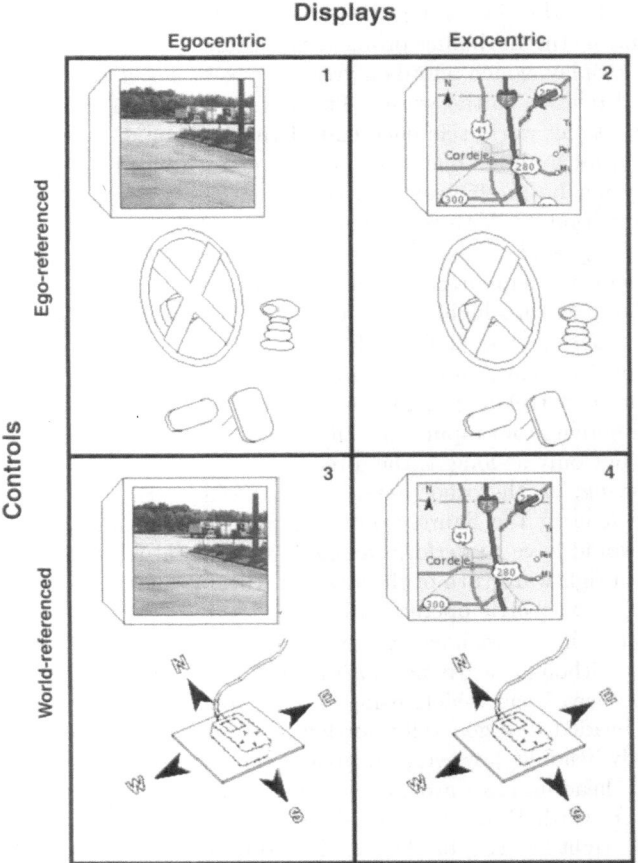

Figure 1.11 2x2 space of control-display mappings for vehicle simulator.

system being controlled, the overall degree of congruence depends not just on the device itself but on the context in which it is used, especially in relation to the associated display. For example, from a control point of view, although both devices influence direction of travel, they could each be programmed differently, as either zero-order or first order input devices.

Considering in turn the overall pros and cons of the cases shown in Figure 1.11 [23] [24] [26], in quadrant 1 we have an egocentric display together with a highly congruent control device, in the sense that a leftward control input will cause the simulated vehicle to turn to the left (as the visual display rotates to the right). Such a setup is very effective for *local guidance*, since it enables the operator/driver readily to follow an established trajectory to get from one point to another. In terms of Figure 1.9, this case lies at the ego-reference frame side of the continuum. This kind of display does not, however, encompass "flying up" above ones vehicle to survey the scene globally,[7] thus increasing the potential for keyholing, as mentioned earlier.

In quadrant 4, the keyholing problem is greatly reduced, because we now have an exocentric display, which affords a great degree of *global awareness* about where one is situated relative to the world. This case is indicated at the right hand side of Figure 1.9, at the world-reference end of the continuum. Although the display shown in quadrant 4 allows one efficiently to plan a route from A to B, traversing the route may not be as efficient as with the out-the-window view at ground level provided in quadrant 1. With respect to the control device shown in quadrant 4, a relatively high degree of control/display congruency is achieved, due to the fact that all mouse movements map directly onto the direction of motion of the vehicle. In this case, mouse input is thus highly consistent with the direct manipulation metaphor for which the mouse is best known [34].

In quadrant 2 we have an exocentric (world-referenced) display combined with an ego-referenced control device. In one sense control with the steering wheel should not be very different from input with the mouse, so this combination should work satisfactorily, but only as long as the direction of travel is generally northward. For southward driving, on the other hand, as indicated by the south-west orientation of the vehicle icon in the figure, confusion will be more likely to occur because the operator would have to perform a significant mental rotation to figure out, for example, that a rightward turn of the steering wheel would cause the vehicle icon to turn towards the exocentric operator's left.

In quadrant 3, finally, we have an egocentric display combined with an exocentric control device. Although clearly not optimal for local guidance, it might nevertheless be possible to "drive" the vehicle using the controller shown – but only as long as the vehicle is headed in a generally northward direction. The larger the deviation from a northerly heading, however, the greater the expected degree of confusion. For example, when heading eastwards, control input would be to the right in order to travel straight forward. Whenever the vehicle is headed in a southerly direction, west would be on the right and east on the left of the display. It would thus be necessary to apply a rightward force on the mouse controller in order to make the vehicle travel eastwards, to the left, and vice versa. The likelihood of control reversals would therefore increase greatly as the heading deviates more and more from a northerly direction.

1.2.4   Summary and Implications of Control-Display Issues

In the preceding discussion a number of tradeoffs among the four cases presented are outlined. These involve tradeoffs between egocentric versus exocentric displays and between ego-referenced versus world-referenced control inputs. Although the cases of quadrants 1 and 4 in Figure 1.11 are clearly more "natural" than quadrants 2 and 3, the latter have been included here not only for the sake of completeness of the

---

[7]Although this statement most often pertains to cases where the information in the display is comprised of real unmodelled image data, as depicted in Figure 1.11, it is nevertheless conceivable that one could also fly above a scene made up of real (3D) data which had been gathered from a number of viewpoints and integrated into a comprehensive 3D database ideally incorporating image interpolation as well. In spite of this, such a fly-through capability would certainly be more tractable with a virtual model rather than using real data.

discussion, but also because they represent realistic cases which can actually occur with Mixed Reality teleoperation systems (such as IIRO). That is, as discussed in Section 1.1, and as illustrated in the discussion of Figure 1.8, Mixed Reality display systems provide users with the opportunity to move back and forth between real world and virtual world scenes. In general, as discussed in Sections 1.2.1 and 1.2.2, real-world images typically provide increased detail, due to higher resolution relative to modelled objects, but they are also often provided only from an egocentric, out-the-window viewpoint. This is especially true for systems which move through their environment, and thus for which an external camera is not usually feasible. As indicated in Figure 1.9, such views are most effective for local guidance, that is, the task of maintaining an accurate trajectory from A to B. Conversely, exocentric displays are more conducive to global situational awareness, which comprises such tasks as landmark recognition, path planning, and obstacle avoidance, since they allow one to view the world from a number of exocentric viewpoints, either tethered or world-referenced. Together with the flexibility which accompanies the capability provided by MR to transit back and forth between real and virtual worlds, however, comes the issue of preserving control-display compatibility, as the advantages of one control scheme with one viewpoint metaphor become disadvantageous with another viewpoint. Because the consequences of such incongruencies include increased mental workload (due to the need to perform more mental transformations) and increased probability of control reversals, as well as other errors, the issues outlined in this section are expected to become increasingly relevant in future MR system research.[8]

1.3   Global Taxonomy of Mixed Reality Display Integration

In this final section we return to our original discussion of the meaning of Mixed Reality in Section 1.1 and discuss how the various issues of centricity and control presented in Section 1.2 pertain to a selection of different types of MR systems. In other words, for each class of MR system, our objective is to place the various factors presented in this paper into a single unified framework, defined within a space determined by its location along the RV continuum, the centricity continuum and the control-display congruency continuum. Our summary is presented in Figure 1.12.

1.3.1   HMD Based AR

In this section we consider the important class of AR displays based on optical or video see-through, that is, all cases in which the user "wears" the display system.

---

[8]It is important to take note of the fact that we have intentionally omitted consideration here of the issue of fixed versus rotating maps in our discussion of the exocentric displays in Figure 1.11. Much of the theory behind the concept of rotating displays, which corresponds to the central range of the centricity continuum depicted in Figure 1.9, can be reviewed in the writings of Wickens and his colleagues [22] [23] [26]. Research in our laboratory is currently focussing on exploring this concept experimentally, as a means of addressing some of the issues introduced in the present section.

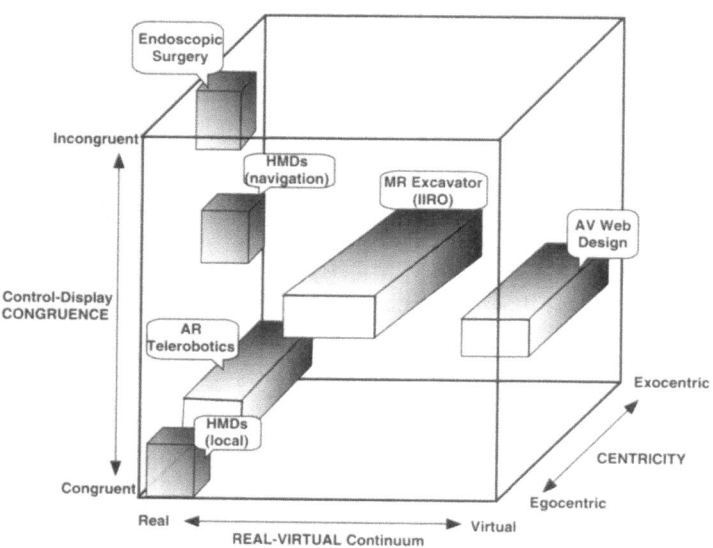

Figure 1.12   Global Taxonomy of MR Display Integration.

As discussed above, however, a distinction must be made between tasks that require information which promotes local task execution and those which require support for global situational awareness. Whereas the former class is the subject of most AR developments, for applications in manufacturing, maintenance, medicine, entertainment, etc., the latter is receiving increasing interest from the military, for example, where HMD's have been proposed for tactical displays for soldiers [8].

Consider first the more conventional AR displays for local task execution, labelled "HMD's (local)" in Figure 1.12. Because such displays are by definition based on a real-world background, this class, must lie very close to the real end of the RV Continuum, as shown. For the same reason, most such displays are also very close to the Egocentric end of the Centricity axis. Finally, for most cases one would expect to encounter good control-display congruency, since such displays are frequently designed for the user to interact directly with his immediate environment.

The classification is different for the second class of see-through AR displays, labelled "HMD's (navigation)" in Figure 1.12. Here we are referring to AR displays which attempt to provide global navigational information in a head-up form [3] [9]. From the point of view of defining AR, such displays also lie very close to the real end of the RV continuum. However, because global navigational is by definition exocentric, such displays will lie closer to the middle of the Centricity axis, in the sense that the graphic information is by nature usually top-down. Finally, for the reasons discussed above, this class of displays should be placed relatively far away from the origin, in the direction of low congruence between the outside world display and the superimposed control related navigation information.

1.3.2 Endoscopic Surgery

In many ways the field of endoscopic surgery resembles the case study presented in conjunction with Figure 1.11. In endoscopic surgery, a video camera and a set of specialised instruments are introduced into a patient's body, in lieu of direct viewing of the surgical field. With this relatively recent capability, however, comes a new set of problems involving control-display compatibility due to variations in the orientation of the intraoperative camera. For example, depending on the circumstances, a movement of the surgeon's hand, which might cause the instrument to move from left to right in her normal visual field, might cause the instrument to move from right to left on the video monitor [35]. Augmented Reality offers a number of potential advantages in this area, both for presenting navigational information and for providing the means to estimate absolute 3D distances and dimensions which would not otherwise be possible [13].

This promising area of application is represented in Figure 1.12, at the Real end of the RV Continuum, in light of the fact that the primary display medium is simple video. It is also depicted as being somewhere in the middle of the Centricity continuum. In using the term Centricity here, we are referring to the user's view of that which is being manipulated, i.e. the instruments and the surgical site, both of which are removed from the camera looking in at them. Furthermore, adding computer generated graphics containing navigational information, for example, could displace the cube even farther from the Egocentric end of the Centricity continuum, similar to the adjacent military HMD's. Finally, in light of the large obstacles which currently remain to be overcome in providing the means to map control movements unambiguously onto corresponding displayed responses, the cube has been placed fairly close to the maximum level of incongruence along the C/D Congruence dimension.

1.3.3 AR Telerobotics

The block labelled "AR Telerobotics" in Figure 1.12 refers to the system illustrated in Figure 1.3. Once again this block lies at the Real end of the RV Continuum. As discussed earlier, depending on the user's nominal viewpoint with respect to the overlaid graphics and the nature of the control laws, the Centricity could be anywhere along that axis. In using the term Centricity this time, we are referring to how the camera is located relative to the objects being manipulated. For the particular resolved control robot shown in Figure 1.3 this is an exocentric view; however for other cases, it could also have been more egocentric. For that reason, the AR Telerobotics block is shown stretching across most of the Centricity axis. However, it is shown fairly close to the highly congruent end of the Congruence axis, by virtue of the fact that the controller is fairly well matched with the display in this particular case [16] [17].

1.3.4 MR Excavator

The block labelled "MR Excavator (IIRO)" in Figure 1.12 refers to the case study presented in Section 1.2.1. Because this is truly an example of many levels of Mixed Reality, it is shown covering a large part of the centre of the RV Continuum. It is also shown stretching across the Centricity dimension, in light of the fact that users are able to alter their view flexibly between egocentric and exocentric. Finally, the IIRO block is shown in the middle of the Congruence axis, since the control display matching can be considered neutral in this case for the reasons discussed earlier.

1.3.5   AV Web Design

The block labelled "AV Web Design" refers to the example shown in Figure 1.4, which is a case of Augmented Virtuality applied towards the design of 3D web pages. Consequently, this block is located at the Virtual end of the RV Continuum. Because users are able to browse such pages either egocentrically, by moving their mouse through the central portion of the image shown in Figure 1.4, as well as exocentrically, by navigating through the plan view at the lower right, we have extended this block also to cover the whole Centricity continuum. Finally, although the control viewpoint is generally consistent with traversing the environment, the only means available for traversing the 3D web page at the present time is a 2D mouse, which can operate in a number of modes, including both position control and velocity control, the C/D Congruence has been rated neutral here.

1.3.6   Conclusion

From the examples presented in this section, a number of observations emerge. It is interesting to note how, in general, the blocks tend to spread out across the Centricity axis as the different systems vary from mostly real (AR) to mostly virtual (AV). This reflects the great flexibility offered by MR displays, where, as with the case of the IIRO excavator, users are able to exploit the advantages of both the real components and the virtual components of their system. Another message to be derived is that current applications of Mixed Reality are not occupying only one corner of the space defined in Figure 1.12, but rather are spread out over most of the taxonomy space. It is important to realise, however, that the classifications presented here are not definitive, but are subject to interpretation, as well as modification in response to changes in specific operational contexts. With this in mind, it is our hope that this graphic representation will serve both as an indicator of the great diversity of activity in the field of MR as well as a useful framework for permitting researchers to understand the similarities and differences characterising their respective endeavours.

Acknowledgements

The authors gratefully acknowledge Spar Aerospace Ltd (under the "Interactive Intelligent Remote Operations (IIRO)" Project) and the Institute for Robotics and Intelligent Systems (IRIS) (under the "Effective Display and Tele-Control Technology Integration for Real and Virtual Environments" project) for their generous support of the work reported here. We would also like to acknowledge the cooperation of the various members of the University of Toronto ETC Lab for their contributions, and especially Dr. Haruo Takemura, on leave from the Nara Institute of Science and Technology, for his generous willingness and exceptional capacity to assist throughout the composition and editing of this paper.

References

[1] R. T. Azuma: "A survey of augmented reality," *Presence*, vol.6, no.4, pp.355–385, 1997.

[2] H. Fuchs and J. Ackerman: "Displays for augmented reality: Historical remarks and future prospects," in (Y. Ohta and H. Tamura, eds.) *Mixed Reality – Merging Real and Virtual Worlds*, Ohmsha(Tokyo)-Springer Verlag(Berlin), 1999.

[3] D. Curtis, D. Mizell, P. Gruenbaum, and A. Janin: "Several devils in the details: Making an AR app work in the airplane factory," *Proc. First Int'l Workshop on Augmented Reality (IWAR'98)*, 1998.

[4] D. Reiners, D. Stricker, G. Klinker, and S. Muller: "Augmented Reality for construction tasks: Doorlock assembly," *Proc. First Int'l Workshop on Augmented Reality (IWAR'98)*, 1998.

[5] J. Molineros, V. Raghavan, and R. Sharma: "AREAS: Augmented Reality for evaluating assembly sequences," *Proc. First Int'l Workshop on Augmented Reality (IWAR'98)*, 1998.

[6] M. Bajura, H. Fuchs, and R. Ohbuchi: "Merging virtual objects with the real world: Seeing ultrasound imagery within the patient," *Computer Graphics (Proc. SIGGRAPH'92)*, vol.26, no.2, pp. 203–210, 1992.

[7] A. Stokes, C. Wickens, and K. Kite: *Display Technology: Human Factors Concepts*, Society of Automotive Engineers (SAE), 1990.

[8] W. O. Blackwood, T. R. Anderson, C. T. Bennett, J. R. Corson, M. R. Endsley, P. A. Hancock, J. Hochberg, J. E. Hoffman, and R. B. Kruk: *Tactical Displays for Soldiers – Human Factors Considerations*, National Academy Press, 1997.

[9] M. M. Glumm, W. P. Marshak, T. A. Branscome, M. McWesler, D. J. Patton, and L. L. Mullins: *A Comparison of Soldier Performance Using Current Land Navigation Equipment With Information Integrated on a Helmet-Mounted Display (ARL-TR-1604)*, Army Research Laboratory, 1998.

[10] P. Milgram and F. Kishino: "A taxonomy of mixed reality visual displays," *IEICE Trans. on Information and Systems (Special Issue on Networked Reality)*, vol.E77-D, no.12, pp.1321–1329, 1994.

[11] P. Milgram, H. Takemura, A. Utsumi, and F. Kishino: "Augmented reality: A class of displays on the reality-virtuality continuum," *Proc. SPIE Conf. Telemanipulator and Telepresence Technologies*,, vol.2351–34, pp.282–292, 1994.

[12] P. Milgram, S. Yin, and J. J. Grodski: "An augmented reality based teleoperation interface for unstructured environments," *Proc. ANS 7th Topical Meeting on Robotics and Remote Systems*, pp.966–973, 1997.

[13] M. Kim, P. Milgram, and J. Drake: "Computer assisted 3D measurements for micro-surgery," *Proc. 41st Annual Meeting of Human Factors and Ergonomics Society*, pp.787–791, 1997.

[14] B. L. K. Davey, R. M. Comeau, P. Munger, L. Pisani, D. Lacerte, A. Olivier, and T. M. Peters: "Multimodal interactive stereoscopic image-guided neurosurgery," *Proc. Visualization in Biomedical Computing (VBC) 94.*, Sep. 1994.

[15] R. T. Azuma: "A survey of augmented reality," *SIGGRAPH'95 Course Notes*, pp.1–38, 1995.

[16] P. Milgram, A. Rastogi, and J. J. Grodski: "Telerobotic control using augmented reality," *Proc. IEEE Int'l Workshop on Robot-Human Communication (RO-MAN'95)*, pp.21–29, Jul. 1995.

[17] A. Rastogi, P. Milgram, D. Drascic, and J. J. Grodski, "Telerobotic control with stereoscopic augmented reality," *Proc. SPIE Conf. Stereoscopic Displays and Virtual Reality Systems III*, vol.2653, pp.115-122, 1996.

[18] G. U. Carraro, J. T. Edmark, and J. R. Ensor: "Techniques for handling video in virtual environments," *Proc. SIGGRAPH'98,* pp.353–360, 1998.

[19] J. Ballantyne, M. Greenspan, and M. Lipsett: "Virtual environments for remote operations." *Proc. ANS 7th Topical Meeting on Robotic and Remote Systems*, pp.545–549, 1997.

[20] P. Milgram and J. Ballantyne: "Real world teleoperation via virtual environment modelling," *Proc. 7th Int'l Conf. on Artificial Reality and Tele-Existence (ICAT'97)*, pp.1–9, 1997.

[21] E. M. Gagnon, M. Greenspan, and J. W. Ballantyne: "Real-time surface following for teleoperation in virtual environments" *Proc. SPIE Conf. Telemanipulator and Telepresence Technologies IV*, vol.3206, pp.66–73, 1997.

[22] C. D. Wickens: *Engineering Psychology and Human Performance (Second Edition)*, Harper-Collins, 1992.

[23] E. L. Faye and C. D. Wickens: "Strategies for display integration in navigational guidance and situation awareness," ARL-95-4/NASA-95-1, Aviation Research Laboratory, Institute of Aviation, University of Illinois at Urbana-Champaign, 1995.

[24] C. D. Wickens: "Integration of navigational information for flight," ARL-95-11/NASA-95-5, Aviation Research Lab, Institute of Aviation, University of Illinois at Urbana-Champaign, 1995.

[25] C. D. Wickens, C. Liang, T. Prevett, and O. Olmos: "Egocentric and exocentric displays for terminal area navigation," ARL-94-1/NASA-94-1, Aviation Research Lab, Institute of Aviation, University of Illinois at Urbana-Champaign, 1994.

[26] C. D. Wickens: "Frame of reference for navigation," in (D. G. A. Koriat, ed.) *Attention and Performance*, Academic Press, vol.16, 1998(in press).

[27] D. D. Woods: "Visual momentum: A concept to improve the cognitive coupling of person and computer," *Int'l J. Man-Machine Studies*, vol.21, pp.229–224, 1984.

[28] S. Roscoe: "Horizon control reversals and the graveyard spiral," *CSERIAC (Crew System Ergonomics Information Analysis Center) Gateway, VII*, pp.1–4, 1997.

[29] S. R. Ellis, M. Tyler, W. S. Kim, and L. Stark: "Three-dimensional tracking with misalignment between display and control axes," *Proc. 21st Int'l Conf. on Environmental Systems*, SAE Technical Paper Series 911390, 1991.

[30] E. P. McCormick, C. D. Wickens, R. Banks, and M. Yeh: "Frame of reference effects on scientific visualization subtasks," *Human Factors*, vol.40, no.3, pp.443–451, 1998.

[31] K. Hinckley, R. Pausch, J. C. Goble, and N. F. Kassell: "A survey of design issues in spatial input," *Proc. ACM Symp. on User Interface Software and Technology (UIST'94)*, pp.213–222, 1994.

[32] D. A. Bowman, D. Koller, and L. F Hodges: "Travel in immersive virtual environments: An evaluation of viewpoint motion control techniques," *Proc. IEEE Virtual Reality Annual Int'l Symp. (VRAIS'97)*, pp.45–52, 1997.

[33] S. Zhai and P. Milgram: "Quantifying coordination in multiple DOF movement and its application to evaluating 6 DOF input devices," *Proc. ACM Conf. on Human Factors in Computing Systems (CHI'98)*, pp.320–327, 1998.

[34] A. Dix, J. Finlay, G. Abowd, and R. Beal: *Human-Computer Interaction*, Prentice Hall Europe, 1998.

[35] J. G. Holden, J. M. Flach, and Y. Donchin: "Perceptual-motor coordination in an endoscopic surgery simulation," *Surgical Endoscopy*, vol.13, 1999(in press).

[36] N. I. Durlach and A. S. Mavor (eds.): *Virtual Reality: Scientific and Technological Challenges*, National Academy Press, 1995.

[37] D. Terzopoulos and K. Fleischer "Deformable models," *The Visual Computer*, vol.4, no.6, pp.306–331, 1988.

[38] S. Fukatsu, Y. Kitamura, T. Masaki, and F. Kishino: "Intuitive control of "Bird's Eye" overview images for navigation in an enormous virtual environment," *Proc. ACM Symp. on Virtual Reality Software and Technology (VRST'98)*, pp.67–76, Nov. 1998.

[39] M. W. McGreevy and S. R. Ellis: "Format and basic geometry of a perspective display of air traffic for the cockpit," NASA Technical Memorandum 86680, Jun. 1991.

Chapter 2

Displays for Augmented Reality: Historical Remarks and Future Prospects

Henry Fuchs
Jeremy Ackerman

The University of North Carolina
at Chapel Hill, U.S.A.

2.1 Introduction

An augmented or "mixed reality" system is one in which computer graphics objects are added into the user's surrounding scene in a meaningful way. Augmented reality includes applications ranging from "heads up" displays for fighter pilots to complex real-time visualizations of multiple data sets in medical applications. Other applications include architectural previewing and engineering design.

Augmented reality systems require the use of specialized display devices. Typically optical see-through or video see-through displays are used. It is important to note that the conception of augmented reality systems should not be limited to applications using head mounted displays. In the following sections the principles of several display technologies are presented and examples of their application to augmented reality are discussed.

2.2 Summary of Augmented Reality Display Technologies and Examples of Their Use

Display devices for augmented reality applications are most easily categorized as head mounted and non-head mounted devices.

2.2.1 Head Mounted Displays

Head mounted displays have been used for augmented reality applications since the 1960s [1], and have been the mainstay of many augmented reality systems to date. Two kinds of head mounted displays have been used: optical see-through and video see-through. As their name implies, optical see-through systems combine the real and synthetic imagery via some optical merging mechanism, most commonly a "half-silvered" mirror. Video see-through systems combine synthetic images with "real" images of the user's surroundings by combining two video streams, one typically generated by computer, the other coming from a video camera mounted on (or near) the user's head.

Optical see-through head mounted displays

Perhaps the first augmented reality system was developed in the late 1960s by Ivan Sutherland [2], first at Harvard University and then at the University of Utah. This display is shown in Figure 2.1.

The display demonstrates the key components and principles needed for optical see-through displays: There is a distinct display for each eye, since each eye's view of the world is from a slightly different vantage point; there is a see-through capability allowing each eye to view simultaneously the user's surroundings and the computer-generated imagery; a half-silvered prism sitting directly in front of each eye allows some of the light of the user's surroundings to pass through to the user's eyes. Also passing through to the user's eyes is some of the light that originates at the face of the CRTs mounted on the sides of the head. Each eye observes the combination of two images superimposed one on top of the other. If light levels in the environment are carefully adjusted, and both the environment and the synthetic imagery are sufficiently simple, both can often be comprehended simultaneously.

More recent implementations of optical see-through displays have utilized newer display technologies and novel optical systems to improve the performance of this type of head mounted display. Other display devices, such as small LCD panels, can be placed directly between the viewer's eyes and the environment, and significantly reduce the complexity of the optics and the weight and cost of the device. Several displays based on this principle are now commercially available.

Optical see-through displays have the very difficult problem of accurate registration of real and synthetic imagery in space and time. Changes in head position or orientation can occur very rapidly during one's normal interaction with the surrounding environment. The combined processes of tracking (orientation and position), and rendering and display of synthetic imagery can easily take close to 100 milliseconds. By the time an image corresponding to the tracking data gathered at the beginning of the rendering cycle can be displayed, the head has already moved considerably.

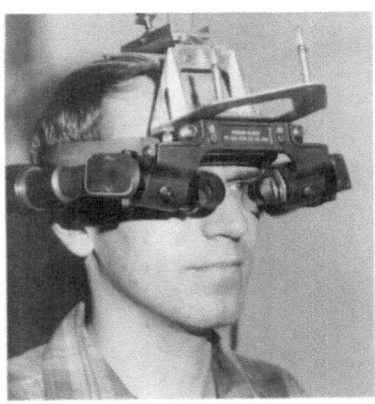 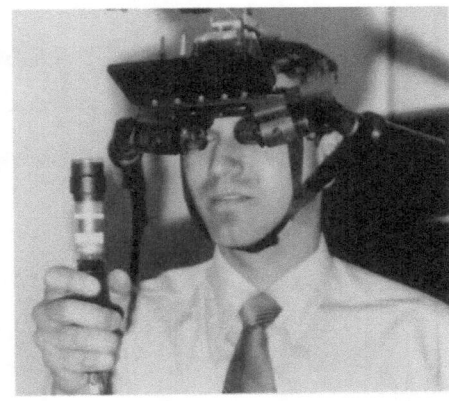

Figure 2.1   Two images of Ivan Sutherland's head-mounted display (circa 1970). One should note the CRTs on either side of the head and the optics in front of the CRTs to put the image in front of the eyes. Images are viewed on half silvered prisms, giving the user a view of both the synthetic imagery and the real world. (Courtesy, Dept. of Computer Science, University of Utah).

The image now displayed is misregistered with the real world, which may lead to confusion, misinterpretation of the scene, and disorientation.

A second difficulty with optical see-through displays is light intensity. If the synthetic imagery is too bright relative to the ambient light, the user's environment will not be visible. If the reverse is true, the synthetic imagery will be either invisible or visible only as a vague shadow in the environment.

A final difficulty with optical see-through displays is that of occulusions. If a synthetic object should appear in front of a real world object, it will (in general) appear to be a semitransparent ghost floating in front of the object. If parts of the real world object are brightly lit, those portions of the synthetic object that are in front of the bright spots will appear more transparent or invisible.

An optical see-through display system manufactured by CAE is used in their simulators and trainers for military aviation. This head mounted display is expensive and heavy, making it impractical for many other applications. This particular head mounted display uses two light-valve type projectors for each eye. The central part of the user's visual field, the area with greatest acuity, is fed with one projector, while the larger surrounding, and lower acuity, region is fed by the second projector. An optical system is used to combine the light from each projector. This system provides greatest display resolution where it is needed most.

In the CAE system, use of a head mounted display reduces the space required for the simulator to the size of the simulated aircraft's cockpit, without the additional space needed for other display techniques used by other manufacturers. A see-through head mounted display is preferable to a virtual reality type display to enable the pilots to see the instruments and their own hands as they learn to operate an unfamiliar cockpit. This environment is ideal for the limitations of optical see-through devices discussed above. There is no reason why synthetic imagery (of the

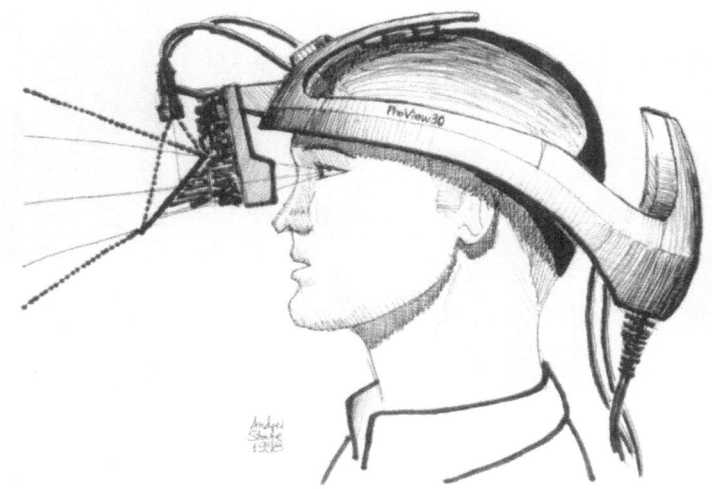

Figure 2.2   This drawing, by Andrei State of UNC-Chapel Hill, shows one method proposed to mount cameras on a Kaiser Electro Optics Proview 30 for use in the UNC augmented reality systems. The image of the world is reflected up to the camera so that the optical path length between world objects and the camera is very close to the path length to the user's eyes. The mirrors are moved with the displays when the displays adjusted for optimal viewing. ((c)1998, Dept. of Computer Science, UNC-Chapel Hill)

environment that the aircraft is flying through) should ever need to be overlaid on the real objects (inside the cockpit). The simulator designers can design the lighting systems inside the cockpit to insure that both the real world and the synthetic imagery are always visible to the user.

Optical see-through displays were also a part of the demonstration of augmented reality air hockey given by Mixed Reality Labs at SIGGRAPH'98. In this demonstration, both optical and video see-through displays were available to conference attendees to play air hockey. The puck and paddles were rendered (registered to a real table) in the head mounted displays of two players. The displays used were relatively light weight and comfortable [3].

Video see-through HMDs

A video see-through display gives the user a view of the world through one or more video cameras mounted on the display. Synthetic imagery is combined with the image captured through the video cameras by the computer, and the combined signal is sent to the display.

No video see-through displays are commercially available at this time; it is, however, relatively simple to mount one or two small video cameras on a standard head mounted display designed for virtual reality (Figure 2.2). Additional hardware and software are needed to combine real and synthetic imagery. One approach is to employ a frame grabber for each camera. Alternatively, techniques like chroma-keying

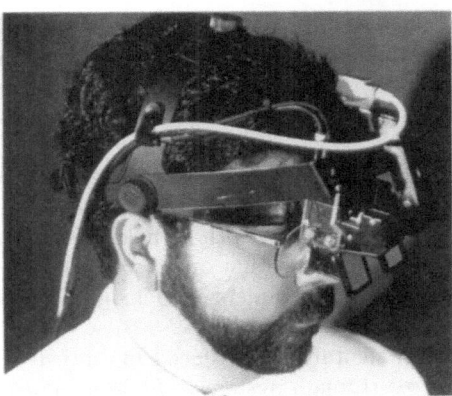

Figure 2.3   This photograph shows David Casalino, MD wearing the video see-through head mounted display, built in collaboration with the UNC-Chapel Hill and the University of Utah. This display features a relatively narrow field of view for the displays, but has an open design allowing unencumbered vision around the display. The displays and cameras are mounted on independently adjustable pods to optimize viewing performance. The camera view is aligned with the user's eyes by optics which fold the optical path. ((c)1998, Dept. of Computer Science, UNC-Chapel Hill)

could be employed to add the video signal into the data stream after the synthetic imagery is rendered.

Video see-through also has a number of underlying difficulties. Among these are camera offset, limited field of view, temporal lag, fixed focus, and limited depth of field. When only a single camera is used, stereopsis is also lost. Some of these difficulties (camera offset and limited field of view) have been overcome in an experimental display built collaboratively by the University of North Carolina and the University of Utah (Figure 2.3).

Without carefully designed optics, it is difficult to align the camera's view with the normal viewing axis of the eye. This means that the image sent to each of the user's eyes is taken from a perspective other than that of the observer's eyes. A significant change in horizontal spacing between the cameras distorts the user's sense of depth because the stereo pairs have an effective interpupillary distance different from that to which the user is accustomed. Typically it is easiest to mount the camera or cameras above the user's eyes vertically. This displacement gives the user a false sense of height, particularly for relatively nearby objects.

Another problem is that of field of view. Care must be taken in the selection of cameras and lenses so that the field of view of the camera and lens combination matches the field of view of the displays. Without this precaution, the user's view of the world will appear noticeably distorted.

Like optical see-through displays, video see-through displays can suffer from temporal lag. However, it is possible to compensate for lag, such that the effects of lag are not noticeable in a given image being displayed by the HMD. This can be achieved by synchronizing video image capture with the report from the tracker.

The synthetic imagery generated from that tracker data must be displayed with the video image captured at the same point in time. This mechanism of dealing with the lag inherent in the system solves the misregistration problem, but may introduce noticeable delays between when movements in the real world occur and when they are seen.

Fixed focus and depth of field are problems inherent in using video cameras. Objects outside the focal distance of the cameras will be out of focus, and typical cameras have a relatively narrow depth of field. While it may be possible to auto-focus type mechanisms to insure that objects seen by the camera are in focus, changing the focus may change the camera model significantly and lead to misregistration of synthetic imagery.

The problem of environmental lighting faced by optical see-through systems is not a concern in video see-through systems. Lighting must be considered when selecting cameras and camera settings, as cameras operate optimally on a much narrower range of illuminance than can the human eye.

Occlusion poses a challenging problem in video see-through head mounted displays. When a real object passes between the user and the location of a virtual object in space, the synthetic object will typically, and incorrectly, fully occlude the real object that it should appear to be behind. This confusion of depth cues can destroy the illusion that the virtual objects are truly existing in the user's environment. In some applications moving objects may be tracked with position sensors, or another method may be employed to determine the three-dimensional geometry of the real world. Without this type of three-dimensional information about real objects, this type of occlusion error is difficult, if not impossible, to solve.

At the University of North Carolina, video see-through head mounted displays have been used extensively for the surgical applications being developed [4]. Video see-through displays were chosen because of the ability to synchronize synthetic imagery with the video image from the real world, thereby reducing the likelihood that any individual image shown to the user would misrepresent the spatial relationship between real and virtual objects. Our system has used a number of displays, both that we were involved in designing (Figure 2.3), and those made from commercially available displays (Figure 2.2).

Our system has used several displays designed for use in virtual reality applications. We have learned that the wide field of view displays that these systems typical have, while helpful for a sense of immersion in virtual environments, do not provide adequate resolution for the surgical site at the center of the user's view. Our medical collaborators have also been unhappy with the inability to use peripheral vision and to see around the display easily. Our own design (Figure 2.3) and our most recently acquired commercially available display (a Kaiser Electro Optics Preview 30, Figure 2.2) solve these two problems by leaving the display open lateral to the display for each eye and by providing relatively narrow field of view displays. These devices also have the ability to flip the displays up and out of view if it is not needed.

2.2.2   Non-head Mounted Displays

There are some environments and situations where wearing a head mounted display is impractical or unfeasible. This may be true of situations requiring long term use, or extensive use of other resources in a non-augmented environment. The following subsections describe a few approaches to other types of displays.

Systems of this type provide the benefit of offering the higher resolution of a conventional monitor for augmented reality applications. The user also does not experience the fatigue and discomfort of wearing a heavy and unwieldy head mounted device.

These devices are limited in their usefulness by their limited working volume and by potential space constraints.

Small displays

Displays the size of a small monitor (around 30cm by 20cm) can be employed to build an augmented reality system. One approach was proposed in 1975 by Knowlton [5]. This display was proposed as a way of creating a programmable keyboard. In his system, a real but unlabeled keypad was to be augmented with button labels drawn by computer graphics. In his approach, a half-silvered mirror was placed between the user and the keypad. The half-silvered mirror provides the user a view of the real objects behind the mirror, and also reflects the image from a conventional display (such as a CRT) to the user. This type of system can be thought of more generally as placing arbitrary real objects behind the mirror and adding whatever synthetic imagery is needed.

Devices of this type can be further enhanced in several ways. If the user's head position is tracked, synthetic imagery can be drawn from the correct perspective. This device can be used to reflect alternating left and right stereo pairs rather than single perspective images. A stereo illusion of dimensionality can be created if the user uses the display wearing shutter glasses [6].

Displays of this type face a problem with illumination similar to the one encountered in optical see-through head mounted devices. Real objects behind the mirror will wash out the synthetic image if they are more brightly lit than the illumination provided by the monitor. The optical see-through problem of not being able to fully occlude real surfaces with synthetic imagery is present with these devices.

These displays have a working volume limited to the space behind the mirror. Generally the mirror needs to be maintained at a fixed geometry relative to the conventional display and is, therefore, not easily moveable to a new position.

In situations where the user may want to move very close to the mirror, the display may fail in two ways. First, the user may obstruct the path of the light from the monitor to the mirror, thereby blocking the light path needed to view the synthetic imagery. Second, as the user approaches the display, each pixel subtends a greater portion of the user's field of view. As a result, the user can gain no better enhancement of their view of the synthetic objects by moving the head closer to where they perceive the object to be.

This type of display is currently being used in surgical simulators (so the surgeon's own hands are visible), and in some telesurgical devices being developed. Figure 2.4 shows how it was proposed to use these types of displays in UNC's surgical augmented reality applications.

In the field of neurosurgery another novel approach to augmented reality is used. Preoperatively acquired images, including three dimensional graphics created during

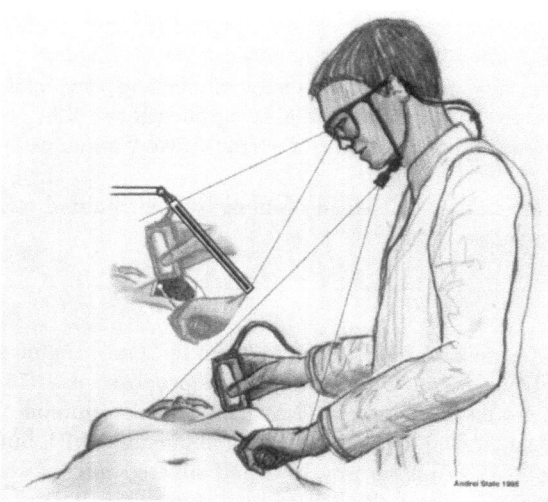

Figure 2.4   This drawing, by Andrei State of UNC-Chapel Hill, presents one vision of how a relatively small display might be employed for use in an augmented reality medical system. This particular drawing shows an alternate implementation in which a strategically placed camera enables use of video see-through. ((c)1998, Dept. of Computer Science, UNC-Chapel Hill)

surgical planning, may be overlayed onto the surgeon's view through the surgical microscope. This system is known as COMPAS and is rapidly gaining acceptance in the field of neurosurgery (Kelly).

Large area displays

Large displays have perhaps received less recognition than they deserve for augmented reality applications. These displays may be employed in many configurations: front projection, back projection, and even conventional CRTs. Use of larger displays in augmented reality is perhaps best explained by example, because they are not usually thought of as augmented reality display devices.

A display like an ImersaDesk or Virtual Workbench is usually a large monitor, set on either a horizontal or tilted work surface. A user wears shutter glasses so that the display can show three-dimensional objects in stereo. A variety of input devices can be employed to interact with the objects being displayed. These displays are limited to displaying objects near their surface. A virtual object can appear no "taller" than the plan from the far edge of the display to the user's eyes [7].

The large monitors and projector systems in civilian flight simulators may also be thought of as augmented reality devices. These displays augment the real environment of a large jet's cockpit with images of what the outside world should appear like.

Projects like UNC's Office of the Future (Figure 2.5) take the previous two concepts a step further. In the Office of the Future, multiple projectors will be used to

provide a virtual window into another Office. In addition to providing a compelling
telepresent experience, workers in an Office will be able to collaborate on three dimensional models both locally and with remote participants. Large area displays,
unlike the head mounted technologies, allow a user to easily switch back and forth
between making use of the virtual objects and taking or reading notes, or using a
conventional monitor.

Figure 2.5   This drawing, by Andrei State of UNC-Chapel Hill, shows a view of the
Office of the Future. Large displays (provided by the projectors mounted in the ceiling),
cover the walls and desk. The user is able to participate in a discussion with his colleagues
who appear to him as sitting on the other side his desk. He is able to use his personal
computer to make notes while working collaboratively on a design for a new head mounted
display, which appears to be hovering over his desk. ((c)1998, Dept. of Computer Science,
UNC-Chapel Hill)

2.3   Conclusion

Augmented reality systems have the advantage of enhancing a user's environment.
The promise of this technology is very large in, among other field, engineering design,
military, and medicine. The difficulties in building augmented reality systems are
equally large, even at the level of the displays. For some of the problems with current
displays there are potential solutions that may soon be available, but whether the
problems can be solved for others is not clear.

While the information provided in this paper is not definitive, we hope that it
will be useful in gaining a better understanding of the abilities of current display
devices. It is important to begin developing augmented reality applications that are
amenable to the current limitations, while also keeping an open mind to new ideas,
which will, in time, reduce these limitations.

Acknowledgements

This work was supported by (1) "Enabling Technologies and Application Demonstrations for Synthetic Environments" (DARPA contract DABT63-93-C-0048), Frederick P. Brooks, Jr. and Henry Fuchs (UNC-CH), Principal Investigators; (2) the "National Tele-Immersion Initiative" (Jaron Lanier, Chief Scientist) of Advanced Network and Services, (Al Weis, President and Chief Executive Officer); (3) "Science and Technology Center for Computer Graphics and Scientific Visualization" (National Science Foundation Cooperative Agreement no. EIA-8920219), Center Director Richard Riesenfeld (University of Utah). Andy van Dam (Brown University), Al Barr (Caltech), Donald Greenberg (Cornell University), and Henry Fuchs (UNC-Chapel Hill), Principal Investigators; (4) "Three-Dimensional Medical Image Display" (National Institutes of Health contract P01 CA47982-09), Stephen Pizer, Program Director. Elizabeth Bullitt MD, Edward Chaney MD, Henry Fuchs, Principal Investigators (UNC-Chapel Hill).

We gratefully acknowledge discussions over many years with close colleagues: Michael Bajura, Gary Bishop, Frederick P. Brooks Jr., David Casalino MD, Vern Chi, D'Nardo Colucci, Kurtis Keller, Mark Livingston, Leonard McMillan, Anthony A. Meyer MD, Ryutarou Ohbuchi, Etta Pisano MD, Stephen Pizer, Jannick Rolland, Michael Rosenthal, Andrei State, Greg Welch. We thank Andrei State in particular for the drawings and Andrew Ade for editorial and production assistance.

References

[1] I. Sutherland: "The ultimate display," *Proc. IFIP Congress*, vol.2, pp.506–508, 1965.

[2] I. Sutherland: "A head-mounted three dimensional display," *Proc. Fal Joint Computer Conference, AFIPS Conf. Proc.*, vol.33, pp.757–764, 1968.

[3] T. Ohshima, K. Satoh, H. Yamamoto, and H. Tamura: "AR^2 Hockey: A case study of collaborative augmented reality," *Proc. VRAIS'98*, pp.268–295, 1998.

[4] A. State et al.: "Technologies for augmented-reality systems: Realizing ultrasound-guided needle biopsies," *Proc. SIGGRAPH 96*, pp.439–446, August 4–9, 1996.

[5] K. Knowlton: "Virtual pushbuttons as a means of person-machine interaction," *Proc. Conf. Computer Graphics, Pattern Recognition and Data Structure*, pp.14–16, May 1975.

[6] K. Knowlton: "Computer displays optically superimposed on input devices," *Bell System Technical Journal*, vol.56, no.3, pp.367–383, 1977.

[7] R. S. Kalawsky: *The Science of Virtual Reality and Virtual Environments*, Addison-Wesley, 1993.

[8] R. Raskar, G. Welch, M. Cutts, A. Lake, L. Stesin, and H. Fuchs: "The office of the future : A unified approach to image-based modeling and spatially immersive displays," *SIGGRAPH 98 Conf. Proc.*, pp.179–188, July 19–24, 1998.

Chapter 3

Virtualized Reality: Digitizing a 3D Time-Varying Event As Is and in Real Time

Takeo Kanade
Peter Rander
Sundar Vedula
Hideo Saito

Carnegie Mellon University, U.S.A.

3.1   Introduction

Virtualized Reality is for 4D digitization - capturing and modeling a time-varying 3D event into a computer. A real event is observed by multiple cameras. The multiple synchronized output video streams are analyzed to produce voxel representations of a scene for each moment. First, image-based stereo is used to compute a range map corresponding to each camera view; thus a single time instant of the dynamic event is modeled as a collection of color-range image pairs from different viewpoints, and the full event is modeled as their sequence. These range images are then fused into a global 3D model - consisting of both voxel and surface representations. Finally, as an appearance model, view-dependent texture information is attached by back-projecting the original color images onto the recovered surfaces.

Unlike other techniques, what is captured and modeled in Virtualized Reality is not a collection of views, but an explicit representation of the "event". The applications of Virtualized Reality include simulation, training, telepresence, and entertainment. For example, by viewing a dynamic event model of a skilled surgeon performing an operation, students could revisit the operation, free to observe from anywhere in the reconstructed operating room. Spectators could watch a basketball game from any stationary or moving point on or off the court. Once modeled, however, many more things are possible, which are not possible in solely view-based approaches. The recovered scene geometry enables the editing - addition, removal, or alteration - of the event. That is, it is possible to manipulate the reality in the computer; the motion of objects might be altered by computing the Newton-Euler dynamic motion equations after the event, and virtual objects can be added to the event or real components can be removed from it at view time.

We first introduced the concept of Virtualized Reality in 1993 [1] [2], based on the work of the CMU video-rate stereo machine [3]. Then we moved on to develop the "3D Dome", consisting of 51 cameras and analog VCRs [4], with which we demonstrated an off-line system for Virtualized Reality [5]–[7]. Now, we have developed a new fully digital "3D Room" [8]. This paper presents an overview of the current status - methods and examples - of Virtualized Reality at CMU Robotics Institute. The technical details are presented in the above papers.

3.2   Modeling Real Events into Virtual Reality

In the real world, we experiment and observe the results by altering the spatial positions of objects, exerting forces, and adding more objects. Very often, one wants to know the outcome before actually exercising those changes to the real world - remodeling one's house is a good example. In some cases, real experiments may be dangerous, costly or even impossible to perform. In these cases, a virtual reality experiment is a convenient alternative. In virtual reality, experimental choices are tested in "simulation", and the results are presented mostly by rendering visual, audio and haptic information to the user (See Figure 3.1). One of the most important, and yet, least developed capability in this scheme is that of modeling the real event for incorporation into virtual reality. While modeling the reality involves many diverse aspects, such as geometrical, material, optical, dynamic, and so on, our focus in this paper is spatial and appearance modeling.

Our goal is to digitize and model a time-varying three-dimensional event of a large scale in its totality. There are a few aspects that separate our current work from most of the past work. First, the task is 4D digitization - a time sequence of three-dimensional shapes and positions is to be acquired. Second, while the 3D sensors provide 2-1/2-D representations of the scene from each view, the desired output must be the full 3D, or whole scene representation of the event - in other words, the set of view dependent data must be converted to a scene-centered description. Finally, rather than modeling a small toy object on a turn table, our interest is modeling an event whose spatial extent is at least that of a room containing multiple people. A large physical space poses many issues in calibration, visibility, and illumination.

We set out to design and develop a dynamic scene modeling system that would realize this vision. First, the video capture system must achieve sustained real-time performance in order to capture meaningful events. Second, the observation system

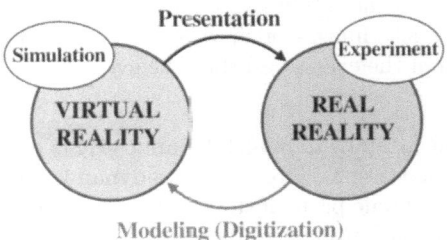

Figure 3.1   Modeling is a critical step for use of virtual reality for simulating on a real world event.

must not interfere with the normal appearance or activity of the event, since the goal is not only geometric modeling, but also modeling of visual appearance. Third, and most importantly, the system must work with minimal human-operator intervention, if at all. The tremendous amount of data generated by multiple video cameras make most of human-interactive approaches unrealistic.

3.3   Related Work

Recent research in both computer vision and graphics has made important steps toward realizing this goal. Work on 3D modeling (e.g., Hilton et.al [9], Curless and Levoy [10], Rander et. al. [6], and Wheeler et. al. [11]) presents volumetric integration of range images for recovering global 3D geometry. Sato et. al. [12] have developed techniques to model object reflectance as well. Note that most of these techniques rely on direct range-scanning hardware, which tends to be too slow and costly for a multi-sensor dynamic modeling system. Debevec et. al. [13] use a human editing system with automatic model refinement to recover 3D geometry and a view-dependent texture mapping scheme to texture the model. This 3D recovery method does not map well to our objectives it relies on human editing.

For the purpose of only re-rendering the event, the image-based rendering approach [14] has been studied extensively. Katayama et. al [15] demonstrated that images from a dense set of viewing positions on a plane can be directly used to generate images for arbitrary viewing positions without the need for correspondences. Levoy and Hanrahan [16] and Gortler et al. [17] extend this concept to construct a four-dimensional field representing all light rays passing through a 3D surface. New view generation is posed as computing a 2D cross section of this field. These approaches require a very large number (typically thousands) of real images to model the scene faithfully, making extension to dynamic scene modeling impractical.

View transform exploits correspondences between images to project pixels in real images into a virtual image plane [18]. View interpolation [19] [20] and view morphing [21] interpolate the correspondences, or flow vectors, to predict intermediate viewpoints. By computing a form of camera calibration directly from the correspondences, these views are guaranteed to be geometrically correct [21]. Virtual objects can be added into the newly rendered images in a visually consistent manner. However, these solely image- or view-based approaches do not construct an explicit representation of the event, and thus the event itself cannot be manipulated in the computer.

The only large-scale attempt to model dynamic events other than our own is Immersive Video [22]. The stationary parts of the dynamic environment are modeled off-line by hand. The dynamic parts of the event are identified in each image by subtracting an image of the background from each input image. The resulting "motion" masks are intersected using an algorithm similar to standard shape-from-silhouette methods, resulting in a volumetric model. The final colored model is acquired by back-projecting the input images onto the geometric model. Because shape is recovered from silhouettes, the final model will be unable to identify cavities in the real scene, which places restrictions on the types of objects that can be modeled.

3.4   Virtualized Reality Studio: From Analog "3D Dome" to Digital "3D Room"

Figure 3.2 (a) is a picture of the 3D Dome - our first virtualized reality studio facility. A 5-meter diameter geodesic dome was equipped with 51 cameras placed at nodes and the centers of the bars of the dome. They provided viewpoints all around the scene. Color cameras with a 3.6mm lens were used for achieving a wide view (about 90o horizontally). All of the cameras were synchronized with a single common sync signal, so that the images taken at the same time instant from different cameras correspond to the same scene. Due to cost, the 3D Dome took the strategy of real time recording and off-line digitization. Each of 51 camera outputs is recorded by each of 51 consumer-grade VCRs. Every field of each camera's video is time stamped with a common Vertical Interval Time Code (VITC). The tapes are digitized individually off-line under the control of a computer program; the computer can identify and capture individual fields of video using the time code.

Recently we have upgraded the studio setup to the "3D Room" - a fully digital system that can digitize all of the video signal in real time while an event occurs [8]. As shown in Figure 3.2 (b), a large number of cameras (at this moment 49 of them) are mounted on the walls and ceiling of the 20 feet × 20 feet × 9 feet room, all of which are synchronized with a common signal. A PC-cluster computer system (consisting of 17 PCs at this moment) can capture all the video signals from the cameras simultaneously in real time in the 4-2-2 format as uncompressed and unlossy full frame images with color (640 × 480 × 2 × 30 byte per seconds).

The processing results contained in this paper, however, are all from the data captured by the old 3D Dome.

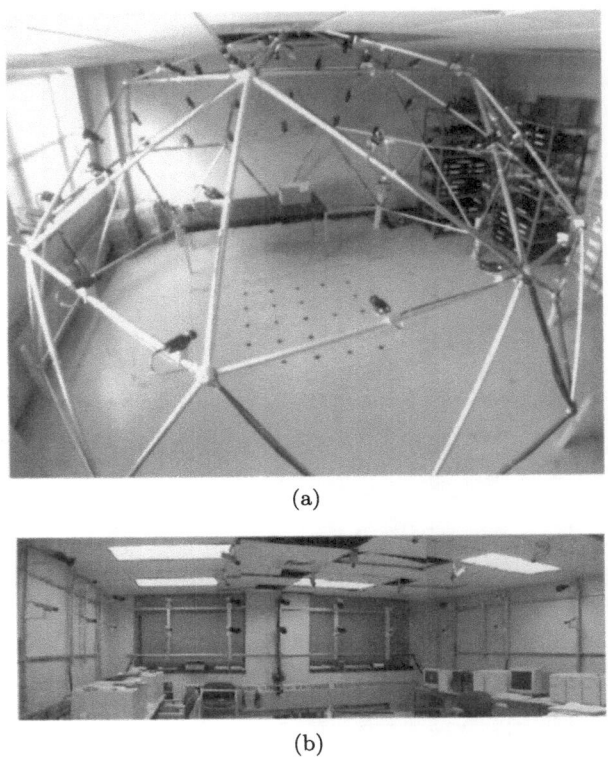

(a)

(b)

Figure 3.2   Studios for Virtualized Reality: (a) 3D dome; (b) 3D room.

3.5   Creation of Three-Dimensional Model

3.5.1   Overview of Processing

Figure 3.3 shows an overview of the current processing method for Virtualized Reality. The input images are passed through an image-based stereo algorithm, generating a range image (a) corresponding to each camera at each time instant. A volumetric integration technique is then applied to recover a single global model by extracting the polygonal approximation to the iso-surface (b) for each time instant. Each model is then reprojected into each camera to create a new range image (c). These range images have much fewer errors than those originally computed by stereo. The stereo process is repeated to obtain the refined range estimate (d) (Section 3.5.5). These are then volumetrically integrated again. Before extracting the final model (f), free space is carved out using foreground-background masks (e).

3.5.2   Stereo Image Matching for Range Image Creation

Stereo algorithms compute depth using triangulation. Strong calibration of the cameras allows computation of the depth in a Euclidean ("world") coordinate system for

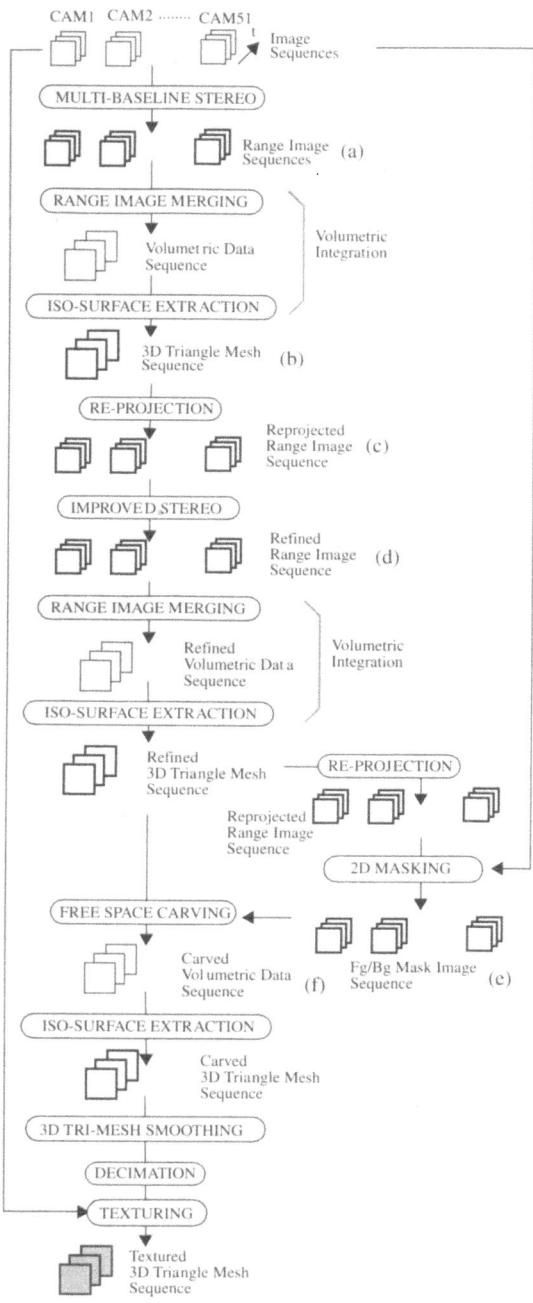

Figure 3.3   Overview of the processing method of creating a Virtualized Reality model.

each pixel, given its correspondence(s) in the other image(s). We adapted the multi-baseline stereo algorithm (MBS) [23] for a general, non-parallel camera configuration by incorporating the Tsai camera model [24] The choice of MBS was motivated primarily by two factors. First, MBS recovers dense depth maps - a depth estimate for every pixel in the intensity image. Second, MBS can take advantage of the large number of cameras to improve the depth estimates.

Basic stereo

We begin by considering a simple stereo system with two parallel cameras. Assume that both cameras are pinhole projectors with the same focal length . The second camera is laterally shifted down the negative X axis of the first camera by distance , which is referred to as the baseline between the two cameras. The goal of stereo is to estimate the depth to a scene point corresponding to each pixel in the reference camera. This requires the determination of a corresponding image point in the second camera for each pixel in the reference camera. For any pair of corresponding points, the difference between these image points gives the disparity :

Stereo searches for the corresponding points along the epipolar line, and selects the disparity yielding the best match. Increasing the baseline amplifies the disparity between corresponding image points, giving better precision for depth recovery. Increasing the baseline, however, also increases the difficulty of finding correct matching points. MBS retains the advantages of a large baseline while reducing the chances of incorrect matches.

Adaptation to MBS

Multi-baseline stereo attempts to improve matching by computing correspondences between multiple pairs of images, each with a different baseline. Since disparities are meaningful only for each pair of images, we rewrite the above equation to derive a parameter that can relate correspondences across multiple image pairs:

The search for correspondences can now be performed with respect to the inverse depth , which has the same meaning for all image pairs with the same reference camera, independent of disparities and baselines. The resulting correspondence search combines the correct correspondence of narrow baselines with the precision of wider baselines.

Robust correspondence detection

A most typical method to compute correspondences between a pair of images is to compare a small window of pixels from one image to corresponding windows in the other image. The matching process for a pair of images involves shifting this window as a function of and computing the degree of matching, usually with normalized correlation or sum of squared differences over the window at each position. The MBS does this computation for all the image pairs, adds the resulting matching scores for each , and finds the value which shows the best match. We use normalized correlation, which is less immune to the image noises due to viewing angles or camera's non-uniformity.

Figure 3.4 (b) shows an example depth map obtained by MBS for a scene shown in Figure 3.4 (a). The farther points in the depth map appear brighter. We apply the MBS stereo to compute a depth map for each of 51 camera views. In doing so, 3 to 6 neighboring cameras provide the baselines required for MBS. Because matching becomes increasingly more difficult as the baseline increases, adding more cameras may not necessarily improve the quality of the depth map.

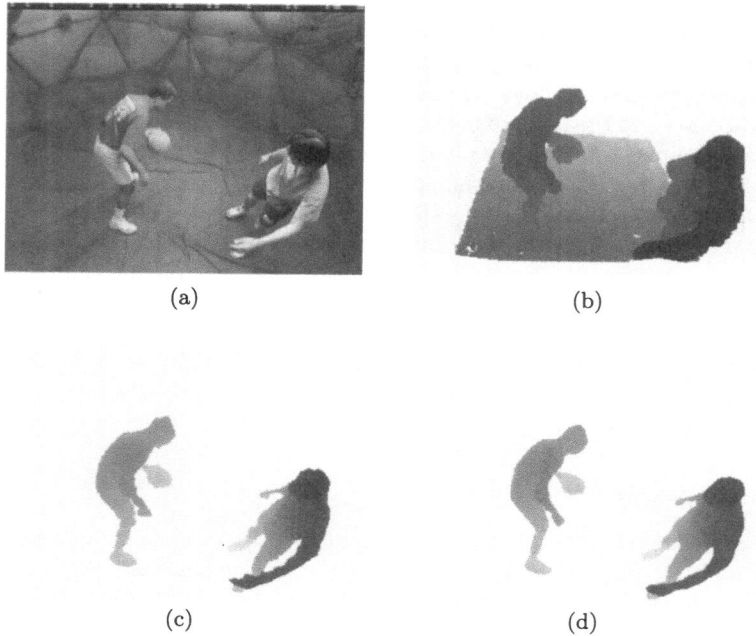

Figure 3.4   (a) Original image (b) Range image from stereo (c) Reprojected range image from first volumetric integration (d) Reprojected from the second volumetric integration.

Any window-based stereo matching has two problems. One is that in image regions with low texture, the depth estimate will have low reliability. A more serious problem is the phenomena of fattening or thinning an object along its boundaries. At the object boundary, a window contains both foreground or background surfaces, for which disparities are different. As a result, the "best" disparity found could be either that of background, foreground, or in-between, depending the strength of their texture patterns. Thus the foreground object becomes either fattened into the background or eaten by the background. Because of these problems, it is unavoidable that the resultant depth map using only stereo includes errors in depth, especially at or near depth discontinuity. Section 3.5.5 will discuss improvements of the MBS stereo results.

3.5.3   Volumetric Merging Depth Maps

The initial range images are then merged, to recover a single volumetric model of each scene. The problem of merging multiple range estimates into a single object model has received much attention recently, but mainly working with high quality range images [9] [10] [25] [26]. The depth maps generated by stereo under normal lighting conditions (i.e., no structured lighting or special textures) suffer from problems inherent in window-based correlation, as described above. Fusion of the multiple depth maps of the same scene can improve the precision of surface localization. We adapted a volumetric integration method, proposed by Curless and Levoy [10], to construct a global 3D surface model of the scene [6].

The scene space to be modeled is divided into small voxels. Each voxel accumulates the signed distance to the surfaces in the range images. To add a range image into the volume, the image is first converted into a set of triangular surfaces by tessellating the image, connecting each pixel to its neighbors. If neighboring pixels have a large difference in depth, no tessellation occurs between the pixels. A weight may also be attached to each range estimate, allowing incorporation of range estimate reliability into the fusion process. Next, each voxel is projected onto the tessellated surface along the line of sight of the sensor providing the current depth map. From this projection, the signed distance from the surface (positive for front and negative for rear, for example) to the voxel is computed, and it is added to the voxel value. The process is repeated for each voxel. After accumulating across all range images, the voxels implicitly represent the surface by the zero crossings of their values. Extraction of an implicit surface, or iso-surface, is done by the marching cubes algorithm [27] [28], which generates 3D triangle meshes representing the implicit surfaces.

In order to overcome the false surfaces (fattening or thinning) generated by stereo, we made one noteworthy change to the original Curless and Levoy algorithm. Instead of limiting the extent of contributions by each tessellated surface only to the neighboring voxels, we allowed the algorithm to adjust the values of all voxels which lies in front of the surface as viewed from the sensor that has generated this surface. At the same time, for voxels far in front of the surface, we clamp the weighted, signed distance contribution of each viewpoint so that this single view does not overwhelm all others in the fusion process. This modification gives significant improvement in the ability of the algorithm to reject the numerous outliers in our range images, while not significantly degrading the recovered shape [6].

3.5.4   Decimation and Texturing

One drawback of the volumetric merging algorithm is the large number of triangles in the resulting models. For example, fusing range images of our dome itself, with no foreground objects, at the 1-cm resolution, created a model with 1,000,000 triangles. The number of triangles in the model is directly related to the resolution of the voxel space, so increasing the voxel resolution will increase the number of triangles in the final model. To reduce the number of triangles, we apply an edge-collapse decimation algorithm [29]. Typically we obtain a reduction of 20:1 without visible degradation of mesh quality.

3.5.5   Stereo Improvement and Foreground/Background Separation

By projecting the triangle mesh obtained by depth map merging into the depth buffers of the original cameras, we obtain an approximate range image, with values for depth at all the foreground pixels. Using this approximate value as much tighter bounds on the range of depth values for the pixel, we perform a second round of stereo matching. This process can significantly improve the stereo results. The improved range images from this iteration are again merged to yield a more accurate 3-D model.

Figure 3.4 shows the results by this refinement process. Compared with the initial range image (Figure 3.4 (b)), one can observe progressive improvements in the reprojected range image from volumetric integration (Figure 3.4 (c)) and the final model obtained from the second round of merging.

The volumetric integration process creates a 3D triangle mesh representing the surface geometry in the scene. To complete the surface model, a texture map is constructed by projecting each intensity (or color) image onto the model and accumulating the results. A simple approach would be to average the intensity from all images in which a given surface triangle is visible, but a more sophisticated method is used to learn a view-dependent texture map. At the time of re-rendering, the contributions of multiple views are weighed so that the most "direct" views dominate the texture computation, and the multiple views are super-resolved for improving the image quality [1].

Figure 3.5 shows a comparison of image quality between an original image (a) and the images rendered from the constructed model (b)-(d). Figure 3.6 (b) shows a rendered textured image from a position close to the original camera. Figures (c) and (d) show rendered images from virtual viewpoints far away from the real camera. We see that good shape and texture recovery ensure that the quality of the rendered images closely match that on the original image, both when the virtual viewpoint is near an original image, and when it is away from it. Note that the original images have been digitized only with a half resolution of NTSC (512×256) in this example.

3.6   Combining Multiple Events

There is often a desire to combine many different kinds of events that occurred separately in space and time. Also, a large environment may be created by modeling its individual components separately. Virtualized Reality allows such spatio-temporal integration of event models that are created separately.

Each virtualized reality model is typically represented by a triangle mesh, with a list of vertex coordinates, texture coordinates, and polygon connectivities. The vertex coordinates of each such model are defined independently with respect to a local coordinate system. To combine different models, a simple rotation and translation is applied to the vertex coordinates alone of each triangle mesh, so that each local origin is mapped to the location of the world origin with the desired orientation.

A little more attention needs to be paid to integrating the temporal components.

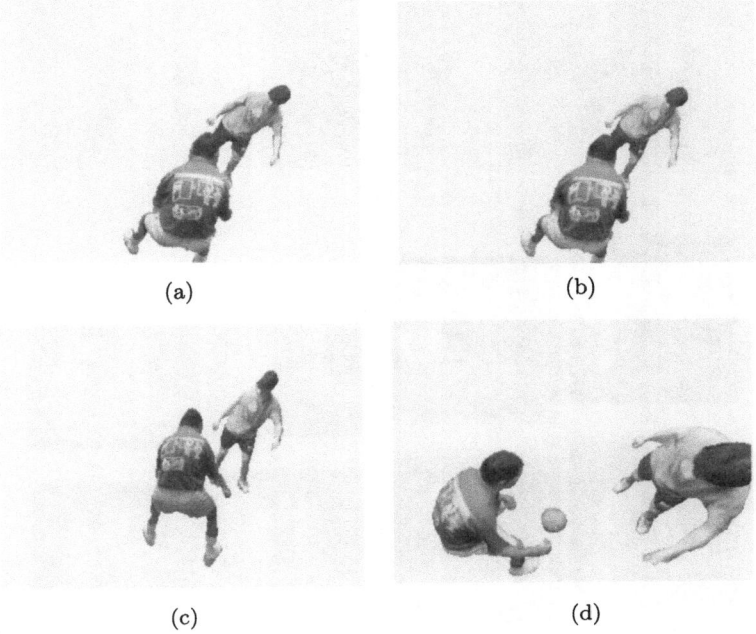

Figure 3.5 (a) Original image (masked); (b) Virtual image rendered from the same view point; (c) and (d) Virtual images from other view points.

Firstly, if one or more sequences are not modeled at the frame rate of the desired virtual sequence, those sequences would need to be subsampled or supersampled. Or in some cases, the event is used in a reverse order, as in the example shown later. Once this is done, each time frame on component event sequence is mapped to a frame on the global time scale.

In addition, since our models are a metric description of the real event, they can be also integrated with traditiona: VR models, such as those often produced by CAD programs.

3.7　Examples

"Basketball One-on-One"
Figure 3.6 shows an example where a single 10-second event was digitized. During the event, two people move and a ball is passed from one person to the other. Figure 3.6 (a) shows two frames of the input image sequence. Figure (b) shows the time-varying 3D model of the event; the video sequence of this model (shown at the presentation) clearly shows that detailed motions, such as motion of the pony tail hair of the woman and flapping of man's suit, are correctly and faithfully digitized.

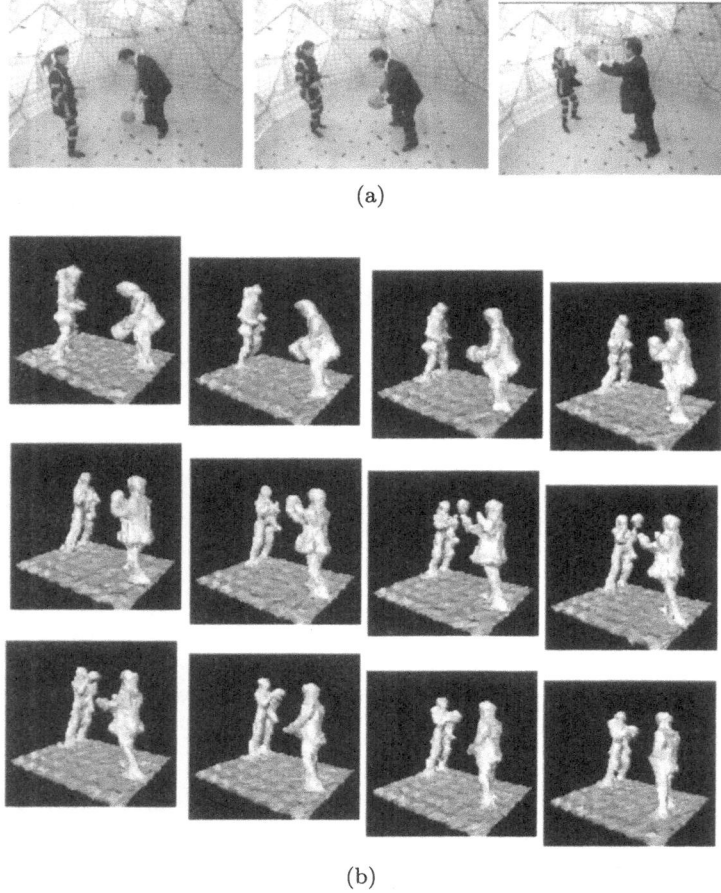

(a)

(b)

Figure 3.6   The Virtualized Reality model of "basketball one-on-one" event. (a) input images; (b) a time-varying 3D model.

"Three-man Basketball"

In our example, two separate events are recorded. The first event involves two players, where one player bounces a basketball and passes it off to the side while the other attempts to block the pass. The second event involves a single player who receives a basketball and dribbles the ball. Since we are free to choose frames in reverse, we actually record the second event by having the player throw the ball to the side. Both these events are recorded separately, so no camera ever sees all three players at once. The volumetric models for the first time frame for each of these models are shown in Figures 3.7 (a) and (b), respectively. The aim is to combine the events, so that the final model contains a motion sequence where the first player passes the ball to the third player, as the second player attempts to block this pass.

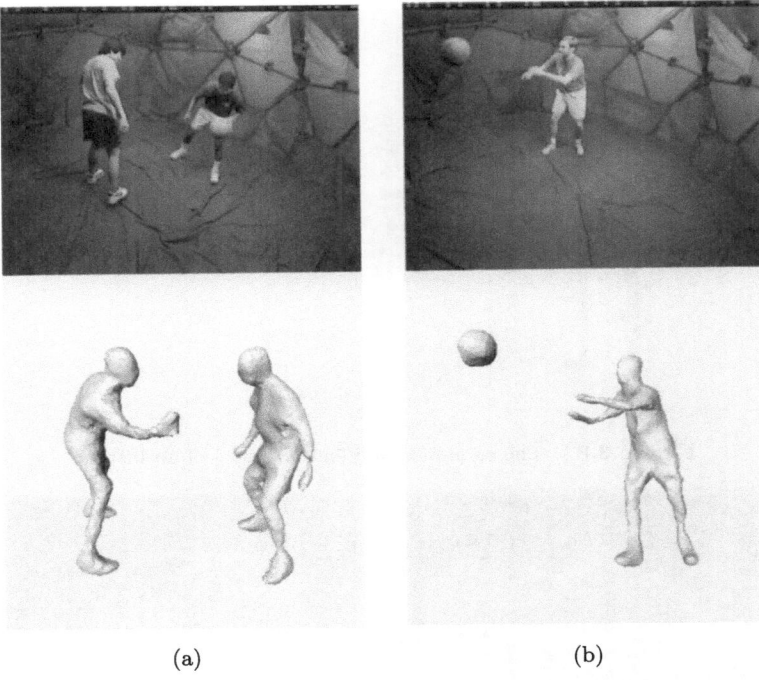

(a)   (b)

Figure 3.7   The volumetric model: (a) one of the frame instant of the first event (b) the second event (see color pages).

The spatial transform is done so that the ball at the end of the last frame of the first event coincides with the ball at the beginning of the second event. The constructed models for the last frame of the first event and the first frame of the second event are modified to remove the ball (this is a simple operation since the ball is physically disjoint from the rest of the scene) and this is used for time instances of the first event after the ball has left the scene, i.e. while it is being dribbled by the third player.

Figure 3.8 shows the volumetric model obtained by combining both events. The spatial and temporal smoothness of the volumetric models generated can be seen in Figure 3.9, which shows a time-lapse volumetric model over all time frames, that is how the space is occupied by the combined event.

The virtualized reality model of the combined event is placed inside a virtual basketball court. Figure 3.10 shows a several images of a flythrough of the event, as the camera flies over and in-between the players - the non-existent views in the original input sequence. In this case, the virtual model is static in time, but it is easy for the virtualized model to be integrated with a time-varying virtual model.

Figure 3.8   The combined volumetric model of an instant.

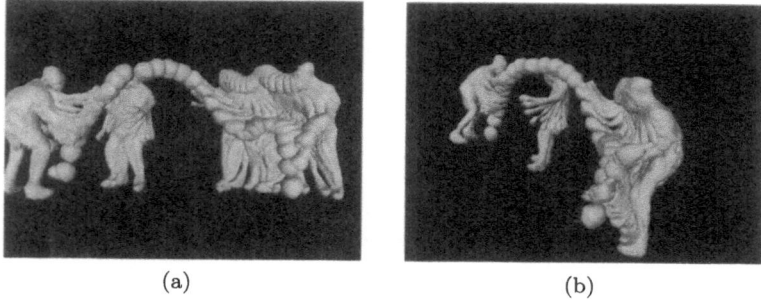

(a)   (b)

Figure 3.9   Time lapse volumetric model of the entire sequence of the combined event. (a) and (b) display the model from two angles.

3.8   Conclusions

Our Virtualized Reality system provides a significant new capability in creating virtual models of time-varying large-scale events involving free-form and large objects such as humans, using multiple video streams. In addition to the modeling, we have the capability to produce synthetic video from a varying virtual viewpoint. There are no restrictions on the positions from which virtual views can be synthesized. The system is complete to go from captured real image sequences to virtual image sequence, with no human input required with regard to knowledge or structure of the scene. In addition, it provides the capability to integrate two or more independent motion sequences with each other, or with an existing static or dynamic VR model.

The model creation process described in this paper is a two-stage process: stereo matching for 2-1/2 D surface extraction from each view and merging them into a single representation. After all, the quality of the results is strictly dependent on

Figure 3.10   Images of a fly-by of the combined event on a basketball court. The virtual camera trajectory includes a spiral motion going up and down around the players and moving in-between the players (see color pages).

the quality of correspondence. Currently we are working on a method to create a 3D model directly into a voxel representation by fusing multiple images in a manner similar to voxel coloring by Seitz and Dyer [30]. Also, it appears that this methods allows us to exploit multi-image information more systematically, as well as allowing the image rendering process into the voxel representation.

References

[1] T. Kanade, P. J. Narayanan, and P. W. Rander: "Virtualized Reality: Concepts and early results," *IEEE Workshop on the Representation of Visual Scenes*, Boston, pp.69–76, Jun. 1995.

[2] T. Kanade, P. J. Narayanan, and P. W. Rander: "Virtualized Reality: Being mobile in a visual scene," *Int'l Conf. on Artificial Reality and Tele-Existence and Conf. on Virtual Reality Software and Technology*, Japan, Nov. 1995.

[3] T. Kanade, A. Yoshida, K. Oda, H. Kano, and M. Tanaka: "A stereo machine for video-rate dense depth mapping and its new applications," *Proc. IEEE CVPR'96*, San Francisco, CA, Jun. 1996.

[4] P. J. Narayanan, P. W. Rander, and T. Kanade: "Synchronizing and capturing every frame from multiple cameras," Robotics Institute Technical Report, CMU-RI-TR-95-25, Carnegie Mellon Univ., 1995.

[5] P. J. Narayanan, P. W. Rander, and T. Kanade: "Constructing virtual worlds using dense stereo," *Proc. Sixth Int'l Conf. on Computer Vision*, Bombay, India, pp.3–10, Jan. 1998.

[6] P. W. Rander, P. J. Narayanan, and T. Kanade: "Recovery of dynamic scene structure from multiple image sequences," *Int'l Conf. on Multisensor Fusion and Integration for Intelligent Systems*, Washington, D.C., pp.305–312, Dec. 1996.

[7] P. W. Rander, P. J. Narayanan, and T. Kanade: "Virtualized Reality: An immersive visual medium," *IEEE Multimedia*, vol.4, no.1, pp.34–47, Jun. 1997.

[8] T. Kanade, H. Saito, and S. Vedula: "The 3D room: Digitizing time-varying 3D events by synchronized multiple video streams," Technical Report, CMU-RI-TR-98-34, Robotics Institute, Carnegie Mellon Univ., Pittsburgh, Dec. 1998.

[9] A. Hilton, J. Stoddart, J. Illingworth, and T. Windeatt: "Reliable surface reconstruction from multiple range images," *Proc. ECCV'96*, pp.117–126, Apr. 1996.

[10] B. Curless and M. Levoy: "A volumetric method for building complex models from range images," *Proc. SIGGRAPH'96*, Aug. 1996.

[11] M. D. Wheeler, Y. Sato, and K. Ikeuchi: "Consensus surfaces for modeling 3D objects from multiple range images," *Proc. DARPA Image Understanding Workshop*, 1997.

[12] Y. Sato, M. D. Wheeler, and K. Ikeuchi: "Object shape and reflectance modeling from observation," *Proc. SIGGRAPH'97*, pp.379–388, 1997.

[13] P. Debevec, C. Taylor, and J. Malik: "Modeling and rendering architecture from photographs: A hybrid geometry- and image-based approach," *Proc. SIGGRAPH'96*, Aug. 1996.

[14] L. McMillan and G. Bishop: "Plenoptic modeling: An image-based rendering system," *Proc. SIGGRAPH'95*, Los Angeles, 1995.

[15] A. Katayama, K. Tanaka, T. Oshino, and H. Tamura: "A viewpoint dependent stereoscopic display using interpolation of multi-viewpoint images," *Proc. SPIE: Stereoscopic Displays and Virtual Reality Systems II*, vol.2409, pp.11–20, 1995.

[16] M. Levoy and P. Hanrahan: "Light field rendering," *Proc. SIGGRAPH'96*, Aug. 1996.

[17] S. J. Gortler, R. Grzeszczuk, R. Szeliski, and M. F. Cohen: "The Lumigraph," *Proc. SIGGRAPH'96*, Aug. 1996.

[18] S. Laveau and O. Faugeras: "3-D scene representation as a collection of images," *Proc. ICPR'94*, 1994.

[19] E. Chen and L. Williams: "View interpolation for image synthesis," *Proc. SIGGRAPH'93*, 1993.

[20] S. Vedula, P. W. Rander, H. Saito, and T. Kanade: "Modeling, combining, and rendering dynamic real-world events from image sequences," *Proc. 4th Conf. on Virtual Systems and Multimedia*, vol.1, pp.326–332, Gifu Japan, Nov. 1998.

[21] S. M. Seitz and C. R. Dyer: "View morphing," *Proc. SIGGRAPH'96*, pp.21–30, 1996.

[22] R. Jain and K. Wakimoto: "Multiple perspective interactive video," *Proc. IEEE Conf. on Multimedia Systems*, May 1995.

[23] M. Okutomi and T. Kanade: "A multiple-baseline stereo," *IEEE Trans. on Pattern Analysis and Machine Intelligence*, vol.15, no.4, pp.353–363, 1993.

[24] R. Tsai: "A versatile camera calibration technique for high-accuracy 3D machine vision metrology using off-the-shelf tv cameras and lenses," *IEEE J. Robotics and Automation*, vol.RA-3, no.4, pp.323–344, 1987.

[25] H. Hoppe, T. DeRose, T. Duchamp, M. Halstead, H. Jin, J. McDonald, J. Schweitzer, and W. Stuetzle: "Piecewise smooth surface reconstruction," *Proc. SIGGRAPH'94*, 1994.

[26] G. Turk and M. Levoy: "Zippered polygon meshes from range images," *Proc. SIGGRAPH'94*, Jul. 1994.

[27] J. Bloomenthal: "An implicit surface polygonizer," in (P. Heckbert, ed.) *Graphics Gems IV*, pp.324–349, 1994. (ftp://ftp-graphics.stanford.edu/pub/Graphics/GraphicsGems/GemsIV/GGemsIV.tar.Z).

[28] W. Lorensen and H. Cline: "Marching cubes: A high resolution 3D surface construction algorithm," *Proc. SIGGRAPH'87*, pp.163–170, Jul. 1987.

[29] A. Johnson: "Control of mesh resolution for 3D computer vision," Robotics Institute Technical Report, CMU-RI-TR-96-20, Carnegie Mellon Univ., 1996.

[30] S. M. Seitz and C. R. Dyer: "Photorealistic scene reconstruction by voxel coloring," *Proc. Computer Vision and Pattern Recognition Conf.*, pp.1067–1073, 1997.

[31] S. B. Kang and R. Szeliski: "3-D scene data recovery using omnidirectional multibaseline stereo," *Proc. IEEE CVPR'96*, San Francisco, CA, Jun. 1996.

[32] T. Werner, R. D. Hersch, and V. Hlavac: "Rendering real-world objects using view interpolation," *IEEE Int'l Conf. on Computer Vision*, Boston, 1995.

Chapter 4

Steps Toward Seamless Mixed Reality

Hideyuki Tamura
Hiroyuki Yamamoto
Akihiro Katayama

Mixed Reality Systems Laboratory, Japan

4.1 Introduction

It was 1984 when William Gibson used a term *cyberspace* in his SF novel "Neuro-mancer" [1] and described a space in which an observer could experience and see visualized data in the globally inter-linked computer network. Five years later in 1989, a company called VPL realized a system in which an observer could actually experience Gibson's cyberspace. They named this technology virtual reality (VR), and it casts a great impact upon people since an observer wearing a stereoscopic head mounted display (HMD) can interact with a virtual space created in a computer in real time.

Now, we can present a far more realistic cyberspace by applying the technical innovations of computers in this decade, especially innovations related to computing power and the resolution of liquid crystal displays. However, reality in this synthetic world is limited by nature. Because of this limitation, people have begun to positively utilize rich information in the real world.

We have been participating in the "Key-Technology Research Project on Mixed Reality Systems" (MR project) in Japan. At the planning stage of this project, it was our obligation to build an innovative information technology and a human interface technology that could be pragmatically utilized in the first decade of the 21st century while going beyond the limitations of traditional VR technology. At that time, people were already using the term "augmented reality" (AR) in reference to a concept to explain augmentation of the real world with electronic information using the power of a computer. The concept of AR is the antithesis of the closed world of virtual spaces. Some problems were already pointed out such as the emotional impediments created by being absorbed in a virtual world, and the physiological influence of HMD covering the entire view field of the observer. The AR using a see-through HMD was evaluated as having the potential of solving these problems since an observer could see the surrounding space through the HMD.

On the other hand, there was a trend toward the effective use of cyberspace. Note that the rapid growth of the Internet made the information space much greater than the general population. Cyberspace on the Internet is neither a scientific calculation result nor a fantastic illusion produced by imagination or hallucination. It is a place where one can perform realistic business or electronically augment such business. Popularization of this type of cyberspace gradually requires a virtual world that can be used without awareness of the border between the real and virtual worlds. It is also obvious that as the data transmission bandwidths of the computer network become wider, the quality of visualization of this type of cyberspace improves.

Given this, we have adopted Paul Milgram's "Mixed Reality" (MR) [2] as a theme of our project. His MR includes "augmented virtuality" (AV), the counterpart of AR, that enhances or augments the virtual environment with raw data from the real world (of course, AR is a subset of MR). He considers that AR and AV are continuous(Figure 4.1). By adopting the relatively broader concept of MR, the goal of our project has been set to develop a technology seamlessly merging the real and virtual worlds.

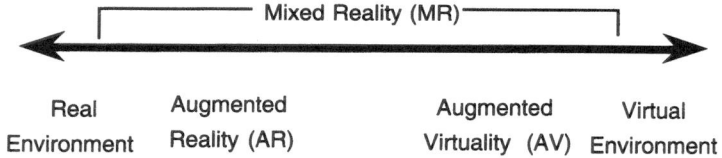

Figure 4.1   Reality-virtuality continuum.

Merging or integration of the real and virtual worlds should not be considered from the point of augmentation, which makes one world primary and the other secondary, but rather should be considered from the point of mixture as in MR technology.

Our Mixed Reality Systems Laboratory Inc. was established to conduct this project in January 1997 using funds provided by the Japanese government and Canon Inc. This national project is relatively short-term one and will be extended to March 2001 with the collaboration of three universities in Japan, the Univ. of Tokyo (Prof. M. Hirose), the Univ. of Tsukuba (Prof. Y. Ohta), and Hokkaido Univ. (Prof. T. Ifukube).

In this paper we first describe the seamless continuum between AR and AV, as well as the seamless fusion of physical and virtual spaces. Then we describe the steps toward seamless MR space by clarifying the problems and potential of MR technology through our research project.

4.2 Outline of the MR Project

4.2.1 Steps Between Augmented Reality and Augmented Virtuality

P. Milgram has pointed out six classes of displays for MR [2]. In our MR project, we have set four steps (classes) from AR to AV as shown below.

Class A : Optical see-through MR

The most prominent AR method invclves seeing the real world through optical see-through glasses on which computer generated images are superimposed. In this system, the two worlds are merged on the retinas of an observer.

Class B : Video see-through MR

This method uses TV cameras attached to an HMD to take scenes appearing to an observer, usually through the naked eye, into the system and electrically merges the video images of the real world and virtual images. The observer feels that the images from the two different worlds are merged evenly.

Class C : On-line tele-presence

By extending the concept of video see-through, a system can provide video images transmitted from a remote site into the HMD. This system can produce an MR world by merging computer generated images into the video images from a remote stereoscopic camera that is synchronized with the tracked head action of an observer.

Class D : Off-line tele-presence

With a little modification of the idea of Class C, it becomes possible to create a system in which images are reconstructed from an image database with various pre-recorded images, eliminating the need to capture remote scenes in real time. The system of this class is developed by simply adding a method to provide images according to the viewpoint of an observer. This kind of method must not be classified as AR, but rather as AV. The system, however, which merges computer generated images into reconstructed realistic images, can be thought of as an MR system.

The four classes can be expressed shown in Table 4.1. It is apparent that the characteristics or properties change gradually from class A to D.

The order of the classes is decided on the basis of how easy it is to figure out; from the simplest optical see-through to a fully virtualized environment. However, this

Table 4.1   Types of MR representation.

A. optical see-through MR	B. video see-through MR	C. on-line tele-presence	D. off-line tele-presence
optically merged	electrically merged		
on site		remote site	
on-line			off-line

order does not imply that any particular class is better or more evolved. Application purpose or usage is what ultimately determines which class of MR system should be used. We believe that many subclasses will be necessary if research is to advance. If research becomes diversified, it may require 2D or 3D mapping to describe the overall figure of MR technology.

4.2.2   Research Themes of the MR Project

The research themes of the MR project, which has been founded by the Ministry of International Trade and Industry, are officially classified as shown below.

(1) Technologies for merging the real and virtual worlds

- To develop technologies for building a mixed environment model from the geometric and radiometric structures of the real world, using 3D image measurement and computer vision
- To develop technologies that enable the seamless and real-time registration of physical space and cyberspace
- To totally evaluate a mixed reality system integrated with 3D image display

(2) 3D image display technologies

- To develop a compact and lightweight head-mounted display, with the aim of achieving a mixed reality system that incorporates state-of-the-art optics design theory
- To develop a high-luminance and wide-angle 3D display without eyeglasses
- To establish methods to quantitatively measure, evaluate, and analyze the impact of 3D display on people, as well as to obtain physiological data to prevent and minimize hazardous effects (such results will be fed back into the design of displays and other equipment to develop imaging and display equipment that reflects the importance of safety and physical comfort considerations.)

One of the characteristics of this project is that it includes the development of new 3D (stereoscopic) displays, as well as research on the methodology or algorithms aimed at a "seamless MR." Research theme (1), technologies for merging real and virtual worlds, is a different type of classification of the four classes of MR given in the previous section. It was thought that the development of a new 3D display was inevitable to reproduce the merged results of the real and virtual worlds as realistically as possible. The goal of this project is not only to write a paper or to obtain a patent, but also to realize prototypes that will work in real time and that will be applicable to the pragmatic or commercial systems of 21st century. Thus it is necessary to develop new 3D image displays conforming to MR technology.

"Seamless" is the final goal or slogan of our project, although we do not really think that it is possible to perfectly fuse the two worlds. A technology or a system of technologies that can be flexibly applied according to pragmatic precision or cost requirements is preferable. The term "seamless" implies a fine balancing between opposite requirements at various levels, granulation of the predefined classes into subclasses, and continuous stacking of the technologies that conform to each subclass.

4.3 Approaches to Seamless Augmented Reality

4.3.1 Three Problems to Be Solved

First, we consider problems related to AR. An AR system based on the real world will be seamless when a virtual object can be placed in the real world without any sense of incongruity. We think the following three problems must be solved for the registration of physical and virtual spaces.

(1) **Resolution of static error:** It is necessary to ensure that the coordinates of physical space and virtual space coincide, also precisely fix a user's position and direction of view. Building up a methodology for advance calibration, we are developing a method for real time registration.

(2) **Resolution of rendering error:** Measures, which prevent rendered images of a virtual space from incongruity caused by some factors such as contrast or tone by merging with physical space, need to be taken here. In addition, it is being made seamless by estimating the position and the direction of light sources in physical space and shading.

(3) **Minimization of dynamic error:** Any method results in a time lag in rendering the virtual environment because of the movement or rotation of a user's viewpoint. Here, we are advancing the measuring speed and precision with several sensors, and consequently minimizing the time lag.

The first static error can be treated as positional misalignment in the geometric registration of two spaces. This is not only the most important problem in AR, but also important in our MR project. We, therefore, did our best to solve this problem, and this will be discussed later.

The second rendering error can be treated as a problem of image quality in photometric registration. This is also a problem in AV, which introduces raw data from the real world into the synthesized virtual world, as well as in AR. In the system of optical see-through (class A), this photometric gap is an inherent problem that can not be solved by research efforts.

Given this problem, the video see-through system is much more advantageous than the optical see-through system, since it is possible to adjust images into any level electronically (by digital processing) after capturing the real world with a video camera. At the beginning, the resolution of the liquid crystal display (LCD) used in the HMD was too low to make an observer feel that the virtual object was real. However, the quality of LCDs these days makes it possible to apply the video see-through system pragmatically.

Figure 4.2 compares the quality of images in the optical see-through system and the video see-through system. Figure 4.2 (a) is taken by a video camera placed in the HMD of the optical see-through system. Note that it is hard for an observer to recognize dark objects. It is possible to outline signs or wireframes superimposed onto the HMD by using bright graphic objects. However, it is quite hard for the optical see-through system to merge an object of natural color as shown in the figure. This problem becomes much worse if it is used outdoors in bright daylight.

(a) Optical see-through   (b) Video see-through

Figure 4.2   Optical see-through vs. video see-through (see color pages).

The video see-through image shown in Figure 4.2 (b) still has some sense of incongruity. The virtual vase placed on the real table looks as though it is floating on the table since it does not cast a shadow. In feature movies with a lot of visual effects, people can solve this problem by manually adjusting images, a process that takes a great deal of time. However, in AR or MR, it should be possible to narrow this photometric gap instantly since these systems require real-time rendering. In order to achieve this real-time rendering, the system must measure the lighting of the real world, model it in a computer, and calculate the shadow cast onto the real world by a virtual object in that lighting environment.

Ikeuchi et al.[3] and Debevec et al.[4] are now struggling with this problem. We are planning to utilize their results in our MR system.

To deal with the third problem of dynamic error or time lag, Azuma has characterized the error and summarized current efforts underway. Reduction of system lag and the prediction of future locations are some of the methods being applied. Time lag is primarily dependent on the position sensors and the geometric registration methods being used. For the optical see-through system, it is very hard to move virtual objects completely by following the motion of an observer.

The video see-through system is also advantageous in dealing with this problem since it can delay the display of images acquired from the real world until the rendering of the virtual objects is completed. In actual experiments, we noticed that the graphic rendering was faster than the video capturing if a fast graphic computer was used (we used SGI ONYX2 InfiniteReality). In any case, the timing involved in merging real and virtual images can be freely adjusted in a computer. Since a slower timing is usually required, the time lag between an observer and a merged image tends to be longer while the time lag between the real and virtual worlds tends to be minimized. This may produce so-called VR sickness.

4.3.2 Geometric Registration in AR Systems

In the optical see-through AR, the problem of positioning is synonymous with the problem of deciding the 3D position of a user's viewpoint. The viewpoint is generally measured by 3D sensors such as magnetic, ultrasonic or gyroscopic sensors. Since these sensors can not always provide the accuracy required, sensor errors cause positional misalignment.

In the video see-through AR, on the other hand, the problem of positioning is synonymous with the problem of seeking the position of a camera, and, therefore, the computer vision approach can be utilized to solve this problem. We can also utilize a method to register the images of virtual objects directly on images captured by the camera [5]. This method, however, has some problems, including the lack of reliability or time lag, although it's easy to realize accurate positioning since we can directly handle the positioning error on the image.

Some trials using both physical 3D sensors and image-based methods to realize accurate positioning have recently been reported. Bajura and Neumann [6] have suggested a method to compensate for positioning errors caused by magnetic sensors by using image information in the video see-through AR. State et al.[7] have expanded this method and have suggested a method to compensate for vagueness in positions obtained through image information by using sensor information. Since we cannot expect a perfect physical 3D sensor for the AR system at present, such sensor fusion seems to be the best approach to ensure practical geometric registration.

Our MR project is now confirming this sensor fusion through the development of working MR systems. So for, we have developed the following two AR systems.

(a) AR² Hockey: A case study of collaborative AR

We have developed the AR AiR Hockey (AR² Hockey) system as a case study of a collaborative AR for human communication. In this study, collaborative AR is a method for establishing an environment in which participants get together and collaborate while sharing physical space and cyberspace simultaneously.

Air hockey is a game in which two players hit a puck with mallets on a table, attempting to shoot it into goals. In our AR² Hockey, a puck is in virtual space. Each player wears an optical see-through HMD and hits a virtual puck on a real table. Figure 4.3 (a) shows the scene of playing AR^2 Hockey and Figure 4.3 (b) is an image seen through the HMD when the system is operating.

(a) Playing scene   (b) Augmented view(optical see-thru)

Figure 4.3   Playing scene of AR^2 Hockey.

Figure 4.4 (a) shows the typical coordinate systems used in a simple AR. Registration is the process that transforms the viewing matrix C_C. In collaborative AR, physical space and virtual space are shared by all the participants. Thus the coordinate systems C_R and C_V exist in the system, and are shared by the participants. At the same time, the coordinate systems C_C and C_D that relate to the viewing transformations exist for each participant. Figure 4.4 (b) illustrates this situation. Thus, the registration algorithm is implemented independently for each participant.

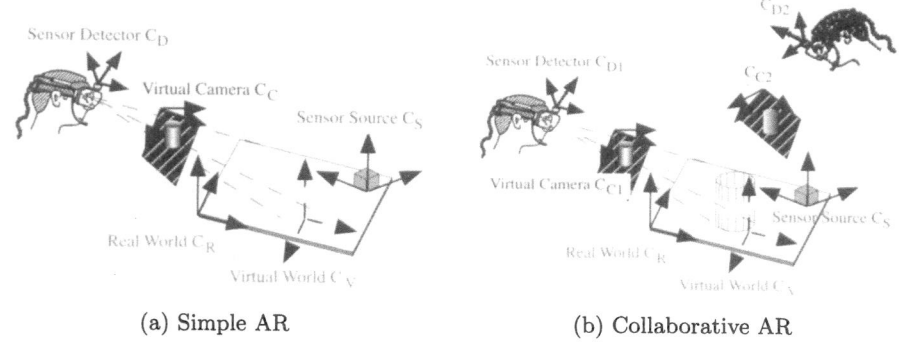

(a) Simple AR   (b) Collaborative AR

Figure 4.4   Coordinate systems in AR.

The optical see-through system is adopted to the AR^2 Hockey system so that players (observers) can easily recognize opponents by their eyes. A Polhemus sensor and a CCD camera are mounted on the HMD of the players. The CCD camera is used not to see outside through the captured video image, but rather to register the virtual object based on the captured image.

The paper [8] explains the first version of our AR^2 Hockey. This system has been modified greatly to be exhibited at the "Enhanced Reality" area of SIGGRAPH 98

[9]. We had to enhance the system throughput in order to operate the system using SGI O2 computers exclusively without any ONYX2 computer. The shape and placement of landmarks were also modified for more accurate registration.

During SIGGRAPH 98, more than 1,000 couples (2,000 players) played the new AR^2 Hockey. One of the most significant characteristics of the SIGGRAPH 98 version of AR^2 Hockey is that anyone could play without any difficulty. They did not have to be developers of this system or trained players. It may be said that the new AR^2 Hockey became the see-through AR system experienced by the largest number of users worldwide.

(b) MR Living Room: A case study of visual simulation with AR

Using the AR^2 Hockey system, we studied mainly static and dynamic registrations, that is, positional misalignment and time lag, by taking a game requiring quick motion as the subject of our research. "MR Living Room" is another experimental AR system for interior simulation. This has been developed using the knowledge obtained from the AR^2 Hockey project while taking technical problems related to image quality consistency into consideration. This section outlines this project.

The MR Living Room has a 2.8 m x 4.3 m floor made of wooden flooring staff half-equipped with a few pieces of furniture and articles. In this space, two observers with see-through HMDs can experience virtual interior simulation such as selecting and placing furniture. Figure 4.5 (a) shows the inside of the experiment space. As shown in Figure 4.5 (b), this room is half-equipped with a few pieces of physical furniture and articles. Virtual furniture and articles are merged into this physical space and presented in real time onto the HMDs. The augmented views are shown in Figure 4.2.

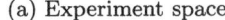

     (a) Experiment space   (b) Non-augmented view

Figure 4.5   MR Living Room.

Geometric registration in this system is performed by fusing sensor-based and image-based methods as in the case of the AR^2 Hockey system. In the MR Living Room system, the sensors for the physical tracker are an ultrasonic sensor to measure the observers' position and a gyroscopic sensor to detect the observers' direction. However, no Polhemus sensor is used because the area in which the observer can move around is much greater than in the AR^2 Hockey system. Since landmarks (fiducials) placed on the table of the AR^2 Hockey system are not as elegant as in the living room, the system uses small devices emitting infrared rays placed on such places as walls or a bookshelf. One of the two CCD cameras mounted on the HMD, shown in Figure 4.6, detects these infrared ray markers and the other is used to obtain a video signal for the video see-through image. Observers can experience this system using the optical see-through mode, although the video see-through mode is suitable for the purpose of this system since the uniformity of the image quality is quite important for this kind of simulation.

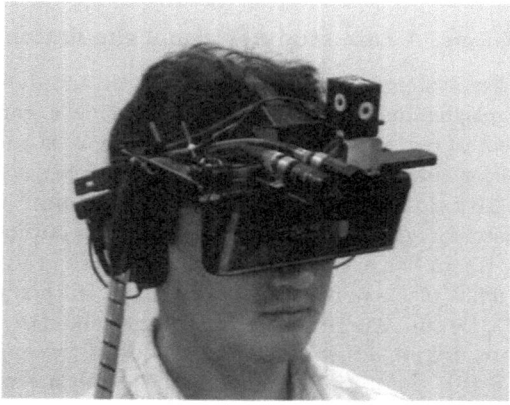

Figure 4.6   See-through HMD.

Since the system requires a greater registration area and a higher registration accuracy than the AR^2 Hockey system, we have developed a new method to detect the position and posture of the observer by using multiple landmarks. In this system, the number of observable landmarks varies depending on the viewing angle of the observer. Therefore, we have to work out an algorithm [10] to adjust misalignment adequately based on the number of observable landmarks. Figure 4.7 shows the case when two or three landmarks can be observed.

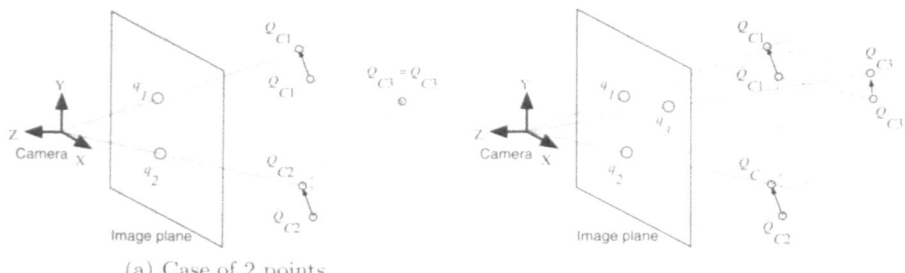

(a) Case of 2 points

Figure 4.7   Compensation of registration errors by landmarks.

This algorithm forms a single framework that can treat both the case that depends only on a physical tracker and that which depends only on landmarks as in the image-based method. It is remarkable that this algorithm can also treat the intermediate state between these two cases (Figure 4.8). It is quite important for a pragmatic system that can treat the intermediate state between two extreme methods seamlessly to cope with various occasions in the actual application.

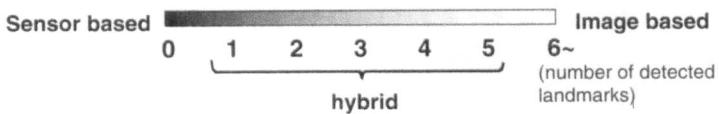

Figure 4.8   Hybrid registration.

4.4 Approaches to Seamless Augmented Virtuality

4.4.1 Background of Image-Based Rendering

A few years before starting MR project, we had studied AV. In that study we had been seeking a way to handle objects and their backgrounds having complex shapes that could not be drawn using conventional computer graphics techniques in virtual space. We tried to reconstruct a scene that coincides with the viewpoint of observers from various real images without expressing virtual environment with data based on the geometric models. At that time, Apple Computer's QuickTimeVR [11] had already been introduced, and this was capable of reproducing a scene as a photographic panorama seen from a limited viewpoint. Note that, in QuickTimeVR, the scene did not follow the motion when an observer moved.

Our goal was to find a method to reconstruct an image that produced motion parallax when an observer moved around it. The method had to reconstruct the required image from images captured by multiple cameras placed evenly on a line by interpolating images from those cameras. The problem eventually became the simpler problem of finding a straight line from an epipolar plane image (EPI) [12] . This was really a technique of computer vision or image processing. Applying this theory, we developed the Holomedia system [13] that gives an observer stereoscopic images through liquid crystal shutter glasses with a head tracker. No geometric data was used in this method. Now, such approaches are called "image-based rendering (IBR)."

By generalizing the method based on the EPI, we advanced to image-based rendering based on the "Ray Space." This method, advocated by H. Harashima and others [14], is one that produces a radiometric representation of an object as a bundle of rays which go through a certain point on a screen at a certain time. Figure 4.9 illustrates this. The theory has the same basis as the Lumigraph [15] or light field rendering [16]. All these methods perform image-based rendering from a lot of pictures captured from the real world.

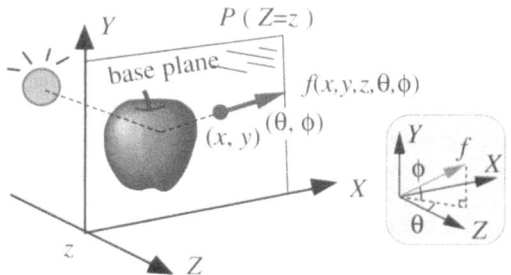

Figure 4.9   Ray space description.

4.4.2   Mixed Rendering of Model-Based Data and Ray-Based Data

The method stated above has realized a procedure to render a photo-realistic scene without describing an explicit geometric shape. Note that an image seen from any point of view can be generated just by collecting necessary raw images. Theoretically, it is proved. However, there is a actual problem involving the image acquisition method and the large amount of data.

As AR and AV are contiguous and involve many intermediate steps, fully modeled and non-modeled methods are also contiguous with various intermediate steps (Figure 4.10) [17].

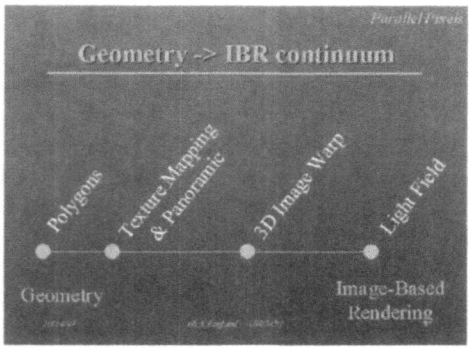

Figure 4.10   Model – non-model continuum(from [17]).

We then tried to draw an image by merging geometric model-based data and ray-based data. Finally, we were able to complete a system in which an observer could walk through MR (or AV) space that is constructed by complex objects represented by ray-based data placed in circumstances of polygon-represented graphic data. Figure 4.11 shows an example of this type of data structure. The system expanded from the VRML viewer is called CyberMirage [18] [19].

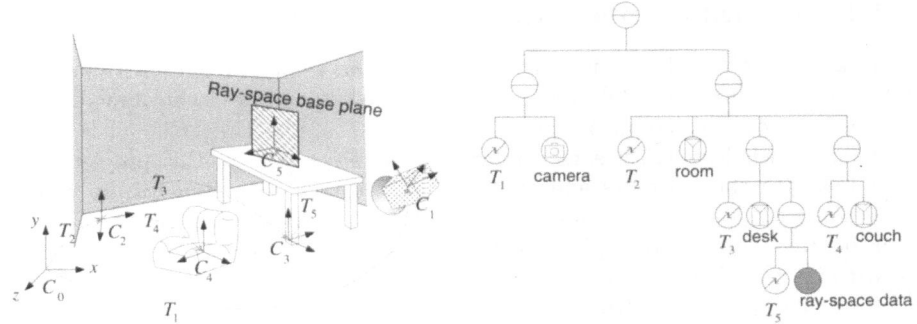

Figure 4.11   Ray-based data embedded in VRML data structure.

Collaborative CyberMirage [20] is an expanded version of the CyberMirage in which multiple remote participants can visit cyberspace on a network and communicate with each other in real time while recognizing other participants as avatars. This system was tested by linking multiple points several dozen kilometers apart from each other with lines of 6 Mbps. The research is mainly reviewed from the point of telecommunication system such as how to compress and transmit huge image-based data such as some megabytes per object.

In the MR project, we studied methods to merge the image-based data in any class of implementation from Class A to D. A series of CyberMirage systems target cybershopping at a virtual mall (Figure 4.12). We have already achieved a certain success for the photo-reality of a single object not affected by circumstances.

Figure 4.12   Virtual mall generated by mixed rendering(see color pages).

The next problem to be solved is the shading of an object placed under lighting. Shading becomes fixed when we reproduce an object from images captured under fixed lighting. In order to achieve a seamless merge, we have to find some measure to alter shading according to lighting conditions in the real world in which the object exists.

4.4.3   Toward Seamless Extension from Near to Far-Range

The further problem is to augment or enhance the whole virtual environment by applying raw images to solid objects and their background or circumstances.

We have been collaborating with Hirose Lab. of the University of Tokyo to virtualize a whole town of some hundreds meters square. In this project, a car mounted with three sensors and eight cameras is used to systematically capture images of a static town. The goal is to build a system where an observer can freely walk or drive through the cybercity generated by these images. This is the class D MR system referred to in Section 4.2.1. Progress in this project is detailed in Chapter 10 of this book [21].

Objects in front of an observer and a wide area such as the cybercity are two opposing extremes. However, we also have to think about the intermediate state. The mid-range, several meters to tens of meters from an observer, is the most difficult range to express in the virtual world.

Objects in front of an observer can be expressed using the model-based method or can be reconstructed using the image-based method. Distant landscape can be texture mapped using image-based data or can be reproduced with ray-based rendering. However, there is no adequate method for dealing with the mid-range. If you map a texture to coarse model data of, for example, a wall and a floor, observers can easily find fault with them if they come close to them. Fine model data are not practical since it takes too much time to create. The ray-based approach is also not practical because of the large amount of data required.

The challenge now is to build an AV system where an observer can seamlessly move through from the short-range to the far-range.

4.5   3D Display Technologies

4.5.1   Head Mounted Display for MR

As we mentioned, two choices are available for the augmentation of the physical world with the virtual world: optical see-through (Class A) and video see-through (Class B). Video see-through HMD works using a closed-view HMD and one or two cameras attached to the HMD. The video images from the cameras are combined with computer-generated images and the combined video is displayed on the HMD. This configuration is often used in applications where accurate registration is necessary.

On the other hand, optical see-through HMD uses optical combiners so that users can see the physical world through glasses and simultaneously look at an image displayed on the HMD monitor. Since the physical world is viewed almost directly, there is no time delay, unlike in the case of the video see-through type where at least one frame time is delayed. In addition, the resolution of the physical space is only limited by the resolution of human fovea, not by the display device of the HMD. The time lag, however, between the physical world and the virtual world is crucial.

Apart from the see-through methods described above, the current HMDs have the following issues.

(1) Although the display image size can be set independently of the device size, increasing the view angle results in an increase in the size of HMD optics as well as the entire HMD.

(2) If the number of pixels of the display device is assumed to be constant, increasing the viewing angle inevitably deteriorates the resolution of the displayed image.

Given this, we decided to adopt a prism optics system to make the HMD compact and lightweight, and to give it a high resolution [22]. The prism optics system is provided with the function of lenses by implementing the reflective surfaces of off-axial surfaces. As shown in Figure 4.13, the light emitted from the LCD is incident to the prism, totally reflected by the reflection surface 1, then reflected again by the reflection surface 2. The light is then output from the prism by being transmitted through the reflection surface 1, and led into the eye to form a virtual image. As shown here, the foldback of the light caused by the reflection surfaces makes it possible to implement a thin optical system, that is, leading to a reduction in the size and weight of the HMD.

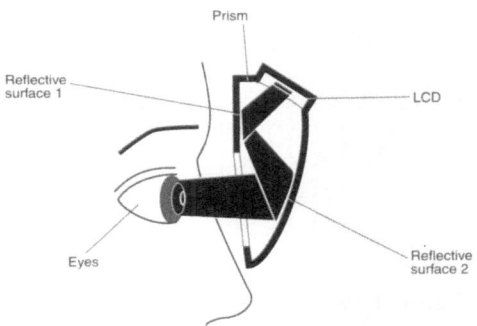

Figure 4.13   Free curve prism optics.

With this kind of prism, all reflective surfaces are off-axial so a large amount of off-axial aberration is generally produced. To reduce this off-axial aberration, we adopted 3D aspherical surfaces without rotational symmetry as the prism surfaces rather than ordinary spherical or aspherical surfaces with rotational symmetry.

The configuration shown in Figure 4.13 is not capable of implementing a display based on the optical see-through. The reason for this is that when the light is transmitted from the outside world to the eye through the prism as shown in Figure 4.14, optical refractivity is produced inside the prism, causing errors in the observation of the outside world, alteration of the magnification, and distortion. Therefore, we implemented an optical see-through HMD by incorporating another prism to compensate for these factors. This is shown in Figure 4.15. This new configuration has succeeded in providing HMD optics of almost identical thickness to the closed HMD with the configuration shown in Figure 4.13.

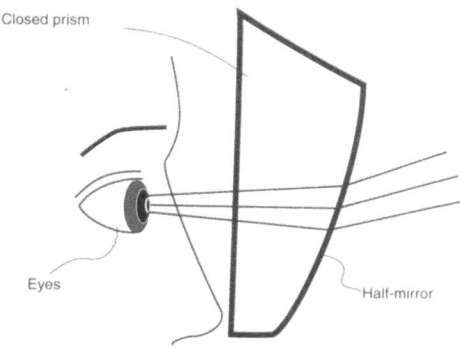

Figure 4.14   See-through problems with free curve prism.

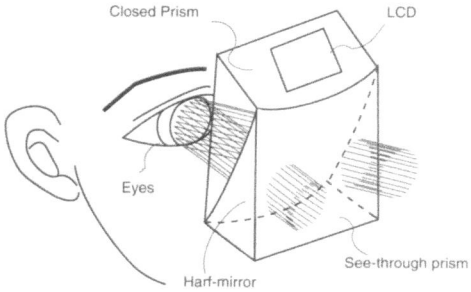

Figure 4.15   Joint free curve prism optics.

Figure 4.16 shows the HMD with the optical see-through configuration described above. This HMD utilizes two TFT LCDs of 920,000 pixels as the display unit, and can present a set of stereoscopic color images with a resolution of 640 x 480 VGA. The horizontal viewing angle is set to 51 degrees with very little distortion. The transparency rate of the outside world is also set to a high value to achieve more realistic see-through feeling.

In order to make our HMD more suitable for AR/MR, we have to embed the CCD chip into the HMD as the LCD chip instead of attaching a camera to the HMD. The problem with attaching a camera is that the optical axis of the camera may not coincide with the eye direction of the observer. The coincidence between the view direction of the observer and the optical axis of a camera is quite an important factor for the seamless AR/MR system. The development of such an HMD is also one of the goals of our MR project.

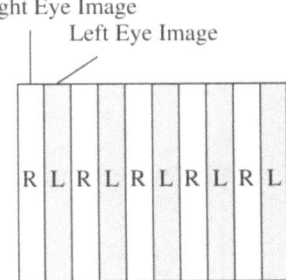

Figure 4.16   Composite images for 3D stereo viewing.

4.5.2   3D Display Without Eyeglasses

The Class C or D tele-presence system does not require a see-through function, and therefore the usual closed-view HMD or stereoscopic display with an LCD shutter or polarizing device will work. Thus, it is preferable for the system not to have bothersome eyeglasses.

The lenticular method and the parallax barrier method are two of the methods of 3D display without eyeglasses that have been known for a long time. An attempt has been made recently to implement 3D displays by combining LCDs, but these methods pose the problems described below.

(1) The arrangement of vertically striped images causes chromatic separation even in the 3D stereo-view area.

(2) Composite image data is not compatible with the field sequential method that is widely used in VR for LCD shutter glasses.

In order to solve these problems, we developed a rear cross lenticular 3D (RCL3D) display [23] [24]. This system displays a composite image consisting of alternately arranged horizontally striped left-eye/right-eye images, and enables 3D stereoscopic viewing by switching the illumination light directivity according to the left and right eyes at every scanning line. In the system, two lenticular lenses (lenticular-H and lenticular-V) and a checkered pattern mask are arranged between the TFT LC panel and the backlight as shown in Figure 4.17. The lenticular-V separates an image into the left and right eyes by alternately switching the direction of illumination from the backlight for each scanning line. The lenticular-H condenses the light from the opening of one of the horizontal lines on the checkered pattern mask to one of the scanning lines of the LC panel. This makes it possible to prevent crosstalk, as well as to expand the 3D stereo visual range in the vertical direction by turning the light through the opening into a flux dispersed toward the observer.

We have already developed prototype systems using this method. Our prototypes are 14.5 inches and 21 inches TFT LCD displays. Now, observers can see a remote scene as a 3D image through these displays while operating a pair of cameras placed on the other end of a network to adjust the viewing angle. An auto-stereoscopic

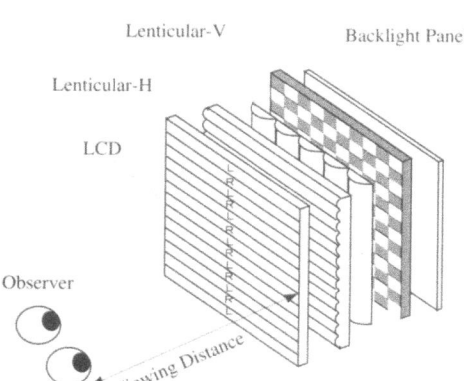

Figure 4.17   Configuration of RCL 3D stereoscopic display.

VRML viewer [25] has also been developed to see a variety of VRML data on various web sites around the world through these displays.

4.6   Concluding Remarks and Future Studies

This paper describes the concept of the MR technology in our project and provides the results to date. Even though we did not talk about it in this paper, we are also studying methodologies to measure and evaluate the influence of these two kinds of 3D displays on people.

Now, research in this project is concentrated on visually merging the real and virtual worlds. However, this does not mean that auditory or haptic stimuli are not applicable to our MR system. Though the original and innovative research in our project is concentrated on visual information, we are going to look into ways of incorporating other sensory data into our total MR system. It should be noted that the enhanced version of AR^2 Hockey has vibrators in the mallets, allowing players to feel vibrations when they hit a puck. In addition, we are going to implement a 3D sound system into the MR Living Room that generates sound from virtual equipment such as a TV set or audio set.

Finally, we would like to mention some objects of future studies, which we have not mentioned in the previous sections. AR^2 Hockey as a collaborative AR system allows only two players. New problems in the geometric registration will arise when we try to support cooperative work by four or five persons. The bandwidth of the magnetic sensor is limited, in that it can support several persons simultaneously. In order to cope with this limitation using the image-based method, a new invention is required to deal with the placement and shape of landmarks. AR^2 Hockey has demonstrated a good future as AR-type amusement. Now, the amusement business is seeking an AR game that will allow several players to play simultaneously. The key to success for such AR games lies in how to increase the number of players while decreasing the cost for each player.

In the area of static registration, our main research theme, we have only studied the case in which a virtual object is superimposed onto the real world, but not the case in which a real object occludes a virtual object. In order to merge the two worlds evenly, we have to solve the problem of mutual occlusion between real and virtual objects, which require us to have geometric models of the real world in the computational form. If the objects in the real world are static, this is easy to do because we can prepare model data for those objects. Moreover, it is also possible to take some time to determine the range of objects. However, if an object moves or the real world is dynamic, we have to develop a new range finder that can capture movement in the real world in real time. This difficult problem is also part of our research theme.

The research theme of our MR project is the development of a real-time, interactive MR system to broaden the application field of MR technology. Such a field is not limited to indoors. We think outdoor applications will give us a far broader field. A variety of wearable computers will be available on the market. This means that the power of outdoor computers will soon increase to the level we require, which means that wearable AR/MRs are feasible. It is not so difficult to build a system to support users work or to guide users by superimposing data onto a real world scene. However, the problem becomes greater if we think about the possibility of geometric or photometric registrations in a natural outdoor field. People have already started research on this difficult theme. We think that our research will naturally lead us to study this problem as well.

Acknowledgments

The authors would like to thank Mr. Susumu Matsumura, Mr. Naosato Taniguchi, Dr. Toshikazu Ohshima, Dr. Shinji Uchiyama, Dr. Kiyohide Sato, and the other members of Mixed Reality Systems Laboratory for their cooperation in this project. The Key-Technology Research Project on Mixed Reality is supported by a grant from the Japan Key-Technology Center.

References

[1] W. Gibson: *Neuromancer,* Ace Books, New York, 1984.

[2] P. Milgram and F. Kishino: "A taxonomy of mixed reality visual display," *IEICE Trans. Inf. & Sys.*, vol.E77–D, no.12, pp.1321–1329, 1994.

[3] K. Ikeuchi, Y. Sato, K. Nishino, and I. Sato: "Photometric Modeling for Mixed Reality," in (Y. Ohta & H. Tamura, eds.) *Mixed Reality– Merging Real and Virtual Worlds*, Chapter 8, Ohmsha-Springer Verlag, 1999.

[4] P. Debevec: "Rendering synthetic objects into real scenes: Bridging traditional and image-based graphics with global illumination and high dynamic range photography," *Proc. SIGGRAPH 98,* pp.189–198, 1998.

[5] K. N. Kutulakos and J. R. Vallino: "Affine object representations for calibration-free augmented reality," *Proc. VRAIS'96*, pp.25–36, 1996.

[6] M. Bajura and U. Neumann: "Dynamic registration correction in video-based augmented reality systems," *Computer Graphics and Applications*, vol.15, no.5, pp.52–60, 1995.

[7] A. State, G. Hirota, D. T. Chen, B. Garrett, and M. Livingston: "Superior augmented reality registration by integrating landmark tracking and magnetic tracking," *Proc. SIGGRAPH 96*, pp.429–438, 1996.

[8] T. Ohshima, K. Satoh, H. Yamamoto, and H. Tamura: "AR^2 Hockey: a case study of collaborative augmented reality," *Proc. VRAIS'98*, pp.268–275, 1998.

[9] T. Ohshima, K. Satoh, H. Yamamoto, and H. Tamura: "AR^2 Hockey system," *SIGGRAPH 98, Conference Abstracts and Applications*, pp.110, 1998.

[10] K. Satoh, T. Ohshima, H. Yamamoto, and H. Tamura: "Case studies of see-through augmentation in Mixed Reality Projects," *Proc. First IEEE Int'l Workshop on Augmented Reality*, 1998.

[11] S. E. Chen: "QuickTime VR - An imaged-based approach to virtual environment navigation," *Proc. SIGGRAPH 95*, pp.29–38, 1995.

[12] R. C. Bolles, H. H. Baker, and D. H. Marimont: "Epipolar-plane image analysis: An approach to determining structure from motion," *Int'l. J. Computer Vision*, vol.1, no.1, pp.7–55, 1987.

[13] A. Katayama, K. Tanaka, T. Oshino, and H. Tamura: "A viewpoint dependent stereoscopic display using interpolation of multi-viewpoint images," *Proc. SPIE*, vol.2409A, pp.11–20, 1995.

[14] T. Naemura, T. Takano, M. Kaneko, and H. Harashima: "Ray-based creation of photo-realistic virtual world," *Proc. VSMM'97*, pp.59–68, 1997.

[15] S. J. Gortler, R. Grzeszczuk, R. Szeliski, and M. F. Cohen: "The Lumigraph," *Proc. SIGGRAPH 96*, pp.43–54, 1996.

[16] M. Levoy and P. Hanrahan: "Light field rendering," *Proc. SIGGRAPH 96*, pp.31–42, 1996.

[17] http://www.cs.unc.edu/~nick/parallel_talk/sld009.htm

[18] S. Uchiyama, A. Katayama, H. Tamura, T. Naemura, M. Kaneko, and H. Harashima: "CyberMirage: Embedding ray based data in VRML world," *Video Proc. VRAIS'97*, 1997.

[19] http://www.x-zone.canon.co.jp/CyberMirage/,
     http://www.hc.t.u-tokyo.ac.jp/3D/CyberMirage/

[20] S. Uchiyama, A. Katayama, A. Kumagai, H. Tamura, T. Naemura, M. Kaneko, and H. Harashima: "Collaborative CyberMirage: A shared cyberspace with mixed reality," *Proc. VSMM'97*, pp.9–18, 1997.

[21] M. Hirose, T. Tanigawa, T. Endc: "Building a Virtual World from the Real World," in (Y. Ohta & H. Tamura, eds.) *Mixed Reality–Merging Real and Virtual Worlds*, Chapter 10, Ohmsha-Springer Verlag, 1999.

[22] H. Hoshi, N. Taniguchi, H. Morishima, T. Akiyama, S. Yamazaki, and A. Okuyama: "Off-axial HMD optical system consisting of aspherical surfaces without rotational symmetry," *Proc. SPIE,* vol.2653, pp.234–242, 1996.

[23] H. Morishima, H. Nose, N. Taniguchi, K. Inoguchi, and S. Matsumura: "Rear cross lenticular 3D display without eyeglasses," *Proc. SPIE,* vol.3295, pp.193–222, 1998.

[24] H. Morishima, H. Nose, N. Taniguchi, K. Inoguchi, and S. Matsumura: "An eyeglass-free rear-cross-lenticular 3D display," *SID Int'l Symp. Digest of Technical Papers,* vol.29, pp.923–926, 1998.

[25] S. Uchiyama, H. Yamamoto, and H. Tamura: "An auto-stereoscopic VRML viewer for 3D data on the world wide web," *Proc. VSMM'98,* vol.2, pp.566–571, 1998.

Part II

Registration and Rendering

In general, the literature on virtual reality (VR) tends to focus on the details of operational systems with relatively little theory. VR researchers are inclined to be comparatively more pragmatic than others, since one of their primary goals is to achieve real-time interaction with as much (synthesized) reality as possible. Researchers in mixed reality (MR), whose work focuses on the real world, are compelled to be equally as pragmatic.

In Part II of this book, Chapters 5 – 10 present a collection of algorithms and method oriented papers. The first three chapters deal with geometric registration of real and virtual worlds.

Recently, computer vision technology has been actively applied to image media. Yuichi Ohta, who was one of the first to pay attention to this idea, has been applying his experience in computer vision to the problem of registration in MR. Chapter 5 introduces a stereo algorithm with which one can acquire sufficient depth information with sharp object profiles for representing the occlusion between virtual and real objects. Ohta and his colleagues also present a method based on affine shapes for reconstructing an object's image seen from a given viewpoint from several pictures. Finally, they also describe in detail a linear method for geometrically registering virtual objects onto the real world by obtaining the posture of an observer from the video captured by their head-worn camera.

Ulrich Neumann, who began his work in augmented reality (AR) at the University of North Carolina at Chapel Hill, was also a pioneer in applying computer vision to AR through his work with Mike Bajura in 1993. After moving to the University of Southern California, he expanded his work in tracking fiducials for camera-pose determination. Robust vision-based tracking is the main theme of his research. Chapter 6 describes several years of his group's work on tracking with fiducials and in natural environments. Their recent work in hybrid tracking systems for both indoor and outdoor settings is based on their pragmatic combination of vision and inertial sensing technologies.

Chapter 7 describes an algorithm developed by Naokazu Yokoya and his colleagues to realize video see-through MR in a relatively simple way. The main feature of this method is its ability to determine occlusion between the viewer's hands and virtual objects, by capturing stereo images using a pair of CCD cameras attached to the HMD. The method is especially attractive, because it can run at near video rates without any special hardware.

It is known that the spatial correspondence between real and virtual world objects is one of the most interesting themes of MR. However, in addition to the challenge of correct geometric alignment of images, it is also important to minimize an observer's sense of incongruity relative to the quality of mixed real and virtual images. The matching procedures for achieving this can be regarded as accomplishing "photometric registration" in contrast to geometric registration.

Chapter 8 introduces the goal of Katsushi Ikeuchi and his colleagues to integrate virtual objects into real scenes such that the net result becomes seamless in photometric properties. In order to achieve this seamless integration, a method is required, among other things, to recognize the illumination of the real environment into which the virtual objects are to be integrated. Ikeuchi and his colleagues, who are experts in the photometric approach in computer vision, have established a method to determine reflectance parameters from images of the real world and utilize these to solve the problem of photometric gaps in mixed reality.

The other two chapters are on augmented virtuality, and focus on methods to introduce complicated real world objects and real world surroundings into the virtual world while maintaining their photoreality.

Hiroshi Harashima's group at the University of Tokyo has been seeking a framework that can comprehensively treat visual information in any combination of real and virtual worlds involving two- or three-dimensional images as well as holograms. Their goal is to create a new flexible visual communication system, free from traditional conventions. Their "ray-based representation of visual cues" has attracted much attention as a method for realizing photorealistic rendering without any explicit geometric model. Although this idea may seem to be quite similar to light field rendering [1] and the Lumigraph [2] presented at *SIGGRAPH 96*, Harashima's group has used the same function in principle, but differing slightly in their representation of light beams. In Chapter 9, their research provides us with a powerful method to attain this important goal of mixed reality. They present a collection of examples demonstrating the potential of their method, including a real-time application, an application for high quality images, and an application to merge real and virtual world objects (CyberMirage).

In Chapter 10, Michitaka Hirose describes the research goal of his group, to build a large-scale virtual environment from the real world using image-based methods. Many readers may recall the well known "Aspen Movie Map" system [3], which simulated the interactivity of a large-scale virtual environment in which an observer could actively drive through a town by switching between multiple video sequences captured from a real town. That system was limited in its ability to transit through the reproduced scene, however, to switching only at intersections corresponding to those of the real town, and to use viewing angles that were the same as in the original captured video. Hirose's group has taken upon itself the challenge of providing observers the relatively restriction free ability to change their direction and look around them as though they walk through a virtual town. To synthesize this environment by computer graphics while maintaining the photoreality of the real world, these researchers have developed a data capturing car equipped with three kinds of sensors and eight video cameras, which are used to capture images of the town systematically. Their methods for executing image-based walk-throughs are also reported in this chapter. In summary, it may be said that the research of Hirose's group seeks a way to expand the breadth of Mixed Reality spatially, as opposed to Kanade's virtualized reality which is advancing the frontiers of the field temporally.

References

[1] M. Levoy and P. Hanrahan: "Light field rendering," *Proc. SIGGRAPH 96*, pp.31–42, 1996.

[2] S. Gortler, R. Grzeszczuk, R. Szeliski, and M. F. Cohen: "The Lumigraph," *Proc. SIGGRAPH 96*, pp.43–54, 1996.

[3] A. Lippman: "Movie-Maps: An application of the optical videodisc to computer

graphics," *Computer Graphics (Proc. SIGGRAPH 80)*, vol.14, no.3, pp.32–42, 1980.

Chapter 5

Vision-Based Geometric Registration of Virtual and Real Worlds

Yuichi Ohta
Goki Inoue
Toshihiro Kobayashi
Long Quan

University of Tsukuba, Japan

5.1 Introduction

It is more than thirty years since computer vision (CV) research first began to involve the creation of artificial eyes for intelligent robots. This kind of use of CV technology is herein referred to as "robot-technology CV." As illustrated in Figure 5.1(a), robot-technology CV involves a computer gathering visual information from the outside world and using this information to engage in some activity with that world. Robot-technology CV may be considered as the technology in which machines are designed to replace humans.

In medium technology, on the other hand, the computer is regarded as a medium, and humans play the central role. Just as a human gains information of the outside world through his sense of vision (arrow-1 in Figure 5.1(b)), so he also gains information from images displayed by a computer (arrow-2), and at the same time a computer tries to observe human behavior through its sense of vision (arrow-3) and to carry out actions accordingly. The system formed by arrow-2 and 3 is known as the intelligent human interface, with CV technology playing the role of recognizing the faces, expressions, body movements, and hand movements of humans who come in contact with computers. Such technology can be considered necessary in order to ensure the ease-to-use of computers.

In a system in which a computer displays to humans information on the external world gained through the vision (arrow-4 in the Figure), the computer plays precisely the role of a medium that gathers and communicates information. The imaging medium that originated from paintings is given a role of communicating visual information by overcoming the restrictions of time and space. In addition, by using a computer as the medium, it is possible to create a novel three-dimensional imaging system that effectively communicates visual information without the usual physical restrictions of sensing equipment and display devices.

It should be noted that a Mixed Reality System requires that all of the visual information flows (arrow-1 through 4) in Figure 5.1(b) be executed in parallel and in real time. In other word, the mixed reality is one of the most severe and most challenging application fields of CV and CG technology. Here, we refer to CV research conducted from this point of view as "medium-technology CV."

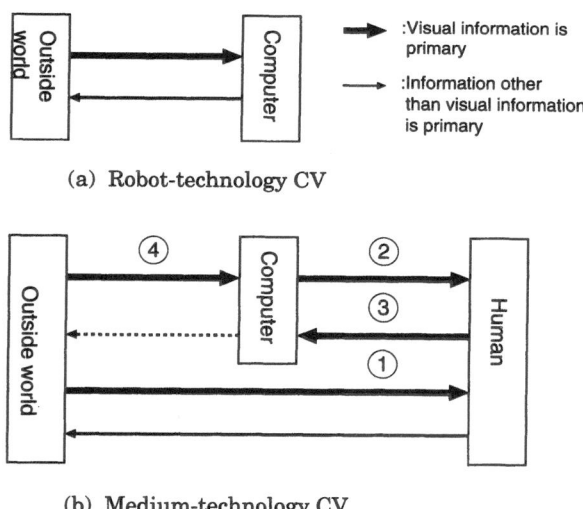

(a) Robot-technology CV

(b) Medium-technology CV

Figure 5.1   CV as medium technology.

The distinction between robot technology and medium technology is fairly straight-forward. Robot technology is aimed at creating machines that can substitute for humans by simulating human behavior. The word *robot* is itself emblematic of a technology aiming at human replacement. By contrast, a medium offers an environment that allows humans to learn, think, and generate ideas by offering them information. A medium does not have the objective of substituting machines for humans, but rather, considers humans the final receivers of information. Medium technology is thus based on the important premise that the role of technology is to maximize human capabilities by assisting, augmenting, and complementing them. The word *medium* denotes technology of cooperation with, and respect for, humans [1].

In the following sections, we introduce our researches on medium-technology CV for realizing geometric registration in mixed reality systems. Geometric registration refers to the generation of virtual objects that show a high degree of coincidence with the real world in terms of appearance, positional arrangement, and even occlusion.

5.2 Depth of Real World by Occlusion Detectable Stereo

In a mixed reality system, and especially in an augmented reality system, virtual objects are often located in a real world. When a virtual object is in front of a real object, the image of the virtual object is simply overlaid on the image of the real one. When the real object is in front of the virtual one, the image of the virtual object should be cropped before the overlay. We need the depth information of the real scene to perform the cropping. In order to generate high quality overlaying, the depth information must have high resolution and sharp object boundaries. Conventional stereo algorithms, however, generate many false correspondences around the occluding boundaries of objects. Such false correspondences seriously blur the depth information of the object profiles and thus affect the quality of the images generated using this depth information.

We solve this problem by using a camera matrix for image capturing. The camera matrix is a set of cameras arranged in a grid on a plane. A polinocular stereo algorithm, called SEA (Stereo by Eye Array), has been developed to fully utilize the characteristics of the camera matrix. In the SEA, a pixel-based correspondence search using the redundancy of polinocular stereo enables us to obtain dense depth information with high spatial resolution. Furthermore, because the arrangement of the camera matrix is identical to that of the pixels on the image, occlusions can be easily detected. This improves the accuracy of the correspondence search around the occluding boundaries, resulting in a drastic improvement in the resolution of object profiles in the depth map.

5.2.1 Coordinate System of the Camera Matrix

Figure 5.2 illustrates the coordinate system of the camera matrix. The origin of the scene coordinates is located at the lens center of the central camera. Other cameras are located at equal intervals to form a grid on the X-Y plane. The location of each camera is labeled as $V^{m,n} = V(mb, nb, 0)$ $(m = -M, \cdots, 0, \cdots, M; n = -N, \cdots, 0, \cdots, N)$ and the image captured by each camera is labeled as $I^{m,n} = I^{V(mb,nb,0)}$. The optical axes of all cameras are set to be parallel to each other. The image coordinates x-y of each camera are set to be parallel to the X-Y axes of the scene coordinates.

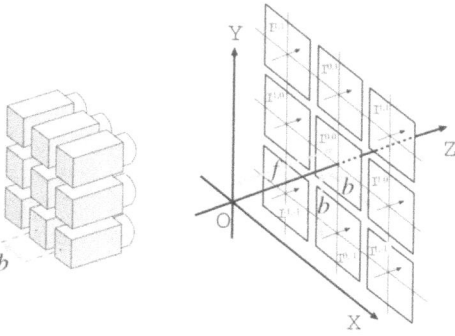

Figure 5.2   Coordinate system of the camera matrix.

5.2.2   Correspondence Search in SEA

SEA uses eight stereo pairs constructed between the central image and the eight peripheral images. First, the dissimilarity values $e^{k,l}(x,y,d)$ between $I^{0,0}(x,y)$ and $I^{k,l}(x - kd, y - ld)$ are computed for each stereo pair, which are the corresponding points on the image pair $I^{0,0}$ and $I^{k,l}$ under the assumption of disparity d. The dissimilarity between two points is evaluated as the summation of RGB distances within the small matching windows whose centers are located at the target points.

$$e^{k,l}(x,y,d) = \sum_{i,j} dis^{k,l}(x+i, y+j, d).$$

Here, $dis^{k,l}(x,y,d)$ denotes the RGB distances between $I^{0,0}(x,y)$ and $I^{k,l}(x-kd, y-ld)$. i and j are the indices which denote the pixel position in the matching window. For example, when the size of the matching window is 5×5, their ranges are $-2 \leq i, j \leq 2$.

Then, all of the eight values, $e^{k,l}(x,y,d)$, are summed up to make the value $e(x,y,d)$ as the penalty for the assumption of disparity d.

$$e(x,y,d) = \sum_{k,l} e^{k,l}(x,y,d).$$

Finally, we can estimate the correct disparity at $I^{0,0}(x,y)$ by choosing the value $\hat{d}$ which satisfies the following condition,

$$\hat{d} = \operatorname*{argmin}_{d}\ e(x,y,d).$$

The disparity at an image point can be estimated using only a pair of images. However, by using a number of image pairs as in the multiple-baseline stereo technique [2], we can greatly reduce the false targets caused by false corresponding points which unexpectedly indicate good similarity. This realizes a dense disparity map with high spatial resolution.

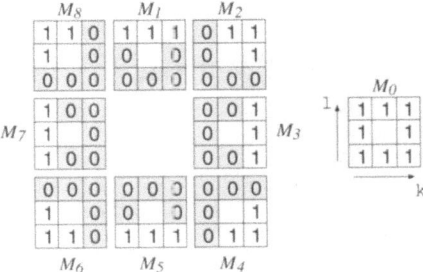

Figure 5.3   Masks representing eight patterns of occlusion.

5.2.3   Occlusion Detection in SEA

A scene point which is visible on the central image $I^{0,0}$ may not be visible on some peripheral images $I^{k,l}$ because it is often occluded by other objects in the scene. This often makes the penalty for the true disparity worse than that for another disparity and causes a false correspondence. If we could distinguish the images on which the target is occluded, it would be possible to estimate the correct disparity, even at an occluding boundary, by utilizing only the images on which the target is observable [3] [4].

By analyzing the real scene statistically [5] [6], we defined the set of occlusion mask patterns illustrated in Figure 5.3, which are able to cover most of the shapes of the occluding boundaries. In this Figure, each mask $M_t(k,l)(t = 1, 2, \cdots, 8)$ represents a pattern of occlusion depending on the orientation of the occluding boundary. The 0 or 1 value on $M_t(k,l)$ indicates whether or not occlusion occurs between the image pair $I^{0,0}$ and $I^{k,l}$; 0 indicates that occlusion occurs and 1 indicates that it does not. A mask M_0, whose values are only 1s, indicates the non-occluding case, *i.e.*, the target scene point is visible on all peripheral images.

All the masks are examined at every point, and the mask that gives the minimum penalty value is chosen. By detecting the occlusion explicitly, it becomes possible to estimate the correct disparity around occluding boundaries. This enables us to acquire a dense disparity map with sharp object profiles which satisfies the requirements for representing the occlusion between virtual and real objects.

5.2.4   Occlusion Masks for Matching Windows

Use of a large matching window is effective for reducing noise and thereby obtaining a stable dissimilarity evaluation. This, in turn, produces a smooth disparity map. However, the large matching window often fails to yield the correct correspondence at the boundary of object, because different disparities may be included in a window.

Figure 5.4 illustrates this problem. Here, the target point is on the farther object near the boundary of the two objects with different disparities. As shown in the Figure, a portion of the farther object is occluded in the matching window on the center image. This part becomes visible on the three images to the right and the pattern in the matching window changes. This creates larger dissimilar values,

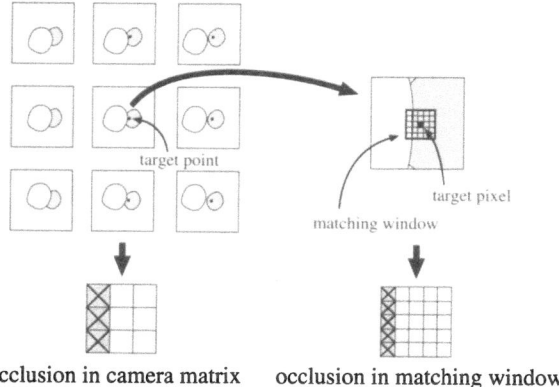

occlusion in camera matrix   occlusion in matching window

Figure 5.4   Occlusion problem in matching window.

even in the case of the true correspondence.

To cope with this problem, we introduce the idea of an occlusion mask for the matching windows as well as for the camera matrix. Because both of the occlusions in the camera matrix and in the matching window depend on the identical shape of the occluding object, it is appropriate to assume that the occlusion pattern in the matching window is similar to that in the camera matrix [7].

5.2.5   Experimental Results

We demonstrate the feasibility of SEA by using real image data. Figure 5.5 shows the scenes used for the experiments. For each scene, a set of 3×3 images were captured on a color video camera mounted on a computer controlled X-Y arm. Only the central images are shown in Figure 5.5. Figure 5.6 shows the disparity maps obtained by SEA for the scenes in Figure 5.5. The disparity values are coded by intensity; larger disparity is indicated with brighter intensity. Here, the window size for the correspondence search was set to 5×5 square, and no smoothing of the disparity maps was performed.

5.3   Synthesis of Novel Views of a Virtual Object from Several Images

In a mixed reality system, the virtual object is often synthesized from the image of the real object. When the observer moves in the real world, the view of the virtual object must change according to the observer's position. The technique for drawing arbitrary views of a 3-D object using its image data is known as Image-Based Rendering. In this section, we describe a framework for synthesizing the image of a face, *i.e.*, a 3-D object, in an arbitrary pose using two photographs. Note that this synthesis is not a morphing. It is based on the Structure-from-Motion theory [8] and is able to generate precise views of 3-D objects.

(a) "Lamp Shade"   (b) "Santa"   (c) "m & m"

Figure 5.5   Three scenes used in the experiment.

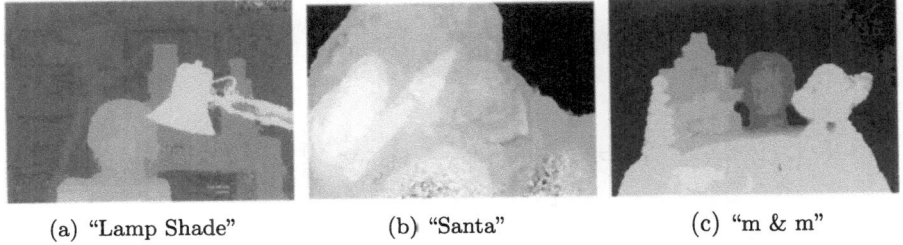

(a) "Lamp Shade"   (b) "Santa"   (c) "m & m"

Figure 5.6   Disparity maps obtained by SEA.

A flow of the process used to synthesize novel views from several images is shown in Figure 5.7. A set of images of an object in various poses is used as an input. First, 2-D coordinate values of the feature points in each new pose are calculated by linear combination of the feature points detected from the input images. Then, the blended texture taken from the input images is mapped onto the triangular patches whose vertices are the feature points. In the following sections, we explain the method used to calculate the coordinate values, and that used for the texture mapping.

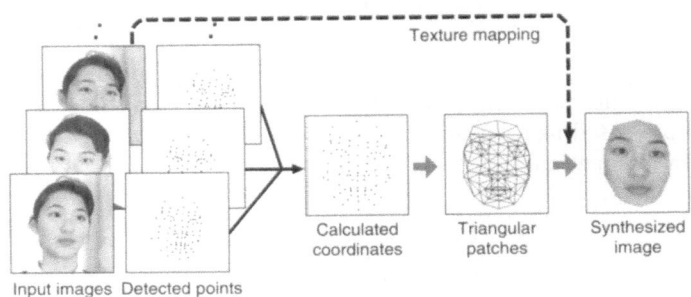

Figure 5.7   Flow of the process.

5.3.1   Calculation of 2-D Coordinate Values

Let B_1 and B_2 be two input images of the same object in different poses. We assume that the feature points are located on the object surface, and that the correspondences of all feature points between two input images are known. Let (x_k^1, y_k^1) and (x_k^2, y_k^2) be 2-D coordinate values of the k-th feature point on images B_1 and B_2, respectively. These 2-D coordinate values are the result of rotating and projecting the points (X_k, Y_k, Z_k) of the 3-D space. The vectors and matrices that indicate these coordinate values are defined as follows.

$$
\begin{aligned}
\boldsymbol{x}^1 &= [x_1^1, x_2^1, \cdots, x_n^1], \\
\boldsymbol{y}^1 &= [y_1^1, y_2^1, \cdots, y_n^1], \\
\boldsymbol{x}^2 &= [x_1^2, x_2^2, \cdots, x_n^2], \\
\boldsymbol{y}^2 &= [y_1^2, y_2^2, \cdots, y_n^2].
\end{aligned}
$$

$$
\boldsymbol{P} = \begin{bmatrix} X_1 & X_2 & \cdots & X_n \\ Y_1 & Y_2 & \cdots & Y_n \\ Z_1 & Z_2 & \cdots & Z_n \end{bmatrix}.
$$

For simplification, we assume rigidity, orthographic projection, and no translation. As shown in the following equations, the vectors which indicate the coordinate values of the feature points of each input image are represented as a product of the 2 × 3 transformation matrix and the 3-D coordinate values of the feature points.

$$
\begin{bmatrix} \boldsymbol{x}^1 \\ \boldsymbol{y}^1 \end{bmatrix} = \begin{bmatrix} \boldsymbol{r}_x^1 \\ \boldsymbol{r}_y^1 \end{bmatrix} \boldsymbol{P},
$$

$$
\begin{bmatrix} \boldsymbol{x}^2 \\ \boldsymbol{y}^2 \end{bmatrix} = \begin{bmatrix} \boldsymbol{r}_x^2 \\ \boldsymbol{r}_y^2 \end{bmatrix} \boldsymbol{P}.
$$

Let $\hat{\boldsymbol{x}}$ and $\hat{\boldsymbol{y}}$ be 2-D coordinate values of the feature points in another view $\hat{B}$. Let $\hat{\boldsymbol{r}}_x$ and $\hat{\boldsymbol{r}}_y$ be the first row vector and the second row vector of the transformation matrix corresponding to $\hat{B}$. If $\boldsymbol{r}_x^1$, $\boldsymbol{r}_x^2$, and $\boldsymbol{r}_y^1$ are linearly independent, a set of coefficients a_{x1}, a_{x2}, and a_{x3} which satisfy equation(5.1) should exist, because the rank of $\hat{\boldsymbol{r}}_x$ is 3. This means that the X-coordinate values of all the feature points on $\hat{B}$ can be represented as a linear combination of 2-D coordinate values on B_1 and B_2, as shown in equation(5.2).

$$
\begin{aligned}
\hat{\boldsymbol{r}}_x &= a_{x1} \boldsymbol{r}_x^1 + a_{x2} \boldsymbol{r}_x^2 + a_{x3} \boldsymbol{r}_y^1. & (5.1) \\
\hat{\boldsymbol{x}} &= a_{x1} \boldsymbol{x}^1 + a_{x2} \boldsymbol{x}^2 + a_{x3} \boldsymbol{y}^1. & (5.2)
\end{aligned}
$$

The base vectors of the linear combination are not always linearly independent. In order to obtain sufficient number of stable base vectors, we apply the principal component analysis of the four base vectors ($\boldsymbol{x}^1$, $\boldsymbol{y}^1$, $\boldsymbol{x}^2$, $\boldsymbol{y}^2$), and we use the first three eigen vectors ($\boldsymbol{p}_1$, $\boldsymbol{p}_2$, $\boldsymbol{p}_3$) as the base vectors. As shown in the following equation, X- and Y- coordinate values can be stably represented as linear combinations of the linearly independent vectors.

$$\hat{\boldsymbol{x}} = a_{x1}\boldsymbol{p}_1 + a_{x2}\boldsymbol{p}_2 + a_{x3}\boldsymbol{p}_3,$$
$$\hat{\boldsymbol{y}} = a_{y1}\boldsymbol{p}_1 + a_{y2}\boldsymbol{p}_2 + a_{y3}\boldsymbol{p}_3.$$

As shown above, the 2-D coordinate values of the feature points can be easily calculated once the coefficients of the linear combination are obtained. In order to determine the coefficients corresponding to the synthesized image, we need at least four representative points in the new view. The coefficients of the linear combination are determined from the representative points, and the 2-D coordinate values of all feature points on the object surface can be calculated [9].

5.3.2 Texture Mapping

Basically, the textures taken from input images which have similar poses and expressions are mapped onto the synthesized image. However, if we take all the textures from one image with the closest facial pose, the synthesized image will be warped unnaturally when the facial pose smoothly changes. This undesirable warping is caused by a drastic deformation of texture. We can solve this problem by texture blending. In our method, facial images are synthesized by mapping the blended texture taken from multiple input images. The larger weight is given for the texture whose pose is nearer to the pose to be synthesized.

Although the object orientation of the input image is unknown, a rough orientation can be estimated. The relative orientation of the new pose is also estimated. Since this value is only used for determining the texture blending weights, small errors are not critical. The weights for texture blending are determined to be inversely proportional to the square of the angle difference of object orientations.

The texture is mapped by using triangular patches whose vertices are the feature points. For each triangular patch, the texture is clipped from multiple input images, and deformed by affine transformation according to the 2-D coordinate values of the feature points. Then, the textures are blended by using the weights and mapped. At this step, each patch is checked to determine whether it is obverse or reverse. If the patch is a reverse patch, the texture is not mapped onto it. This simple rule is adequate for facial image synthesis, because complicated occlusion never occurs.

5.3.3 Experimental Results

We chose 86 feature points on a face. This number of points is relatively small compared to that for ordinary facial models used in other applications. The feature points were located not only on such facial organs as the eyes and mouth, but also on movable areas such as cheeks. The 2-D coordinate values of the points were determined manually by referring to the marks drawn on the face. 156 triangular patches were created, as shown in Figure 5.8(a). Two input images are shown in Figure 5.8(b). Five points (outside corners of both eyes, tip of the nose, and bottoms of both ears) were chosen as the control points with known 3-D coordinates. The

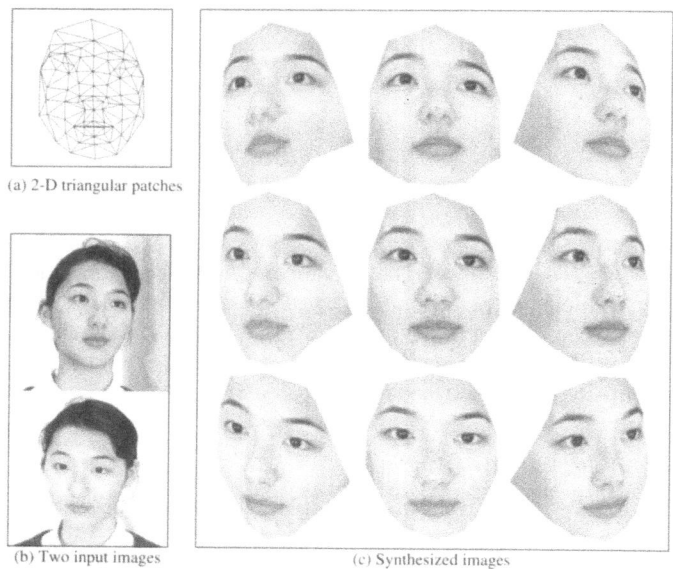

(a) 2-D triangular patches

(b) Two input images

(c) Synthesized images

Figure 5.8   Synthesized images of various poses using two input images.

synthesized images with various poses are shown in Figure 5.8(c). We have extended
the system for arbitrary expression [10] and lip motion [11].

5.4   A Linear Method for Euclidean Motion/ Structure in Real Time

We propose a linear algorithm that uses 9 linear parameters to encode the 8 Euclidean motion parameters of 3 weak perspective views [12]. It can handle in a
single framework the novel view synthesis described in the previous section and the
recovery of camera pose that is necessary for the base image selection and texture
blending. Because it is a linear algorithm, it is stable and works in real time.

5.4.1   Affine Camera Model

For a restricted class of camera models, by setting the third row of the perspective camera $P_{3\times4}$ to $(0,0,0,\lambda)$, we obtain the affine camera initially introduced by
Mundy and Zisserman [13].

$$A_{3\times4} = \begin{pmatrix} p_{11} & p_{12} & p_{13} & p_{14} \\ p_{21} & p_{22} & p_{23} & p_{24} \\ 0 & 0 & 0 & p_{34} \end{pmatrix} = \begin{pmatrix} M_{2\times3} & t_{3\times1} \\ 0_{1\times3} & \end{pmatrix}.$$

The affine camera $A_{3\times4}$ subsumes the orthographic, weak perspective and para-
perspective. Finite points in affine spaces $\mathcal{R}^n$ are naturally embedded into $\mathcal{P}^n$ by

the mapping $\boldsymbol{u}_a \mapsto \boldsymbol{u} = (\boldsymbol{u}_a, 1)^T$ and $\boldsymbol{x}_a \mapsto \boldsymbol{x} = (\boldsymbol{x}_a, 1)^T$. We have, therefore, $\boldsymbol{u}_a = \boldsymbol{M}_{2\times 3}\boldsymbol{x}_a + \boldsymbol{t}_0$, where $\boldsymbol{t}_0 = (t_1/t_3 \ t_2/t_3)^T = (p_{14}/p_{34}, p_{24}/p_{34})^T$. If we further use relative coordinates of the points with respect to a given reference point (for instance, the centroid of a set of points), the vector $\boldsymbol{t}_0$ is canceled, and we obtain the following linear mapping between relative space points and relative image points.

$$\Delta \boldsymbol{u} = \boldsymbol{M}_{2\times 3}\Delta \boldsymbol{x}. \tag{5.3}$$

Equation (5.3) is the basic projection equation for points in an affine camera using relative coordinates.

5.4.2 Unification of 2-View and 3-View Geometry

For notational simplicity, we rewrite Equation (5.3) as $\boldsymbol{u} = \boldsymbol{M}_{2\times 3}\boldsymbol{x}$. We can now examine the matching constraints between multiple views of the same point. Let the three views of the same point $\boldsymbol{x}$ be given as follows.

$$\begin{cases} \boldsymbol{u} &= \boldsymbol{M}\boldsymbol{x}, \\ \boldsymbol{u}' &= \boldsymbol{M}'\boldsymbol{x}, \\ \boldsymbol{u}'' &= \boldsymbol{M}''\boldsymbol{x}. \end{cases}$$

These can be rewritten together in matrix form as

$$\begin{pmatrix} \boldsymbol{M} & \boldsymbol{u} \\ \boldsymbol{M}' & \boldsymbol{u}' \\ \boldsymbol{M}'' & \boldsymbol{u}'' \end{pmatrix} \begin{pmatrix} \boldsymbol{x} \\ \lambda \end{pmatrix} = 0,$$

where $\lambda \neq 0$ encodes the (unrecoverable) global scale factor of the reconstruction.

Because the vector $(\boldsymbol{x}, \lambda)^T$ cannot be zero, the rank of the coefficient matrix is at most 3, so all of its 4×4 minors vanish. Each expansion of these 4×4 minors is linear in the image coordinates $\boldsymbol{u}$, $\boldsymbol{u}'$ and $\boldsymbol{u}''$, with the coefficients t_i coming from the 3×3 minors of the following 6×3 joint projection matrix: $\begin{pmatrix} \boldsymbol{M} \\ \boldsymbol{M}' \\ \boldsymbol{M}'' \end{pmatrix} = (1\ 2\ 1'\ 2'\ 1''\ 2'')^T$.

There are in total $C_6^3 = 20 = 8 + 4 + 8$ such minors. As will be shown later, 4 of the 20 minors are common to the 2-view and 3-view constraints. All these minors provide a linear coordinate system that spans the joint projection matrix.

Two-view constraints

There are three 2-view constraints corresponding to the 3 pairs of the 3 views, namely, the vanishing of the determinants $[121'2']$, $[121''2'']$ and $[1'2'1''2'']$:

$$\begin{aligned} t_{13}u &+ t_{14}v &+ t_{10}u' &+ t_9\ v' &= 0, \\ t_{15}u' &+ t_{16}v' &+ t_{17}u'' &+ t_{18}v'' &= 0, \\ t_{19}u &+ t_{20}v &+ t_{12}u'' &+ t_{11}v'' &= 0. \end{aligned}$$

These are the affine epipolar geometry. The set of $3 \times 4 = 12$ coefficients t_i for $i = 9, \ldots, 20$ are 12 of the 20 minors of the joint projection matrix.

Three-view constraints

There are four 3-view constraints from the vanishing of the determinants $[121'1'']$, $[122'1'']$, $[121'2'']$ and $[122'2'']$. By careful inspection of the minors, we have

$$
\begin{aligned}
t_4\, u &+ t_8\, v &+ t_{11}u' &+ t_9\, u'' &= 0, \\
t_2\, u &+ t_6\, v &+ t_{11}v' &+ t_{10}u'' &= 0, \\
t_3\, u &+ t_7\, v &+ t_{12}u' &+ t_9\, v'' &= 0, \\
t_1\, u &+ t_5\, v &+ t_{12}v' &+ t_{10}v'' &= 0.
\end{aligned}
$$

Among the 12 minors, 8 of them are new (t_1 to t_8) and 4 of them are common with 2-view constraints (t_9 to t_{12}).

Since each point correspondence gives 4 linearly independent 3-view constraints, 4 points give $(4-1) \times 4 = 12$ linear equations for solving these minors.

The appearance of common minors between 2-view and 3-view constraints is not accidental, since we have 3+4=7 constraints, each with 4 coefficients, for a total of $4 \times 7 = 28$. Since there are only 20 minors, 8 of them should appear more than once.

5.4.3   Euclidean Motion/Structure from 3 Calibrated Affine Views

So far, the linear estimates of the 2-view and 3-view constraints yield, directly but implicitly, the affine motion/structure. To obtain Euclidean motion/structure, we need at least 3 calibrated affine images. Here, we use the unified formulation introduced in [14] for calibrated affine cameras, so that the method developed will be valid for all calibrated orthographic, weak-perspective and para-perspective models.

Each projection matrix $M_{2\times 3}$ can be decomposed into $M = sKR$, where s is a scaling factor of the whole image, K is the intrinsic parameter matrix (for instance, $K = \begin{pmatrix} \xi & 0 \\ 0 & 1 \end{pmatrix}$, where ξ is the aspect ratio for the weak-perspective case), and $R_{2\times 3}$ represents 2 rows of a 3-D rotation matrix. Since we are assuming calibrated cameras, the intrinsic parameter matrix K is known and its inverse can be directly applied to the image points so that its effect is removed completely. Thus the projection matrix M for normalized image coordinates becomes $M = sR$, i.e., a scaled 2×3 rotation matrix. There are in total $8 = 2 \times 3 + 2$ Euclidean parameters for a set of 3 views: two relative 3-D rotations R and G each having 3 d.o.f. and 2 relative scale factors s and s' between the first and the remaining views.

Any linear algorithm will consist of first estimating linearly the minors using multiple view constraints, then extracting the 8 Euclidean parameters from these minors by identifying the projection matrices M and M' with the scaled rotation matrices sR and $s'G$.

Combining 2-view and 3-view constraints   We should keep in mind that, although all 7 constraints are linearly independent, only $3 = (6-3) \times (4-3)$ of them are algebraically independent due to Grassmanian relations. There are in total 20 homogeneous coefficients—minors of the joint projection matrix. How to choose the most appropriate constraints is of primary importance. The selected constraints should be algebraically independent and should contain as few coefficients as possible.

Taking only the three 2-view constraints is a poor choice, since the third one is partially dependent on the first two by the composition rule on the rigid motions, and each one is completely separate from the others. Taking only the 3-view constraints leads to a complicated algebraic manipulation for the extraction of Euclidean parameters. Hence we combine the 2-view and 3-view constraints.

The key observation is that there are common coefficients between the 2-view and 3- view constraints: two of the three 2-view constraints share 4 minors, t_9, t_{10}, t_{11} and t_{12}, with the 3-view constraints. This allows us to use the following combination: two 2-view constraints plus one of the 3-view constraints.

$$
\begin{aligned}
t_4\, u \;\; + t_8\, v \;\; + t_{11} u' \; + t_9\, u'' &= 0, \\
t_{13} u \;\; + t_{14} v \;\; + t_{10} u' \; + t_9\, v' &= 0, \\
t_{15} u \;\; + t_{16} v \;\; + t_{12} u'' + t_{11} v'' &= 0.
\end{aligned}
$$

These 10 unknown minors can be solved as a single homogeneous vector under the constraint $\|t_{10}\| = 1$:

$$
\begin{pmatrix}
u & v & u'' & 0 & u' & 0 & 0 & 0 & 0 & 0 \\
0 & 0 & v' & u' & 0 & 0 & u & v & 0 & 0 \\
0 & 0 & 0 & 0 & v'' & u'' & 0 & 0 & u & v
\end{pmatrix} t = 0.
$$

Any ratio t_i/t_j of the minors can therefore be obtained.

Obtaining partial solutions from 2-view constraints First, from the estimated minors of the 2-view constraint, we can easily obtain the partial Euclidean solution based on Shapiro et al.'s reformulation [15] of Koenderink and Van Doorn's rotation representation [16]. Koenderink and Van Doorn's representation is probably the most appropriate parameterization, since it distinguishes clearly between the entities that can be obtained from two views and those that cannot.

Assume that the affine epipolar geometry of two views is estimated as (a, b, c, d) $(u, v, u', v')^T = 0$. From the 3 ratios, $a : b : c : d$, exactly 3 Euclidean parameters can be extracted. Using a scaled rotation matrix instead of M, the following relation holds: $a : b : c : d = sr_{32} : -sr_{31} : r_{23} : -r_{13}$. Therefore, the relative scale factor between the two views is immediately given as $s = \frac{a^2+b^2}{c^2+d^2}$.

According to Koenderink and Van Doorn, the entire rotation can be decomposed into a rotation in the image plane (assume this rotation angle to be θ) and a rotation through an angle ρ about an axis (angled at ϕ to the positive x axis) in a frontoparallel plane. The rotation matrix in terms of Koenderink and Van Doorn's $\theta - \phi - \rho$ representation can be recomposed as $R_{3\times 3}$:

$$
\begin{pmatrix}
(1 - c(\rho))c(\phi)c(\phi - \theta) + c(\rho)c(\theta) & (1 - c(\rho))c(\phi)s(\phi - \theta) - c(\rho)s(\theta) & s(\phi)s(\rho) \\
(1 - c(\rho))s(\phi)c(\phi - \theta) + c(\rho)s(\theta) & (1 - c(\rho))s(\phi)s(\phi - \theta) + c(\rho)c(\theta) & -c(\phi)s(\rho) \\
-s(\rho)s(\phi - \theta) & s(\rho)c(\phi - \theta) & c(\rho)
\end{pmatrix}.
$$

Therefore,

$$
a : b : c : d = s\sin\rho\cos(\phi - \theta) : s\sin\rho\sin(\phi - \theta) : -\cos\phi\sin\rho : -\sin\phi\sin\rho,
$$

hence, the rotation angle in the image plane is easily determined by $\tan\phi = \frac{d}{c}$, and the rotation axis modulo π out of the image plane by $\tan(\phi - \theta) = \frac{b}{a}$.

Obtaining the full solution with the 3-view constraint   Up to this point, the only unknown is the rotation angle out of the image plane ρ, which is the only component that generates depth information. The one-parameter family of solutions for the rotation matrix between the two views is

$$\boldsymbol{R}_{3\times3}(\rho) = \begin{pmatrix} \boldsymbol{D}_{2\times2} + \cos\rho\,\boldsymbol{E}_{2\times2} & \sin\rho\,\boldsymbol{F}_{2\times1} \\ \sin\rho\,\boldsymbol{G}_{1\times2} & \cos\rho \end{pmatrix},$$

where $\boldsymbol{D}_{2\times2}, \boldsymbol{E}_{2\times2}, \boldsymbol{F}_{2\times1}$ and $\boldsymbol{G}_{1\times2}$ are the known quantities.

Similarly, with the second 2-view constraint, we get another one-parameter family of solutions for the rotation matrix $\boldsymbol{G}_{3\times3}(\rho')$ of the other 2 views in terms of the unknown rotation angle out of the image plane ρ'.

Now, we can use the 3-view constraint to fully determine the motion/structure. It can be easily verified that

$$t_4 : t_8 : t_{11} : t_9 = ss'(r_{11}g_{13} - g_{11}r_{13}) : ss'(r_{12}g_{13} - g_{12}r_{13}) : -s'g_{13} : sr_{13}.$$

Substituting the ratio t_9/t_{11} into t_4/t_{11} and t_8/t_{11}, we get exactly 2 linear equations in $\cos\rho$ and $\cos\rho'$,

$$\begin{aligned} \frac{t_4}{t_{11}} &= a\,\cos\rho + b\,\cos\rho' + c\,, \\ \frac{t_8}{t_{11}} &= a'\cos\rho + b'\cos\rho' + c', \end{aligned} \qquad i.e., \quad \boldsymbol{A}_{2\times2}\begin{pmatrix}\cos\rho \\ \cos\rho'\end{pmatrix} = \boldsymbol{B}_{2\times1}.$$

5.4.4   Experimental System

We have developed a mixed reality system, Table KINGYO, to demonstrate the feasibility of the proposed algorithm ("KINGYO" means goldfish in Japanese). As illustrated in Figure 5.9, a synthesized bowl of goldfish having a geometrically correct shape and position appears on a real table. The observer wears an HMD with cameras. Four feature points (LEDs) near the table are tracked in video rate by means of a special PCI board (Tracking Vision, Fujitsu Ltd.), and the view of the goldfish bowl is synthesized based on their position. Since the textures used for the bowl are taken from either live or recorded videos, the observer sees a virtual bowl with swimming goldfishes on a real table. The synthesis of the bowl itself does not require the recovery of the observer's motion. That is, the bowl can be synthesized using only the 3-view constraints described in section 5.4.2. The motion information is used to select the appropriate textures to generate natural-looking virtual objects.

5.5   Conclusion

In the computer vision field of technology, where R&D work has traditionally been carried out with robotic applications in mind, an emphasis on media, rather than robots, constitutes a new unexplored continent. There a vast space is found, along with dreams and possibilities. At the same time, no guideposts have been established

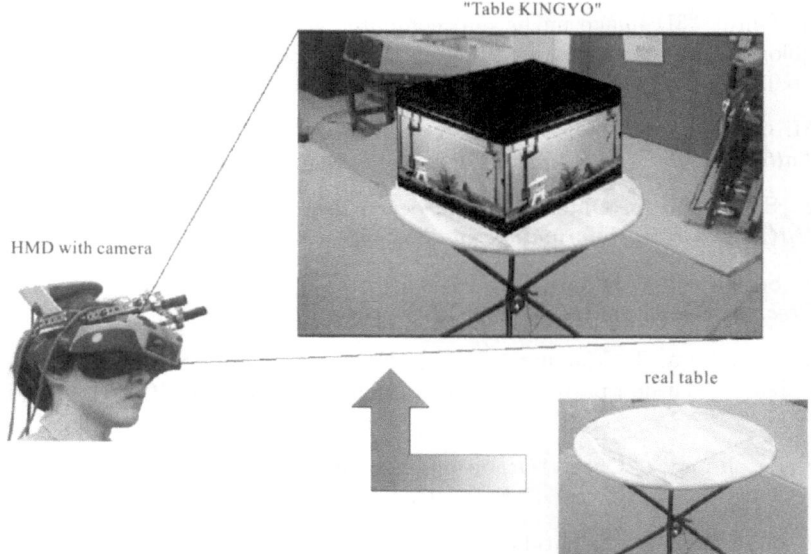

Figure 5.9 "Table KINGYO". A synthesized bowl of goldfish having a geometrically correct shape and position appears on a real table. Since the textures used for the bowl are taken from either live or recorded videos, the observer sees a virtual bowl with swimming goldfishes on a real table.

to direct these new researches. Among these new CV application fields, mixed reality is perhaps the most dramatic and challenging. It requires that multiple visual information flows among humans, computers, and the real world be executed in parallel, in real time, and with stability. Accordingly, many problems remain to be overcome in the development of CV algorithms for use in mixed reality systems.

Acknowledgments

A part of this work was supported by Mixed Reality Systems Lab. and by Ministry of Education, Science, Sports and Culture. The occlusion detectable stereo SEA is a part of the Ph.D. thesis of Kiyohide Satoh, who is currently with Mixed Reality Systems Lab., and the face synthesis work is a part of the Ph.D. thesis of Yasuhiro Mukaigawa, who is currently with Okayama University. We gratefully acknowledge discussions with both of them and also with Yuichi Nakamura, Tomohiko Matsuura, Hiroki Igarashi, and Yasuyuki Sugaya, all are the members of the Computer Vision and Image Media Laboratory, University of Tsukuba.

References

[1] Y. Ohta: "3D image media and computer vision — From CV as robot technology to CV as media technology," *J. Robotics and Mechatronics*, vol.9, no.2, pp.92–97, 1997.

[2] M. Okutomi and T. Kanade: "A multiple-baseline stereo," *IEEE Trans. on Pattern Analysis and Machine Intelligence*, vol.15, no.4, pp.353–363, 1993.

[3] K. Satoh and Y. Ohta: "Passive depth acquisition for 3D image displays," *IEICE Trans. on Information and Systems*, vol.E77-D, no.9, pp.949–957, 1994.

[4] K. Satoh and Y. Ohta: "Occlusion detectable stereo using a camera matrix," *Proc. ACCV'95*, pp.II–331–335, 1996.

[5] Y. Nakamura, T. Matsuura, K. Satoh, and Y. Ohta: "Occlusion detectable stereo – Occlusion patterns in camera matrix –," *Proc. CVPR'96*, pp.371–378, 1996.

[6] K. Satoh and Y. Ohta: "Occlusion detectable stereo – Systematic comparison of detection algorithms –," *Proc. ICPR'96*, pp.280–286, 1996.

[7] Y. Ohta, K. Satoh, I. Kitahara, and T. Matsuura: "Displaying motion parallax by occlusion detectable," *Proc. CVVRHC'98*, pp.35–42, 1998.

[8] S. Ullman, and R. Basri: "Recognition by linear combinations of models," *IEEE Trans. on Pattern Analysis and Machine Intelligence*, vol.13, no.10, pp. 992–1006, 1991.

[9] Y. Mukaigawa, Y. Nakamura, and Y. Ohta: "Synthesis of arbitrarily oriented face views from two images," *Proc. ACCV'95* , pp.III–718–722, 1995.

[10] Y. Mukaigawa, Y. Nakamura, and Y. Ohta: "Face synthesis of arbitrary pose and expression from several images," *Proc. ACCV'98* , pp.680–687, 1998.

[11] L. Gao, Y. Mukaigawa, and Y. Ohta: "Synthesis of facial images with lip motion from several real views," *Proc. 3rd Int'l Conf. on Automatic Face and Gesture Recognition*, pp.181–186, 1998.

[12] L. Quan and Y. Ohta: "A new linear method for Euclidean motion/structure from three calibrated affine views," *Proc. CVPR'98*, pp.172–177, 1998.

[13] J. L. Mundy and A. Zisserman (eds.): *Geometric invariance in computer vision*, The MIT Press, Cambridge, MA, 1992.

[14] L. Quan: "Self-calibration of an affine camera from multiple views," *Int'l J. Computer Vision*, vol.19, no.1, pp.93–105, 1996.

[15] L. S. Shapiro, A. Zisserman, and M. Brady: "3D motion recovery via affine epipolar geometry," *Int'l J. Computer Vision*, vol.16, no.2, pp.147–182, 1995.

[16] J. Koenderink and A. van Doorn: "Affine structure from motion," *J. Optical Society of America*, vol.8, no.2, pp.377–385, 1991.

Chapter 6

Augmented Reality Tracking in Natural Environments

Ulrich Neumann
Suya You
Youngkwan Cho
Jongweon Lee
Jun Park

University of Southern California, U.S.A.

6.1 Introduction

Tracking, or camera pose determination, is the main technical challenge in creating augmented realities. Limiting the degree to which the environment may be modified to support tracking heightens this challenge. The USC Computer Graphics and Immersive Technologies (CGIT) laboratory strives to develop self-contained, minimally intrusive tracking systems for use in both indoor and outdoor settings. Intrusion, in the tracking context, is any type of environment modification, calibration, or constraint upon the environment (*e.g.*, placing calibrated targets or active beacons in the setting).

The goal of a self contained system leads to vision and inertial sensing subsystems that include video cameras, gyros, and accelerometers [1]. Other technologies of interest for future work include RF (GPS, Impulse Radio), compass, pedometer, and altimeter sensors. No single technology is sufficiently robust and accurate (at this time) to satisfy the diversity of needs, so hybrid approaches are a rational goal [2] [3]. An ideal tracking system reports perfect instantaneous six-degree-of-freedom (6DOF) measurements of the sensor pose (*i.e.*, position and orientation), in any environment, under any motion. Perfect tracking is never achieved since temporal delays and accuracy limits are inherent in all measurements. The challenge in tracking research and development is to approach the ideal [2] [4]–[7].

6.2   Indoor Tracking

Indoor and outdoor environments present different tradeoffs. Building structures block signals used by many sensors (*e.g.* GPS or compass). At the same time, structures provide the power and controlled environment that facilitates the installation and use of calibrated targets or beacons. For indoor applications, a minimal set of calibrated passive targets (fiducials) appear practical and prudent if they, in turn, facilitate robust and accurate tracking. Many systems adopt this approach [8]–[16].

From a practical and engineering standpoint, colored-circle stickers make good fiducials for several reasons. They are inexpensive to produce on color inkjet or laser printers. A 2D ellipse models their projection from any viewpoint. The circles represent points, so small stickers suffice, facilitating arbitrary placement, even on small objects. Precise edge measurements are not necessary to compute an accurate centroid, reducing the influence of sensor noise and lighting variations. Color can facilitate the detection of fiducials in a scene image as well as their unique identification among the set of all fiducials.

6.2.1   Fiducial Detection

A practical indoor tracker system needs a robust method to detect fiducials [17]–[22]. We developed two multiscale detection methods. The first approach uses calibrated color region detection and segmentation [23]. The detection is straightforward, however, many variables affect the calibration including fiducial printing; camera and digitizer color response; and lighting. This approach imposes severe calibration requirements upon the user and constrains the deployment of the system. The second detection approach, based on fuzzy membership functions, addresses these deficiencies.

The fuzzy detection approach uses multiscale relationships between neighboring pixel groups to segment fiducials and their backgrounds [24]. There are no threshold values to calibrate and the relationships remain stable under variations in lighting and color. After segmentation, shape and color tests distinguish fiducials from background clutter. Figure 6.1 shows six examples of the colored sticker fiducials.

A fiducial is modeled as two transitions along a line, from background to fiducial color and then to background again. The transitions occur within an expectation interval determined by the range of camera-to-fiducial distances, the fiducial size, and the camera parameters. A minimum spacing between fiducials is assumed, as well as a uniform background around the fiducial. Figure 6.2 illustrates the fiducial detection model. The best edge position is the pixel whose intensity is the average value of its left and right segments. Pixels are grouped into R, C, and L regions according to the following restrictions and membership functions.

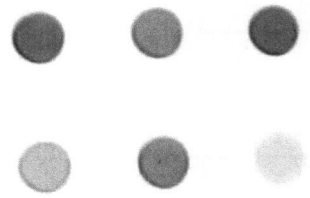

Figure 6.1   Fiducial examples based on primary and secondary colors (see color pages).

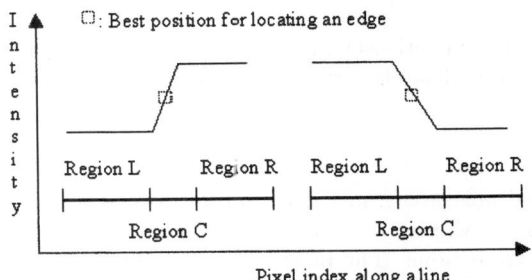

Figure 6.2   Fiducial transition model.

<u>Rule 1:</u> Avg(R) > Avg(C) > Avg(L), or Avg(L) > Avg(C) > Avg(R), where Avg(J) is the average intensity value of region J.

<u>Rule 2:</u> Max(C) - Min(C) > Max(R) - Min(R) & Max(C) - Min(C) > Max(L) - Min(L), where Max(J) and Min(J) are the maximum and minimum intensity values of region J.

<u>Rule 3:</u> The intensity distribution of regions R and L may not overlap.

Membership functions are defined by the following equations.

$$\mu = \mu 1 \times \mu 2$$
$$\mu_1 = \frac{2 \times MIN(|Avg(R) - Avg(T)|, |Avg(L) - Avg(T)|)}{|Avg(R) - Avg(L)|}$$
$$\mu_2 = \frac{Min(R) - Max(L)}{Max(R) - Min(L)}, Avg(R) > Avg(L)$$

or

$$\mu_2 = \frac{Min(L) - Max(R)}{Max(L) - Min(R)}, Avg(L) > Avg(R)$$

μ_1 indicates the similarity of region T to the median intensity value of its two neighboring regions. μ_2 measures the distribution similarity on both sides of a

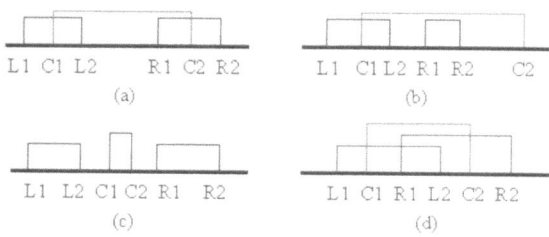

Note. C is a transition region
L & R are left and right homogeneous regions
1 & 2 indicate minimum and maximum intensity values of a region

Figure 6.3   Examples of relationships among transitions and left,& right regions (a)valid fiducial (b-d)invalid cases violating first, second, and third rules, respectively.

transition. An ideal noiseless transition exhibits (Min(R) - Max(L)) = (Max(R) - Min(L)) or (Min(L) - Max(R)) = (Max(L) - Min(R)), and $\mu_2 = 1$. $\mu_2 > 1$ for non ideal or noisy transitions. The peak of membership function μ locates a point closest to an ideal edge. Figure 6.3 shows examples of region groupings that may arise. Only the case shown in Figure 6.3(a) is a valid fiducial detection.

Grouping contiguous pixels that pass all restrictions identifies transitions. The transition pixel with the highest membership value is at the transition position. Connected transitions create line segments. Fiducials have solid color, so line segments with the same color should be grouped. Unfortunately, image noise invokes the need for a similarity measure. The minimum, average, and maximum pixel-intensity values of a region define a region distribution. To avoid the need for thresholds, two line segments are considered as the "same color" when their minimum and maximum intensity distributions overlap. This simple grouping function is efficient and performs well in our experiments.

Regions that pass the color and shape tests described in [24] are taken as valid fiducials. The following summarizes some test results for twelve fiducials on white and black backgrounds. Test images are captured with a Sony DXC-151A color video camera digitized at 640 × 480 resolution. Two different lighting sources, daylight and fluorescent light, and a range of aperture settings ($f = 1.8 \sim 8.0$) of the camera create lighting variations. The algorithm detects all fiducials under every aperture settings, except for the yellow fiducial on a white background at $f = 1.8$ and a green fiducial on a black background at $f = 8.0$ and 5.6 (Table 6.1). The human eye does not easily perceive these undetected fiducials either.

Figures 6.4 (a) and (b) show two images from the Table 6.1 tests. Figure 6.4 (c) and (d) show real time video images of a test scene with a more complex background and shadow patterns. In all cases the white crosses indicate a detected fiducial that also passes the shape and color tests.

Table 6.1 Results of fiducial detection algorithm. Figure 6.4 shows example images of the experiment. Y indicates detection and N indicates no detection.

Number of detected fiducials		11	12	12	12	12	11	11	12	12	12	11	11
White background	Red	Y	Y	Y	Y	Y	Y	Y	Y	Y	Y	Y	Y
	Green	Y	Y	Y	Y	Y	Y	Y	Y	Y	Y	Y	Y
	Blue	Y	Y	Y	Y	Y	Y	Y	Y	Y	Y	Y	Y
	Yellow	N	Y	Y	Y	Y	Y	N	Y	Y	Y	Y	Y
	Cyan	Y	Y	Y	Y	Y	Y	Y	Y	Y	Y	Y	Y
	Magenta	Y	Y	Y	Y	Y	Y	Y	Y	Y	Y	Y	Y
Black background	Red	Y	Y	Y	Y	Y	Y	Y	Y	Y	Y	Y	Y
	Green	Y	Y	Y	Y	Y	N	Y	Y	Y	Y	N	N
	Blue	Y	Y	Y	Y	Y	Y	Y	Y	Y	Y	Y	Y
	Yellow	Y	Y	Y	Y	Y	Y	Y	Y	Y	Y	Y	Y
	Cyan	Y	Y	Y	Y	Y	Y	Y	Y	Y	Y	Y	Y
	Magenta	Y	Y	Y	Y	Y	Y	Y	Y	Y	Y	Y	Y
Aperture(f)		1.8	2.0	2.8	4.0	5.6	8.0	1.8	2.0	2.8	4.0	5.6	8.0
Lighting source		Day light						Fluorescent light					

6.2.2 Scalable Fiducials

Multi-ring color fiducial systems allow for scalable numbers of fiducials and scalable tracking areas. A set of design rules describes fiducial parameter relationships that satisfy a given AR application requirement [25]. In addition, the fiducial detection method converts a complex 5-DOF shape test to a series of simple problems.

When a camera is very far (or close) to a fiducial, the projected fiducial size is too small (or large) to detect it correctly. Therefore, AR systems with single-size fiducials have limited tracking range. The major axis length d of a projected fiducial ellipse is $d = Df/w$, where D is the diameter of the fiducial, $f (= f_u = f_v)$ is the effective focal length, and w is the depth of the fiducial in the camera coordinate system. The tracking range could be extended by combining the detection ranges of different size (D) concentric rings.

Multi-ring color fiducials have varying numbers of rings, each of different sizes (Figure 6.5). The first-level fiducial has one core circle and one outer ring. With increasing levels, additional rings surround the previous level fiducial. The number of rings specifies the fiducial level. Colored rings (*e.g.*, red, green, blue, yellow, magenta, and cyan) create unique patterns that facilitate fiducial identification. Figure 6.4 shows three size levels of multi-ring color fiducials with two regular rules for the ring width: constant width and proportional width.

The proportional width system leads to scalable tracking range. The size ratio c between adjacent levels is constant

$$
\begin{aligned}
D_i &= cD_{i-1} \quad (c > 1) \\
&= c^{i-1}D_1
\end{aligned}
$$

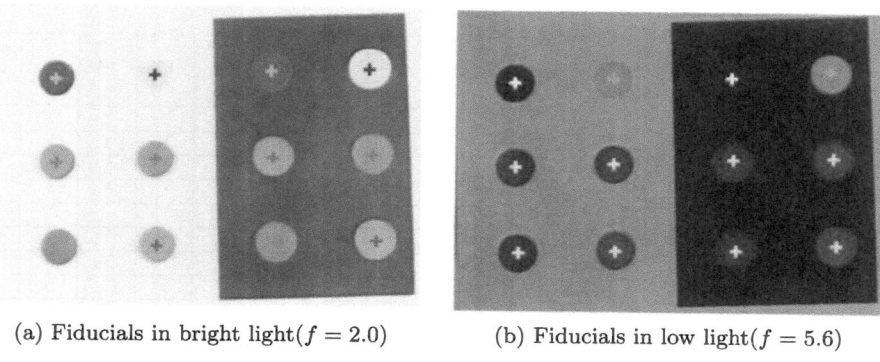

(a) Fiducials in bright light($f = 2.0$)   (b) Fiducials in low light($f = 5.6$)

(c) Truck scene with bright uneven   (d) Truck scene with low uneven
lighting($f = 4.0$)   lighting($f = 8.0$)

Figure 6.4   Fiducial detection under varying light conditions and backgrounds. The green rectangle without a white cross on right side of (d) indicates a detected fiducial that failed color test (see color pages).

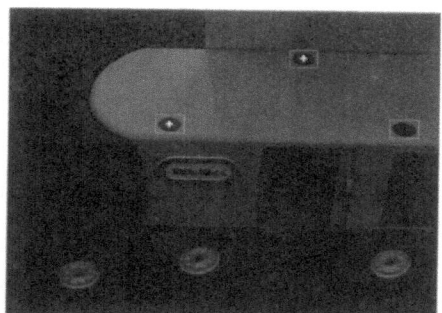

	First level	Second level	Third level
Proportional width ring fiducials			
Constant width ring fiducials			

Figure 6.5   Concentric ring fiducials allow multiple levels and unique size relationships for each ring (see color pages).

and D_i is the diameter of the level i fiducial. The outer rings are more detectable from a distance, so higher level fiducials have a greater detectable range. By combining the detectable ranges of many level fiducials, we extend the system tracking range.

Let the desired tracking range be $Z_{near} \sim Z_{far}$, and the camera focal length be f. Let w be the minimum detectable ring width in an input image. w depends on the camera, the digitizer, and fiducial detection algorithm. Let the tracking range of a level i fiducial be $Z_{near,i} \sim Z_{far,i}$ with the conditions $Z_{near} = Z_{near,i}$ and $Z_{far,n} = Z_{far}$. The largest detectable fiducial size in an image is d_{near} ($\geq D_i f/Z_{near,i}$), and the smallest detectable fiducial size is d_{far} ($\leq D_i f/Z_{far,i}$). To combine the detectable ranges smoothly, there should be no gaps between adjacent work ranges.

$$0 \leq Z_{far,i} - Z_{near,i+1} \leq \frac{D_i f}{d_{far}} - \frac{D_{i+1} f}{d_{near}} = \frac{D_i f}{d_{far}} - \frac{c D_i f}{d_{near}} = \frac{D_i f}{d_{near}}\left(\frac{d_{near}}{d_{far}} - c\right)$$

$$c \leq \frac{d_{near}}{d_{far}} \; and \; c \leq \frac{Z_{far,i}}{Z_{near,i}}$$

The required levels of fiducials can be expressed as a function of the size ratio c for a given camera working range (Figure 6.6).

$$D_n = \frac{Z_{far}}{f} d_{far}, D_1 = \frac{Z_{near}}{f} d_{near}$$

$$D_n = C^{n-1} D_1$$

$$\frac{Z_{far}}{f} d_{far} = c^{n-1} \frac{Z_{near}}{f} d_{near}$$

$$\frac{Z_{far}}{Z_{near}} = C^{n-1} \frac{d_{near}}{d_{far}} \geq c^n$$

$$n(c) = \frac{log(Z_{far}/Z_{near})}{log c}$$

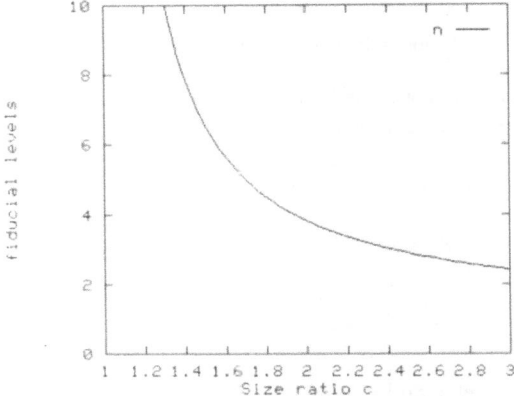

Figure 6.6 Fiducials levels required for an example working range.

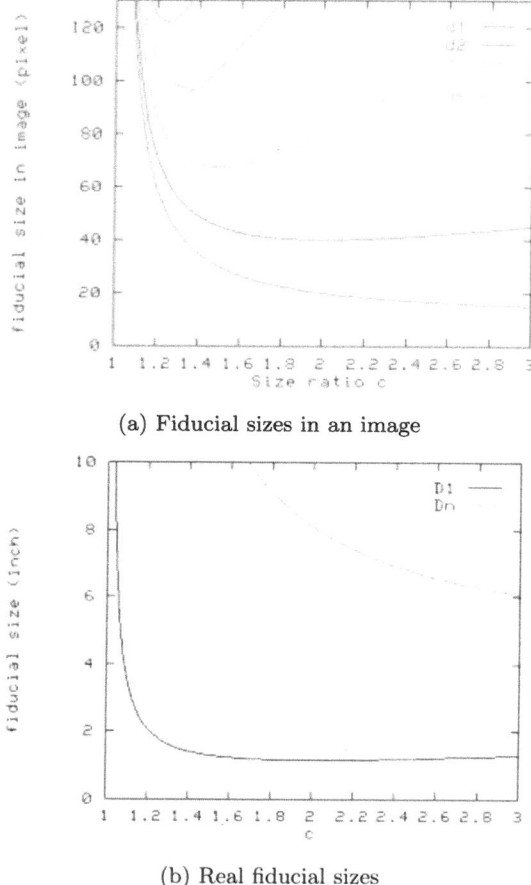

(a) Fiducial sizes in an image

(b) Real fiducial sizes

Figure 6.7   Image and actual fiducial sizes as a function of c.

With the camera at $Z_{far,1}$ from a level n fiducial, the major axis length of the level 1 , 2, and j rings in an input image are

$$d_{far,1} = \frac{2w}{c-1}c \geq d_{far}$$

$$d_2 = \frac{2w}{c-1}c^2 \leq d_{near,2} \leq d_{near}$$

$$d_j = \frac{2w}{c-1}c^j, (1 \leq j \leq n)$$

The diameters of the level 1 and n fiducials are

$$D_n = \frac{Z_{far}}{f}d_{far} = \frac{Z_{far}}{f}\frac{2w}{c-1}c$$

$$D_1 = \frac{Z_{near}}{f} d_{near} \geq \frac{Z_{near}}{f} \frac{2w}{c-1} c^2$$

$$= c^{1-n} D_n = \frac{Z_{far}}{f} \frac{2w}{c-1} c^{2-n}$$

Figure 6.7 shows the major axis lengths of some fiducial levels, and the minimum and maximum fiducial sizes as a function of c.

Additional design parameters, such as fiducial spacing constraints are detailed in [25]. An example design is given for tracking in a 20 × 20 foot room. The camera $FOV_u = 41.7°$ and $FOV_v = 32°$. Let the minimum detectable ring width w be five pixels. The closest operating distance is approximately arms-length (two feet) and the farthest distance is the corner-to-corner length $20\sqrt{2}$ feet. The required number of fiducial levels is $log(20\sqrt{2}/2)/log2 \approx 4$. The fiducial sizes in each level are $D_1 = 1.01"$, $D_2 = 2.03"$, $D_3 = 4.06"$. and $D_4 = 8.1"$. The inter-fiducial distances are $L_1 = 5.6"$, $L_2 = 11.2"$, $L_3 = 22.3"$, and $L_4 = 44.5"$.

A laboratory demonstration uses an SGI Indy (MIPS4400@200MHz with 24-bit graphics) and a SONY DXC-151A color-video camera with S-video output digitized at 640 × 480 resolution. The lens produces a 31.4° horizontal and 24.37° vertical FOV. A three-level proportional-width fiducial set with six colors (red, green, blue, yellow, cyan, and magenta) is printed on a laser printer. The diameters of the fiducials at each level are: first level = 0.8", second level = 1.6", and third level = 3.2". Rings of 20 - 50 pixels in diameter are detected. The detection range of the first level fiducial is $1.5' - 3.7'$, the second level $3.0' - 7.4'$, the third level $5.9 - 14.8'$. The combined system detection range is $1.5' - 14.8'$. Figure 6.8 shows three images of detection results at varied distances from the three level fiducials. Detected fiducials are marked with a white cross-hair at their center. The system performance depends on the number and size of potential fiducials in the image. There is no prediction of fiducial positions, the whole image is processed every frame. The refresh rates are 6.3 frames per second (FPS) at Figure 6.8(a), 7.3 FPS at Figure 6.8(b), 8.1FPS at Figure 6.8(c).

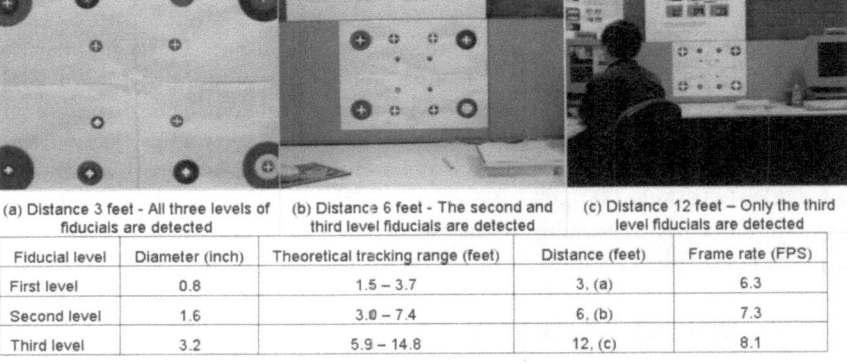

(a) Distance 3 feet - All three levels of fiducials are detected (b) Distance 6 feet - The second and third level fiducials are detected (c) Distance 12 feet – Only the third level fiducials are detected

Fiducial level	Diameter (inch)	Theoretical tracking range (feet)	Distance (feet)	Frame rate (FPS)
First level	0.8	1.5 – 3.7	3, (a)	6.3
Second level	1.6	3.0 – 7.4	6, (b)	7.3
Third level	3.2	5.9 – 14.8	12, (c)	8.1

Figure 6.8 Detection results. The detected fiducials have a white cross hair at their center (see color pages).

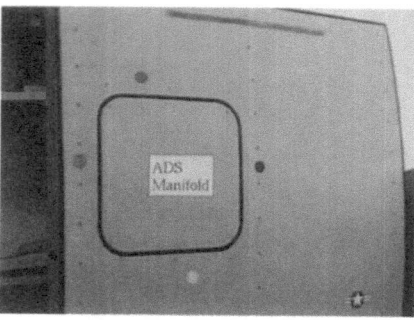

(a) Identification of ADS manifold access panel

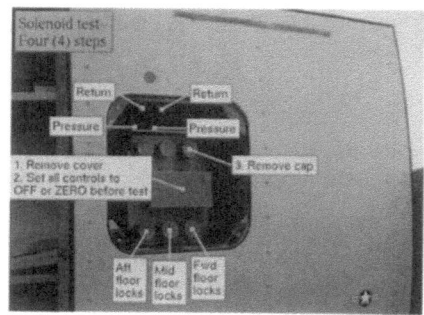

(b) Recognition of open panel causes identification of components and instruction to remove cap

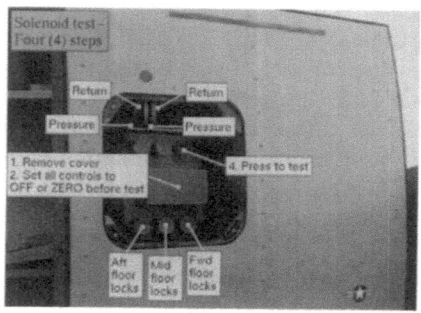

(c) Recognition of cap removal causes new instruction to press button

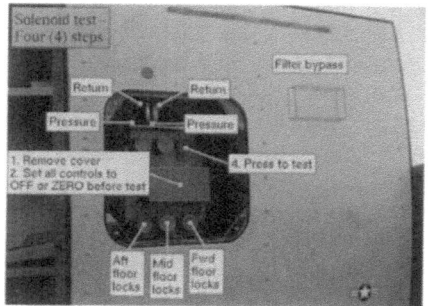

(d) Additional information about position of occluded filter bypass structure in response to user's query

Figure 6.9   Sequence of automatic instructions and information for maintenance task.

6.2.3   Pose Estimation

Three of more detected fiducials facilitate pose calculation. Three points provide up to four possible solutions [26]–[29], four non-planar points provide an analytic solution [30]. Recursive filters [31]–[33] estimate pose from one or more points sensed at any instant in time. Projective invariants provide a relative projection for five points under full perspective [34] or four points under weak-perspective [13]. Model-based methods include [35] [36]. We started with the three-point approach in [26] and we are currently developing robust recursive filters that use all visible features with position confidence values.

6.2.4   Indoor AR Applications

The combination of feature detection and pose estimation provides the basic capability for vision-tracked AR applications [28] [37] [38]. Human factors and cognitive issues for AR manufacturing and maintenance applications are detailed in [39]. Figure 6.9 illustrates a sequence of AR images from a simulated maintenance application.

Annotation alignment with the structure is based on detecting dot fiducials and computing camera pose as described above. This example illustrates the simplicity and minimal intrusion offered by sticker fiducials and the flexibility of the approach. The software library developed for this application supports our laboratory research as well as continued experimental application development.

6.2.5 Extendible Tracking

A drawback of tracking from fiducials is the limited range of camera viewpoints from which the fiducials are visible. Camera pan or zoom can quickly move fiducials out of view, resulting in a loss of camera tracking. Allowing users to interactively place new fiducials in the scene, when and where needed, can extend the range of tracked camera positions. The new fiducials are calibrated from the initial fiducials [40] by using recursive filters that estimate the feature positions as the system is used. Once calibrated to sufficient accuracy and confidence, the new fiducials can support pose calculation.

In [40], two recursive filters are used to estimate the 3D positions of new fiducials. The EKF (Extended Kalman Filter) has the 3D position of a new fiducial as its state. The equations for the filter are simple and fast to compute. The EKF is composed of a *predictor* (time update) and *corrector* (measurement update). The state prediction is simply the prior state since the scene is assumed to be rigid and static.

$$
\begin{aligned}
\hat{x}_k^- &= \hat{x}_{k-1} \\
P_k^- &= P_{k-1} + Q \\
\hat{z}_k &= \boldsymbol{h}(\hat{x}_k^-, c_k, p_c)
\end{aligned}
$$

The corrector equations correct the predicted state value $\hat{x}_k^-$ based on the residual difference between the actual measurement z_k and the measurement estimate $\hat{z}_k$. The Jacobian matrix linearizes the measurement function.

$$
\begin{aligned}
K_k &= P_k^- H_k^T (H_k P_k^- H_k^T + R)^{-1} \\
\tilde{z}_k &= z_k - \hat{z}_k \\
\hat{x}_k &= \hat{x}_k^- + K_k \cdot \tilde{z}_k \\
P_k &= (I - K_k H_k) P_k^-
\end{aligned}
$$

More detail about Kalman filters can be found in references such as [31]–[33].

The RAC (Recursive Average of Covariances) filter projects the screen measurement into the 3D space occupied by the fiducials. Both filters are stable in practice. The EKF has optimal characteristics under certain conditions [33], however the RAC filter gives comparable results, and it is simpler, operating completely in 3D-world space with 3D lines as measurements. The RAC approach eliminates the Jacobian matrix linearization required in the EKF.

Extendable tracking can make use of fiducials as described above or natural scene features such as corners or textures. The latter approach requires methods for natural feature (NF) detection and tracking (correspondence) in the 2D-image sequences produced by a moving camera. A closed-loop NF tracking method [41] is described in the next section.

Figure 6.10 illustrates how extendible tracking is achieved with NF tracking in combination with fiducial tracking and new point calibration. Newly detected natural features are tracked in consecutive images. Feature image coordinates are input to a recursive filter that estimates the 3D feature positions. These 3D positions and their corresponding image coordinates are used to track camera pose in the event that any or all of the fiducials become occluded or undetectable.

Experiment with NF tracking are done in off-line but automatically (no user intervention). Natural feature tracking is about an order of magnitude too slow ($\sim$1Hz) for real-time applications. Optimizations, dedicated hardware, and DSP processor implementations of the algorithms may lead to real-time implementations. The sequence shown in Figure 6.10 demonstrates the automatic detection and calibration of natural features during camera motion, and the automatic use of the natural features to perform camera tracking when fiducials become obscured.

6.3 Outdoor Tracking

The challenge for AR in outdoor settings is to track without any modification or preparation of the environment. Visual sensing is still useful, but there are often no recognizable or predictable targets to support tracking. Whatever type of features are in the environment, they must be detected and tracked (corresponded) in 2D image sequences before any camera motion (tracking) is computed. Corner features or points can be detected in many outdoor scenes, however, tracking them is difficult since corners and points are locally similar to each other. Information about the arrangement of groups of features is useful for establishing correspondences. Region tracking can provide both this neighborhood context to aid tracking, as well as a motion prediction for the individual features that helps minimize the search for the best feature correspondence.

6.3.1 Closed-Loop Feature Tracking

A closed-loop approach provides accurate and robust natural feature detection and tracking in natural environments [41]. The system integrates three main motion-analysis functions, feature selection, motion tracking, and estimate verification, in a closed-loop cooperative manner to track through complex natural imaging conditions. Point features and region features are selected, tracked, and evaluated for their suitability and reliability for computing motions. Continued tracking of either feature type is dependent on the tracking confidence C that is derived from a suitability metric (λ) and the confidence of tracking (δ).

$$C = k_1\lambda + k_2\delta$$

where k_i is a weighting coefficient for each component.

Region tracking is a differential-based local optical flow estimation [42]–[47]. A multi-scale estimation strategy iteratively fits region and point motion estimates to an affine motion model until they agree. The three iterative components (Figure

(a) Tags show positions of automatically detected, tracked,
and calibrated natural features

(b) Initial image with fiducial camera tracking

(c) Frame 295 with camera tracking based on tracked and calibrated natural features

Figure 6.10   The sequence starts by tracking camera pose from fiducials (b), while natural features are automatically detected and calibrated. Note the annotation indicating the blue fiducial and the side door of the truck. As the camera drops low to the ground at the end of the sequence (c), the fiducials are no longer usable for tracking since their aspect is extreme, and natural features automatically support continued tracking.

6.11) of the method are image warping, motion residual estimation, and motion refinement. The method operates independently on selected regions of the image. Every region motion estimate is fit to an affine model. A verification and evaluation measures the confidence of the estimation and the model fit. If the estimate error is large, an iterative refinement is done until the error converges or the region is discarded as unreliable for tracking.

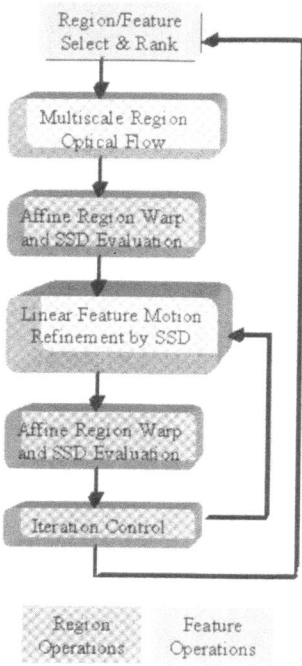

Figure 6.11   Feedback from tracking evaluation controls the iterative refinement of motion estimates within each region.

An affine model of region motion accounts for the geometric distortions that occur with large view variations and long sequence tracking. A simple translation models point feature motion. These models are the basis of the motion verification and evaluation processing. In regions, the optical-flow motion estimate determines the parameters of an affine motion model that specifies a warp of the region into an confidence evaluation frame $(x_c, y_c)^T$.

$$\begin{bmatrix} x_c \\ y_c \end{bmatrix} = \begin{bmatrix} v_2 x_{t0} + v_3 y_{t0} + v_1 \\ v_4 x_{t0} + v_5 y_{t0} + v_6 \end{bmatrix}$$

The confidence frame is compared to the true target image to obtain a measure of tracking error

$$\varepsilon = \frac{\|R_t(\mathbf{x}, t) - R_c(\mathbf{x}, t)\|^2}{max\|R_t(\mathbf{x}, t)\|^2, \|R_c(\mathbf{x}, t)\|^2}$$

Table 6.2 Performace comparison for various optical flow appproaches with synthetic data sequence.

Technique	Average Angle Error	Standard Deviation
Horn and Schunck	11.26	16.41
Lucas and Kanade	4.10	9.58
Anandan	15.84	13.46
Fleet and Jepson	4.29	11.24
Closed-loop approach	2.84	7.69

where $R_t(\mathbf{x}, t)$ and $R_c(\mathbf{x}, t)$ are the true target frame and confidence evaluation frame, respectively. The affine parameters predict the translation of point features within a region. The predictions are refined by a local search for a correlation peak. The refined point motions determine new parameters for their region motion model. The new parameters determine a new confidence frame and a new assessment of tracking accuracy. The process of region and point motion modeling and accuracy assessment is iterated until the estimates converge or features are discarded.

Tracking accuracy feedback is an essential component of the tracking system. As described above, the error information is used by the tracking module for motion correction. The error also used in the feature detection module for continuous feature re-evaluation. Re-evaluation keeps the system working in an "optimum" state by automatically selecting and maintaining only the most reliable features. The tracking accuracy feedback (or tracking confidence) is defined as

$$\delta = \frac{1}{1 + \varepsilon}$$

The closed-loop stabilization of the tracking system is inspired by the use of feedback for correcting errors in non-linear control system. The process make it possible to discriminate between good and poor estimation features, and maximizes the quality of the final motion estimation.

Accuracy comparison

Our approach compares favorably to other published methods. Figure 6.12 is from a synthetic image sequence. Table 6.2 gives numerical comparisons of the motion errors for this sequence that has camera moving along its view axis towards the mountain and valley. The camera motion generates diverging motion flow around the upper right of the mountain, producing one pixel per-frame translation motion in the cloud area and about four pixels per-frame of motion in the lower-left area. For this test, the original image size is (256×256).

An example of tracking natural features in an outdoor scene is shown in Figure 6.13 where the camera pans and translates simultaneously. The tracking method automatically chooses the "best" features, and these are highlighted with motion vectors in Figure 6.13.

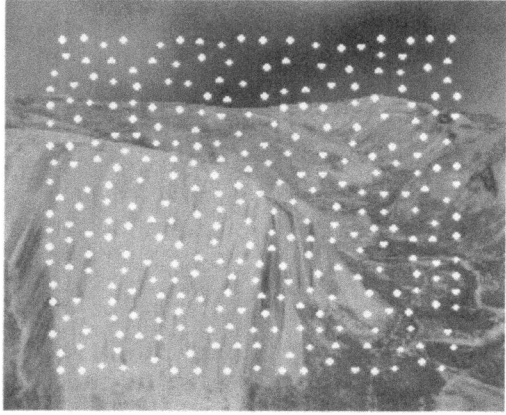

(a) detected tracking features

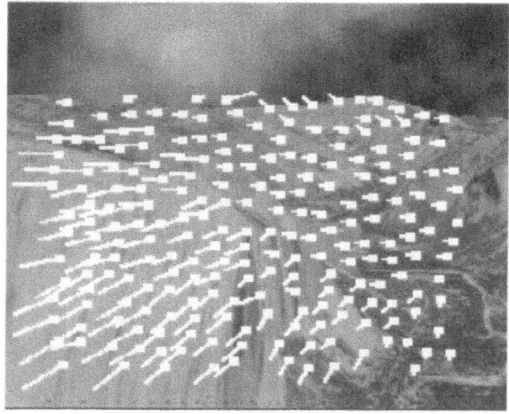

(b) estimated motion field

Figure 6.12   Synthetic image sequence(Yosemite-Fly-Through) for performance evaluation.

Figure 6.13   Tracking result for an outdoor natural scene.

Angle error measure

The angle error measure computed for Table 6.2 treats image velocity as a spatio-temporal vector $\mathbf{v} = (u, v, 1)$ in units of (pixel, pixel, frame). The angular error between the correct velocity $\mathbf{v}_c$ and the estimate $\mathbf{v}_e$ is $Error_{angle} = arccos(\mathbf{v}_c, \mathbf{v}_e)$, where $\mathbf{v} = \frac{(u,v,1)^T}{\sqrt{u+v+1}}$.

This angle error measure represents large and small velocities without the amplifications inherent in a relative measure of small vector differences. The measure has potential bias however; directional errors for small velocities do not give as large an angular error as similar directional errors with large velocities. For this reason, we also used the RMS measure

$$Error_{rms} = \sqrt{\frac{\sum_{\mathbf{x} \in \Omega} (I_c(\mathbf{x}, t) - I_e(\mathbf{x}, t))^2}{MN}}$$

where $I_c(\mathbf{x}, t)$ is a size $M \times N$ region of a real image sequence at time t, and $I_e(\mathbf{x}, t)$ is the reconstructed region based on the estimated motion field. Note that this error measure is similar to the motion residual measure we use to assess tracking accuracy.

Direct image annotation

Robust natural feature tracking enables the direct application of annotation to image sequences. Figure 6.14 illustrates an application where annotations are interactively placed in an initial frame. The tracking system automatically aligns the annotation in the remaining frames of the sequence. This example portrays information that guides astronauts through a tedious equipment installation.

6.3.2 Hybrid Inertial and Video Tracking

Vision tracking is computationally demanding and susceptible to occlusion or numerical instability. Automatic feature calibration (or extendible tracking) addresses this to some degree, however inertial sensors provide a compelling option. Inertial sensors are completely passive, requiring no external devices or targets, however, their drift rates in portable strapdown configurations are too great for practical use by themselves [48]–[50]. A hybrid approach, integrating inertial and vision-based technologies, can exploit the complementary nature of the two technologies and compensate for their respective weaknesses [51].

The basic principles behind inertial sensors for determining orientation and position rest on Newton's laws. Accelerometers measure linear acceleration vectors with respect to the inertial reference frame. In order to subtract the acceleration component due to gravity, the orientation of the linear accelerometers must be accurately determined. Rate gyros sense changes in orientation. A time integration of the gyro outputs computes the orientation changes. The integration of gyro signals and errors gives rise to a linear growth in orientation error. Correction techniques may include magnetic compass measurements, however; compass signals are also noisy and especially subject to errors induced by ferrous materials. Indoor and urban compass data are unreliable. Vision-based corrections, on the other hand, may generalize to a wide range of environments. Since accelerometers require orientation

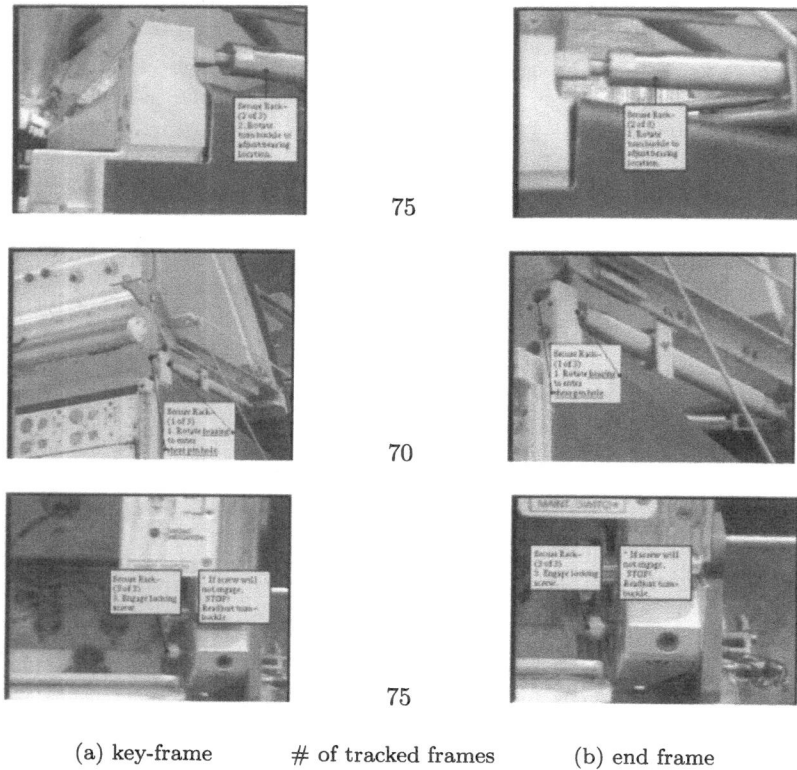

75

70

75

(a) key-frame   # of tracked frames   (b) end frame

Figure 6.14   Direct scene annotation (a) initial frames used to interactively place annotations, (b) later frames in the same sequences showing the automatic tracking of the selected features after the indicated number of frames of hand-held camera motion.

for subtracting the gravity vector, we start our inertial sensor investigations with a focus on rate gyros.

Error sensitivity of inertial tracking

We experimented with a three-degree of freedom (3DOF) orientation tracker produced by InterSense (Model IS-300). This device incorporates three orthogonal gyroscopes to sense angular rates of rotation along its three perpendicular axes. It also has sensors for the gravity (down) vector and a compass [50] to compensate for gyro drift. The measured angular rates are integrated to obtain the three orientation measurements (Yaw, Pitch, and Roll). This system is specified as achieving approximately 1° RMS static orientation accuracy and 3° RMS dynamic accuracy, with 150Hz maximum update rate. Although adequate for interactive applications in virtual reality, this accuracy is inadequate for AR tracking. To demonstrate this, map the specified error into the 2D image domain.

Let (f_x, f_y) be the effective horizontal and vertical focal lengths of a video camera (in pixels), and (L_x, L_y) represent the horizontal and vertical image resolutions, respectively. The field-of-view (FOV) of the camera is calculated as

$$\theta_x = 2tan^{-1}(\frac{L_x}{2f_x})$$

$$\theta_y = 2tan^{-1}(\frac{L_y}{2f_y})$$

If pixels sample the rotation angles uniformly (Yaw and Pitch), the ratio of image pixel motion to the rotation angles (pixel/degree) is given as

$$L_x/\theta_x = \frac{L_x}{2tan^{-1}(L_x/2f_x)}$$

$$L_y/\theta_y = \frac{L_y}{2tan^{-1}(L_y/2f_y)}$$

To illustrate a concrete example of this relationship, consider the Sony XP-999 CCD video camera. Through calibration, we determine the effective horizontal and vertical focal lengths as $f_x = 614.059$ pixels, and $f_y = 608.094$ pixels, when digitized at a 640×480 image resolution. The ratios are $L_x/\theta_x = 11.625$ pixel/degree, and $L_y/\theta_y = 11.143$ pixel/degree. That is, each degree of orientation-angle error results in about 11-pixels of image alignment error. In actual use, the error of the inertial tracker may become larger than the one-degree specified. Increasing the FOV of the camera with a wide-angle lens reduces the pixel error proportionately, however wide-angle lenses produce significant radial distortions that also contribute to pixel error [3].

Figure 6.15 illustrates the measured dynamic alignment error obtained with the inertial gyro. In this experiment, the gyro sensor is attached to a video camera to continually report its orientation. There is no measure of ground-truth absolute pose of the sensor/camera, rather the visual feature motions are tracked to evaluate the gyro sensor accuracy relative to the image. By back-projecting the 3D orientation changes reported by the inertial sensor, the gyro motion estimates can be compared with the observed feature motions in the image plane. Changes in the image-space distances are proportional to any errors accumulated by the inertial system. The error measure is appropriate since the ultimate metric of any augmented reality is the perceived image.

As the camera pans around its azimuth axis, video images are captured in real time, and the inertial tracker continually reports the camera orientation. At the start of each test, ten distinct visual features are selected interactively. These features are automatically tracked in subsequent frames, and their positions are compared to the back projected positions computed from the gyro motion data. The average distance between the vision-tracked feature positions and their corresponding backprojected positions is the accuracy metric. Two 500 frame sequences, a far-view (>100 feet) and near-view * scene, are used for the test. Figure 6.15 illustrates the average error distributions for the two scenes, and it clearly shows the dynamic drifts between the gyro data and tracked features.

---

*We only consider pure rotation of the camera. Although we carefully pan the camera to avoid translations, a minor translation is injected by the offset between the rotation axis and the optical center of the camera. For completeness, we consider both a far-view scene with feature ranges of over 100 feet and a near-view (12×12 foot office) scene that is more sensitive to minor translation.

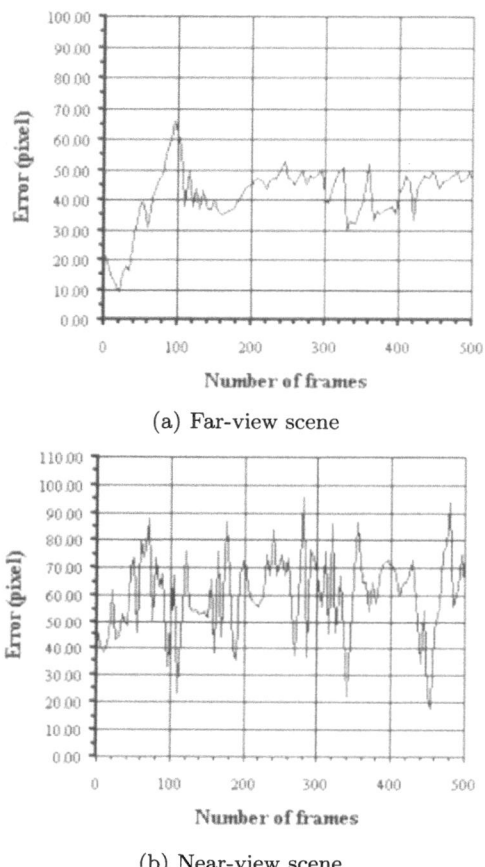

(a) Far-view scene

(b) Near-view scene

Figure 6.15   Average pixel differences between tracked features and backprojected features for (a) distant and (b) near scenes.

Hybrid tracking approach

Our prototype hybrid system fuses inertial orientation (3DOF) data with vision feature tracking to improve vision robustness and correct inertial drift. We treat the fusion as a 2D image stabilization problem. Approximate 2D feature-motion is derived from the inertial data, and vision tracking corrects and refines these estimates in the image domain. Furthermore, the inertial data is a prediction that reduces the correspondence search space and provides tolerance to vision tracking interruptions.

Camera model and coordinates

A CCD video camera has a rigidly mounted 3DOF inertial sensor. There are four principal coordinate systems, as illustrated in Figure 6.16, the world coordinate system $\mathbf{W} : (x_w, y_w, z_z)$, the camera-centered coordinate system $\mathbf{C} : (x_c, y_c, z_c)$, the inertial-centered coordinate system $\mathbf{I} : (x_I, y_I, z_I)$, and the 2D image coordinate system $\mathbf{U} : (x_u, y_u)$.

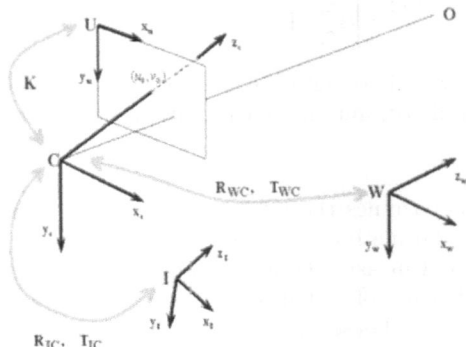

Figure 6.16   Camera model and the related coordinate systems of the hybrid system.

A pinhole camera models the imaging process. The origin of $\mathbf{C}$ is at the projection center of camera. The transformation from $\mathbf{W}$ to $\mathbf{C}$ is

$$
\mathbf{W} : \rightarrow \mathbf{C} : \quad
\begin{bmatrix} x_c \\ y_c \\ z_c \end{bmatrix}
= [\mathbf{R}_{wc} \mid -\mathbf{R}_{wc}\mathbf{T}_{wc}]
\begin{bmatrix} x_w \\ y_w \\ z_w \\ 1 \end{bmatrix}
$$

where the rotation matrix $\mathbf{R}_{wc}$ and the translation vector $\mathbf{T}_{wc}$ characterize the orientation and position of the camera with respect to the world coordinate frame. Under perspective projection, the transformation from $\mathbf{W}$ to $\mathbf{U}$ is

$$
\mathbf{W} : \rightarrow \mathbf{U} : \quad
\begin{bmatrix} x_u \\ y_u \\ z_u \end{bmatrix}
= [\mathbf{K}][\mathbf{R}_{wc} \mid -\mathbf{R}_{wc}\mathbf{T}_{wc}]
\begin{bmatrix} x_w \\ y_w \\ z_w \\ 1 \end{bmatrix}
$$

where the matrix $\mathbf{K}$

$$
\mathbf{K} =
\begin{bmatrix}
\alpha_x f & 0 & u_0 \\
0 & \alpha_y f & v_0 \\
0 & 0 & 1
\end{bmatrix}
$$

contains the *intrinsic* parameters of the camera*, f is the focal length of camera, α_x, α_y are the horizontal and vertical pixel sizes on the imaging plane, and (u_0, v_0) is the projection of camera center (principal point) on the image plane. The intrinsic parameters are calibrated offline.

---

*For simplicity we omitted the lens distortion parameters from the equation.

Camera orientation changes are reported by the inertial tracker, so the transformation between the **C** and **I** is needed to relate the inertial and camera motions. For rotation $\mathbf{R}_{Ic}$ and translation $\mathbf{T}_{Ic}$ the transformation is

$$\mathbf{I} : \rightarrow \mathbf{C} : \quad \begin{bmatrix} x_c \\ y_c \\ z_c \end{bmatrix} = [\mathbf{R}_{Ic}] \begin{bmatrix} x_I \\ y_I \\ z_I \end{bmatrix} + [\mathbf{T}_{Ic}]$$

Since we only use a 3DOF orientation tracker, only the rotation transformation needs to be determined. An automatic calibration method is detailed below.

Static calibration

Camera calibration determines the intrinsic parameters **K** and the lens distortion parameters. We use the method described in [9]. A planar target with a known grid pattern is imaged at measured offsets along the viewing direction. The intrinsic parameters and coefficients of radial lens distortion are computed by an iterative least-squares estimation. These parameters remain constant during our tracking experiments.

The transformation between the inertial and the camera coordinate systems relates the inertial data to the camera motion, and hence to the image feature motions. Measuring this transformation can be difficult, especially with optical see-through display systems. We describe a motion-based calibration, as opposed to the boresight techniques presented in [1]. As discussed above, only the rotation component of the transformation is determined.

Rewriting the relationship between the inertial tracker frame and the camera coordinate frame as

$$\mathbf{x}_C = \lfloor \mathbf{R}_{Ic} \rfloor \mathbf{x}_I + \lfloor \mathbf{T}_{Ic} \rfloor$$

If we consider the vector $\mathbf{x}_c = [x_c, y_c, z_c]^T$ and $\mathbf{x}_I = [x_I, y_I, z_I]^T$ as directions to a point in a scene relative to **C** and **I**, the rotation motion relationship between the two coordinates can be derived

$$\varpi_C = \lfloor \mathbf{R}_{Ic} \rfloor \omega_{\mathbf{I}}$$

$\omega_C = [\omega_{Cx}, \omega_{Cy}, \omega_{Cz}]^T$ and $\omega_I = [\omega_{Ix}, \omega_{Iy}, \omega_{Iz}]^T$ denote the angular velocity of scene points, relative to the camera coordinate frame and the inertial coordinate frame, respectively.

The angular motion ω_I, relative to the inertial coordinate system, is obtained from the inertial tracker output. We need to compute the camera's angular velocity in some way, in order to determine the transformation matrix $\mathbf{R}_{Ic}$. General camera motion can be decomposed into a linear translation $\mathbf{V}_C = [V_{Cx}, V_{Cy}, V_{Cz}]^T$ and an angular motion $\omega_C = [\omega_{Cx}, \omega_{Cy}, \omega_{Cz}]^T$. Under perspective projection, the 2D-image motion resulting from camera motion can be written as

$$\dot{x}_u = \left[\frac{-fV_{Cx} + x_u V_{Cz}}{z_C} + \frac{x_u y_u}{f}\omega_{Cx} - f(1 + \frac{x_u^2}{f^2})\omega_{Cy} + y_u\omega_{Cz} \right]$$

$$\dot{y}_u = \left[\frac{-fV_{Cy} + y_u V_{Cz}}{z_C} + f(1 + \frac{y_u^2}{f^2})\omega_{Cx} - \frac{x_u y_u}{f}\omega_{Cy} + x_u\omega_{Cz} \right]$$

where $(\dot{x}_u, \dot{y}_u)$ denotes the image velocity of point (x_u, y_u) in the image plane, z_C is the range to that point, and f is the focal length of camera. Eliminating the translation term and substituting for ω_c, we have $\dot{\mathbf{x}}_u = \Lambda \lfloor \mathbf{R}_{Ic} \rfloor \omega_{\mathbf{I}}$, where

$$\Lambda = \left[\begin{array}{ccc} \frac{x_u y_u}{f} & -f(1 + \frac{x_u^2}{f^2}) & y_u \\ f(1 + \frac{y_u^2}{f^2}) & -\frac{x_u y_u}{f} & -x_u \end{array} \right]$$

In words, given knowledge of the internal camera parameters, the inertial tracking data $\omega_{\mathbf{I}}$, and the related 2D motions $\lfloor \dot{x}_u, \dot{y}_y \rfloor$ of (at least 5) image features, the transformation $\mathbf{R}_{Ic}$ between the camera and the inertial coordinate systems can be determined.

Dynamic registration

Static registration establishes an initial calibration, however the inertial system accumulates drift and errors with motion. The distribution of drift and error (Figure 6.15) is difficult to model for analytic correction. Our strategy is to minimize the tracking error in the image plane, relative to the visually-perceived image.

Suppose N points are annotated in the scene. Their projections in the image are (x_i, y_i), $i = 1, 2 \cdots N$. Our goal is to automatically track these features as the camera moves in the subsequent frames. Inertial data provides a tracking prediction, and vision tracking provides a correction.

Let $\omega_{\mathbf{C}} = [\omega_{Cx}, \omega_{Cy}, \omega_{Cz}]^T$ be the camera rotation from frame $I(\mathbf{x}, t-1)$ to frame $I(\mathbf{x}, t)$. For the scene points O_i, their 2D positions in the image frame $t-1$ are $\mathbf{x}_{it-1} = [x_{it-1}, y_{it-1}]^T$. The positions of these points in the frame t, due to the related motion (rotation) between the camera and the scene, can be estimated

$$\mathbf{x}_{it} = \mathbf{x}_{it-1} + \Delta \mathbf{x}_{it}$$
$$\Delta \mathbf{x}_{it} = \Lambda \omega_{\mathbf{C}}$$

where $i = 1, 2 \cdots N$, and *Lambda* is determined as above.

The predicted image positions are refined by local searches for the true features. A normalized correlation is used as the feature matching metric. The feature located at $\mathbf{x}_{it-1}$ in frame $t-1$ is correlated with frame t around the location $\mathbf{x}_{it}$ predicated by the inertial tracker. The result of the search forms a correlation surface, and the location of the peak value of the surface is the correct (best) match. Correlation is defined as

$$\varepsilon = \frac{(2n+1)^2 \mu_{t-1t} - \mu_{t-1}\mu_t}{\sqrt{(2n+1)^2 \sigma_{t-1} - \mu_{t-1}^2}\sqrt{(2n+1)^2 \sigma_t - \mu_t^2}}$$

To minimize the effect of pixel position quantization, 2D bilinear interpolation refines the estimate of the peak to subpixel accuracy.

Experiment

Figure 6.17 shows two images from far-view and near-view video sequences. In these frames, black dots identify the feature points that we track and annotate. The yellow boxes are annotation text banners positioned only from inertial data, while the red boxes denote the vision-corrected positions. The resolution of the images is 640×480.

(a) far-view

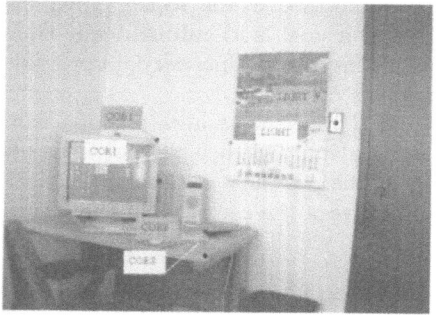

(b) near-view

Figure 6.17   Images from two video sequences showing vision(red) and gyro (yellow) tracking results.

Inertial-only tracking

This test uses only inertial data for tracking. Ten distinct features are manually selected in the initial frame to establish visual reference points. The selected features are backprojected in each frame based on the camera orientation change reported by the inertial tracker. The average differences between the backprojected image positions and the observed (vision-tracked) feature positions are the measure of tracking accuracy in each frame. Figure 6.18 shows the average error distributions for the two scenes confirming that substantial errors occur.

Hybrid inertial-vision: case 1

This test performs inertial tracking and vision-based correction of the integrated gyro error. The predication of 2D image motion is based on the integrated inertial orientation. This approach allows the inertial drift to accumulate, unaffected by any

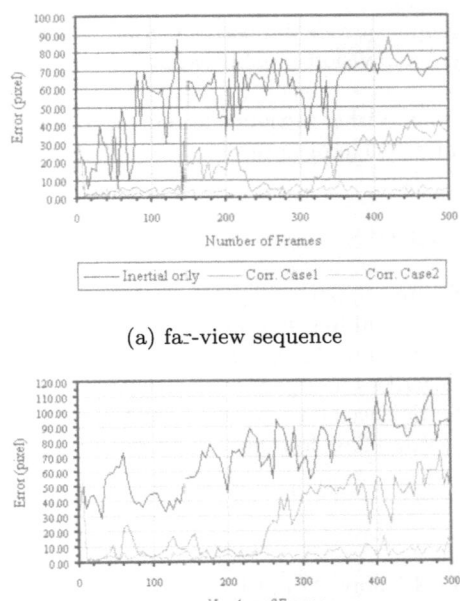

(a) far-view sequence

(b) near-view sequence

Figure 6.18 Hybrid alignment error for (a)far-view and (b) near-view scenes with inertial only, hybrid case 1, and hybrid case 2 methods (see color pages).

vision-based corrections. This simulates the effect of prolonged occlusion or tracking loss in the vision system. Figure 6.18 illustrates the results for both test scenes.

Hybrid inertial-vision: case 2

An alternative correction approach is incremental. Each vision correction results in an adjustment of the gyro state, consequently, the gyro error accumulation (assuming ideal corrections) is limited to periods between corrections. (This is similar to the "ZUPT'ing" method used for linear motion sensing.) The reduced drift integration periods result in better prediction and better registration as illustrated in Figure 6.18. A drawback of this approach is the possibility that a spurious correction error produces a lingering bias in the result.

6.4 Summary and Conclusions

We described our views and strategies to track in (almost) natural environments. Robust fiducial detection, scalable fiducials, and extendible tracking enable vision tracking for indoor settings. The closed-loop approach to natural feature tracking provides the robust long-sequence 2D correspondences needed for natural settings. A fusion of inertial gyro and vision capabilities produce stable and robust orientation measures.

Much remains undone. Other sensors and representations of data can be brought to bear on this critical problem of natural environment tracking.

Acknowledgements

Portions of this work are supported by the Defense Advanced Research Project Agency (DARPA) "Geospatial Registration of Information for Dismounted Soldiers." The Integrated Media Systems Center provides support and facilities. Recognition for much of the work described here goes to the research members of the AR Tracking Group in the Computer Science Department at the University of Southern California, Dr. Suya You, Youngkwan Cho, Jongweon Lee, Jun Park, and Bolan Jian.

References

[1] R. Azuma and G. Bishop: "Improved static and dynamic registration in an optical see-through HMD," *Proc. SIGGRAPH'94*, pp.197–203, 1994.

[2] R. Azuma: "A survey of augmented reality," *SIGGRAPH'95 course #9 notes*, August 1995.

[3] M. Bajura and U. Neumann: "Dynamic registration correction in augmented reality systems," *Proc. VRAIS'95*, pp.189–196, 1995.

[4] I. Sutherland: "A head-mounted three-dimensional display," *Proc. Fall Joint Computer Conf.*, pp.757–775, 1968.

[5] K. Meyer, H. L. Applewhite, and F. A. Biocca: "A survey of position trackers," *Presence: Teleoperators and Virtual Environments*, vol.1, no.2, pp.173–200, 1992.

[6] M. Ward, R. Azuma, R. Bennett, S. Gottschalk, and H. Fuchs: "A demonstrated optical tracker with scalable work area for head-mounted display systems," *Proc. Symp. on Interactive 3D Graphics*, pp.43–52, 1992.

[7] M. Ghazisadedy, D. Adamczyk, D. J. Sandlin, R. V. Kenyon, and T. A. DeFanti: "Ultrasonic calibration of a magnetic tracker in a virtual reality space," *Proc. VRAIS'95*, pp.179–188, 1995.

[8] D. Kim, S. W. Richards, and T. P. Caudell: "An optical tracker for augmented reality and wearable computers," *Proc. VRAIS'97*, pp.146–150, 1997.

[9] U. Neumann and Y. Cho: "A self-tracking augmented reality system," *Proc. VRST'96*, pp.109–115, 1996.

[10] A. State, G. Hirota, D. T. Chen, B. Garrett, and M. Livingston: "Superior augmented reality registration by integrating landmark tracking and magnetic tracking," *Proc. SIGGRAPH'96*, pp.429–438, 1996.

[11] F. Madritsch, F. Leberl, and M. Gervautz: "Camera based beacon tracking: Accuracy and applications," *Proc. VRST'96*, pp.101–108, 1996.

[12] J. P. Mellor: "Enhanced reality visualization in a surgical environment," AI Technical Report 1544, 1995.

[13] K. Kutulakos and J. Vallino: "Affine object representations for calibration-free augmented reality," *Proc. VRAIS'96*, pp.25–36, 1996.

[14] J. Rekimoto: "NaviCam: A magnifying glass approach to augmented reality," *Presence: Teleoperator and Virtual Environments*, vol.6, no.4, pp.399–412, August 1997.

[15] G. Klinker, K. Ahlers, D. Breen, P. Chevalier, C. Crampton, D. Greer, D. Koller, A. Kramer, E. Rose, M. Tuceryan, and R. Whitaker: "Confluence of computer vision and interactive graphics for augmented reality," *Presence: Teleoperator and Virtual Environments*, vol.6, no.4, pp.433–451, August 1997.

[16] D. Koller, G. Klinker, E. Rose, D. Breen, R. Whitaker, and M. Tuceryan: "Real-time vision-based camera tracking for augmented reality applications," *Proc. VRST'97*, Lausanne, Switzerland, pp.87–94, September 1997,

[17] G. D. Hager and P. N. Belhumeur: "Real-time tracking of image regions with changes in geometry and illumination," *Proc. IEEE CVPR*, 1996.

[18] C. G. Healey and J. T. Enns: "A perceptual colour segmentation algorithm," Technical Report, University of British Columbia, 1997.

[19] A. Tremeau and N. Borel: "A region growing and merging algorithm to color segmentation," *Pattern Recognition*, vol.30, no.7, pp.1191–1203, 1997.

[20] J. Kender: "Saturation, hue, and normalized color: Calculation, digitization effects, and use," CMU Computer Science Dept., November 1976.

[21] T. Starner, S. Mann, B. Rhodes, J. Levine, J. Healey, D. Kirsh, R. Picard, and A. Pentland: "Augmented reality through wearable computing," *Presence: Teleoperator and Virtual Environments*, vol.6, no.4, pp.386–398, August 1997.

[22] J. P. Oakley and R. T. Shann: "Efficient method for finding the position of object boundaries to sub-pixel precision," *IVC*, vol.9, pp.262–272, 1991.

[23] Y. Cho, J. Park, and U. Neumann: "Fast color fiducial detection and dynamic workspace extension in video see-through augmented reality," *Proc. 5th Pacific Conf. on Graphics and Applications*, pp.168–177, October 1997.

[24] J. Lee and U. Neumann: "Fuzzy and rule-based fiducial detection for vision-based augmented reality tracking," *Proc. First Int'l Workshop on Augmented Reality*, 1998.

[25] Y. Cho and U. Neumann: "Multi-ring color fiducial systems and a detection method for scalable fiducial-tracking augmented reality," *Proc. First Int'l Workshop on Augmented Reality*, 1998.

[26] M. A. Fischler and R. C. Bolles: "Random sample consensus: A paradigm for model fitting with applications to image analysis and automated cartography," *Computer Graphics and Image Processing*, vol.24, no.6, pp.381–395 1981.

[27] R. Haralick, C. Lee, K. Ottenberg, and M. Nolle: "Review and analysis of solutions of the three point perspective pose estimation problem," *Proc. IJCV*, vol.13, no.3, pp.331–356, 1994.

[28] R. Sharma and J. Molineros: "Computer vision-based augmented reality for guiding manual assembly," *Presence: Teleoperators and Virtual Environments*, vol.6, no.3, pp.292–317, June 1997.

[29] S. Linnainmaa, D. Harwood, and L. S. Davis: "Pose determination of a three-dimensional object using triangle pairs," *IEEE Trans. on PAMI*, vol.10, no.5, pp.634–647, September 1988.

[30] R. Horaud, B. Conio, and O. Leboulleux: "An analytic solution for the perspective 4-point problem," *CVGIP*, vol.47, pp.33–44, 1989.

[31] J. M. Mendel: *Lessons in Estimation Theory for Signal Processing, Communications, and Control*, Prentice Hall PTR, 1995.

[32] G. Welch and G. Bishop: "SCAAT: Incremental tracking with incomplete information," *Proc. SIGGRAPH'97*, pp.333–34, 1997.

[33] T. J. Broida, S. Chandrashekhar, and R. Chellappa: "Recursive estimation from a monocular image sequence," *IEEE Trans. on Aerospace and Electronic Systems*, vol.26, no.4, pp.639–655, July 1990 .

[34] M. Uenohara and T. Kanade: "Vision-based object registration for real-time image overlay," *Proc. Computer Vision, Virtual Reality, and Robotics in Medicine*, pp.13–22, 1995.

[35] T. Tomasi and T. Kanade: "Shape and motion from image streams: a factorization method," Technical Report, Carnegie Mellon University, Pittsburgh, PA, September 1990.

[36] E. Natonek, Th. Zimmerman, and L. Fluckiger: "Model based vision as feedback for virtual reality robotics environments," *Proc. VRAIS'95*, pp.110–117, 1995.

[37] T. P. Caudell and D. M. Mizell: "Augmented reality: An application of heads-up display technology to manual manufacturing processes," *Proc. Hawaii Int'l Conf. on Systems Sciences*, pp.659–669, 1992.

[38] S. Feiner, B. MacIntyre, and D. Seligmann: "Knowledge-based augmented reality," *Comm. ACM*, vol.36, no.7, pp.52–62, July 1993.

[39] U. Neumann and A. Majoros: "Cognitive, performance, and systems issues for augmented reality applications in manufacturing and maintenance," *Proc. VRAIS'98*, pp.4–11, 1998.

[40] U. Neumann and J. Park: "Extendible object-centric tracking for augmented reality," *Proc. VRAIS'98*, pp.148–155, 1998.

[41] U. Neumann and S. You: "Integration of region tracking and optical flow for image motion estimation," *Proc. IEEE Int'l Conf. on Image Processing*, 1998.

[42] P. Anandan: "A Computational framework and an algorithm for the measurement of visual motion," *Int'l J. Computer Vision*, vol.2, pp.283–310, 1989.

[43] D. H. Ballard and C. M. Brown: *Computer vision,* Prentice Hall, ISBN 0-13-165316-4, 1981.

[44] S. S. Beauchemin and J. L. Barron: "The computation of optical flow," *ACM Computing Surveys*, vol.27, no.3, pp.433–466, 1995.

[45] J. R. Bergen and E. H. Adelson: "Hierarchical, computationally efficient motion estimation algorithm," *J. Opt. Soc. Am.*, vol.4, no.35, 1987.

[46] B. K. P. Horn and B. G. Schunk: "Determining optical flow," *Artificial Intelligence*, vol.17, pp.185–203, 1984.

[47] E. C. Hildreth: "Computation underlying the measurement of visual motion," *Artificial Intelligence*, vol.23, pp.309–354, 1984.

[48] K. Britting: *Inertial Navigation System Analysis,* Wiley Interscience, New York, 1971.

[49] D. H. Tittertion and J. L. Weston: "Strapdown inertial navigation technology," *IEE Radar, Sonar, Navigation and Avionics Series 5*, Peter Peregrinus Ltd. UK, 1997.

[50] E. Foxlin: "Inertial head-tracker sensor fusion by a complementary separate-bias Kalman filter," *Proc. VRAIS'96*, pp.184–194, 1996.

[51] E. Foxlin, M. Harrington, and G. Pfeifer: "Constellation: A wide-range wireless motion-tracking system for augmented reality and virtual set applications," *Proc. GRAPHICS 98*, 1998.

[52] "Eagle Eye," Kinetic sciences, http://www.kinetic.bc.ca/eagle_eye.html

[53] D. J. Fleet and A. D. Jeson: "Computation of component image velocity from local phase information," *Int'l J. Computer Vision*, vol.5, pp.77–104, 1990.

[54] B. R. Hunt: "Superresolution of images: algorithms, principles, performance," *IJIST*, vol.6, no.4, pp.297–304, 1995.

[55] B. Lucas and T. Kanade: "An iterative image registration technique with an application to stereo vision," *Proc. DARPA IU Workshop*, pp.121–130, 1981.

[56] H. H. Nagel: "On a constraint equation for the estimation of displacement rates in image sequences," *IEEE Trans. on PAMI*, vol.11, no.1, pp.13–30, 1989.

[57] A. Papoulis: *Probability, Random Variables, and Stochastic Processes, Third Edition*, McGraw-Hill, New York, 1991.

[58] W. K. Pratt: *Digital Image Processing*, A Wiley-Interscience Publication, 1978.

[59] H. Sowizral and J. Barnes: "Tracking position and orientation in a large volume," *Proc. VRAIS'93*, pp.132–139, 1993.

[60] M. Tuceryan, D. S. Greer, P. T. Whitaker, D. Breen, C. Crampton, E. Rose, and K. H. Ahlers: "Calibration requirements and procedures for a monitor-based augmented reality system," *IEEE Trans. on Visualization and Computer Graphics*, vol.1, no.3, pp.255–273, September 1995.

Chapter 7

Stereo Vision Based Video See-through Mixed Reality

Naokazu Yokoya
Haruo Takemura
Takashi Okuma
Masayuki Kanbara

Nara Institute of Science and Technology, Japan

7.1 Introduction

Seamless composition of the real and virtual environments or a technique to put a photo realistic three-dimensional (3-D) model of a real object into a virtual environment is called a mixed reality (MR) technology [1]. Mixed reality enables us to add information on a real environment or to increase the reality of virtual environments. It also draws an attention as one of new promising methods of performing an interaction between real and virtual environments. A number of applications have already been proposed and demonstrated [2]–[5]. The environment that is created by using the mixed reality technology is called a mixed environment.

In order to realize the mixed reality environment, 3-D coordinates of the real and virtual environments must be aligned properly with respect to each other. This is sometimes referred to as geometric registration of real and virtual worlds. For this purpose the user's viewing position and orientation must be precisely measured; see [6] for discussion about registration error. In order to implement a mixed environment with highly realistic sensations, all the processes from acquiring the user's viewing position and orientation to displaying the composed image according to obtained viewing position and orientation information must be done in real-time.

In general, there exist the following two major methods to acquire the user's viewing position and orientation.

- A method that uses 3-D position sensors, such as magnetic, ultra sonic or mechanical sensors; for example, see [3] [7] [8].

- A method that estimates the user's viewpoint by using camera images taken at the viewpoint; for example, see [9]–[12]. This type of method is sometimes referred to as vision-based tracking or registration.

The hybrid of above two is also used to improve the accuracy and the stability of geometric registration [8] [13].

3-D position sensors used in the former method can directly acquire 3-D position and orientation information. However, the drawbacks are that the system requires special equipments and its measuring area is rather limited to a relatively narrow space. On the other hand, the latter estimates the viewing position from a captured image and there is potentially no limitation in measuring area. When the relationship between the camera position and the viewing position is known, the viewing position can be obtained by calculating the camera parameters. This means that conventional techniques studied in the computer vision field can be used to measure the viewing position and orientation [14]–[16].

There are basically two choices in composing real and virtual environments for interactive mixed reality systems: optical see-through and video see-through [2]. Optical see-through uses a standard see-through HMD and is usually accomplished by placing a partially transmissive and reflective mirror on the HMD in front of the user's eyes. The user can see the real world through the mirror and simultaneously can see the virtual world reflected by the mirror. On the other hand, video see-through is accomplished by combining a closed-view HMD with one or two video cameras. Video stream from the camera is composed with virtual objects and then sent to monitors of the HMD in front of the user's eyes. These two approaches have both advantages and disadvantages; see [17] for detailed discussion.

In this chapter, we focus on the combination of vision-based geometric registration and video see-through mixed reality. As suggested above, the vision-based registration does not require any special equipments other than cameras and does not limit the measuring area. Methods to estimate camera parameters from a camera image which contains feature points in the real environment are proposed by several research groups [9]–[12] [18]. The authors of this chapter have already built a prototype system of creating a mixed reality environment that uses four colored markers as feature points [19]. In this system, the position of the viewing point is calculated from the positions of the feature points in a single camera image. The system uses a video see-through HMD as a display device. Thus the displaying-timing of the virtual environment and that of the real environment are synchronized and the alignment error is reduced. Figure 7.1 illustrates the hardware configuration of this system. Figure 7.2 shows examples of composite images produced by the system. In this figure the virtual object, a sake cask, is displayed on a table based on the projection matrix that is recovered from the four markers in the scene. However, a monocular vision system does not allow users to view objects with rich 3-D sensations. Moreover, a virtual object that should be behind real objects occludes them and causes an occlusion conflict. Based on the experience with the monocular video see-through HMD [19], we have developed a new prototype system, a stereoscopic video see-through HMD, that employs a pair of stereo cameras mounted on a standard closed-view mode HMD.

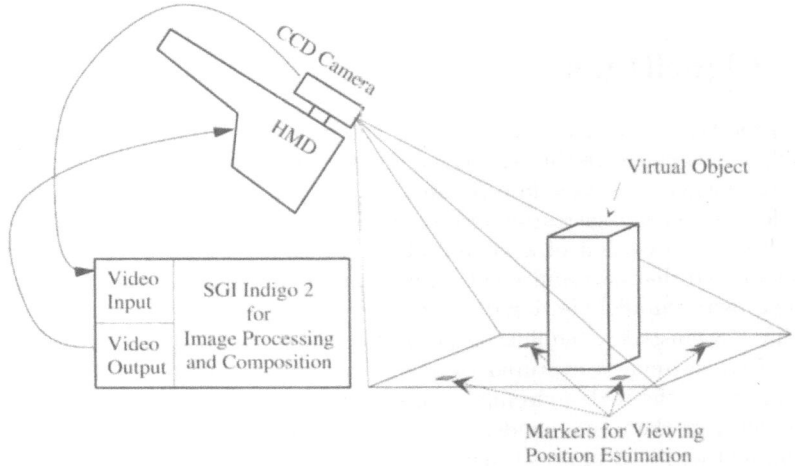

Figure 7.1 Hardware configuration of one camera MR system.

Figure 7.2 Composed images of real and virtual by one camera MR system [19] (see color pages).

This new prototype system recovers camera parameters using three markers whose 3-D positions are not known. The user's hands are regarded as real objects which may occlude virtual objects and the depth of hands is estimated to resolve occlusions properly. When the user's hands are closer to the user than a virtual object, the object is occluded by the hands.

This chapter is structured as follows. In the following section, Section 7.2, algorithms to calculate camera parameters and to estimate the hand depth are described. Section 7.3 then describes experimental results with the proposed methods and the discussion about the prototype system. Finally we summarize the present work in Section 7.4 .

7.2   Algorithms

A new prototype system of mixed reality has been built, in order to evaluate the feasibility of the video see-through mixed reality environment that is based on binocular stereo vision. The system estimates camera parameters using a pair of stereo images for geometric registration of real and virtual world coordinates. The system also realizes an occlusion of a virtual object by the user's hand (real object). It should be noted that two equivalent cameras are mounted on a HMD in front of the user's eyes with parallel viewing directions.

Figure 7.3 illustrates the flow chart of algorithms used in the prototype system. A pair of stereo images captured by stereo cameras are inputted to a computer. These images of the real environment are used to estimate the camera parameters and the depth of the user's hands. Then, the stereo images of the real environment are composed with CG images (images of 3-D virtual objects) and are outputted to the HMD that is worn by the user.

In the following sections, the details of the camera parameter estimation, hand depth estimation and image composition are described.

7.2.1   Camera Parameter Estimation

In order to reduce the image processing complexity, blue markers are used as feature points (fiducials) in a real environment. Note that blue is the complementary color of the typical Japanese skin color. The following three steps are used to estimate camera parameters.

Step 1 [Marker tracking] From input stereo images, blue areas are extracted and the makers' positions in the input images are obtained. The markers are then automatically tracked for further processing.

Step 2 [Marker's 3-D position estimation] The markers' 3-D positions in the camera coordinate system are recovered by applying stereo matching.

Step 3 [Computation of model-view matrix] A projection matrix (model-view matrix) which represents the relationship between the real and virtual world coordinates is calculated using markers' 3-D positions.

Let us describe these three steps in more detail in the following.

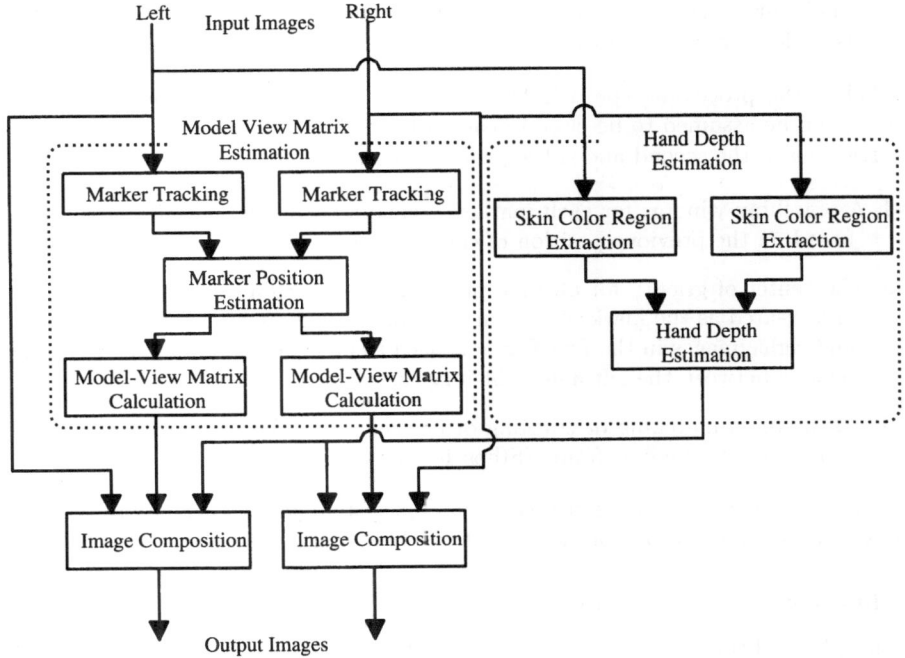

Figure 7.3 Flow diagram of the algorithms used in stereoscopic video see-through MR system.

Tracking of markers

The extraction of blue areas from the entire image frame takes considerable amount of processing time. Therefore, only in the first frame of the image sequence, the blue region extraction over the entire image frame is performed. In the subsequent frames, extracted regions are tracked by using the results in the previous frame. The extracted and tracked regions are used to calculate the screen coordinate of each marker.

In the first frame of the image sequence, the following steps are performed to extract blue marker regions.

1-1 The entire frames in the first stereo pair are scanned to extract blue pixels.

1-2 Blue pixels are segmented into connected components (regions) by applying a component labeling.

1-3 The regions with small numbers of blue pixels are regarded as noises and are eliminated. The other regions are used in the following steps.

1-4 The center of gravity of each blue region is treated as the screen coordinate of each marker.

1-5 When three markers are found in both left and right images, stereo matching is performed based on the epipolar constraint. All markers are then labeled.

When the processing speed is fast enough, the marker's positions in adjacent frames can be assumed to be close to each other. Thus the following steps are used for tracking in the second and subsequent frames in the image sequence.

2-1 A searching window for each marker is generated on the current image frame based on the previous position of each marker.

2-2 The center of gravity for all blue pixels in the searching window is calculated and treated as the marker's screen coordinate. The matching between the left and right images in the first frame is used to determine stereo pairs of marker regions between the left and right images.

2-3 When stereo matching or tracking of markers fails, the system starts over the procedure for the first frame (Steps 1-1 through 1-5).

By using the algorithm described above, the system keeps tracking blue markers and calculates the screen coordinates of three blue markers for the next step.

Estimation of three-dimensional positions of makers

In this phase of the process, the system estimates the 3-D position of each marker in the camera coordinate system. The relationship between two cameras and markers is illustrated in Figure 7.4. The origin of the camera coordinate system is placed at the middle of the left and right lens centers of CCD cameras (O_c in Figure 7.4). X-axis is set along the baseline of the cameras. Z-axis is set to be parallel to the direction of the optical axes of the cameras.

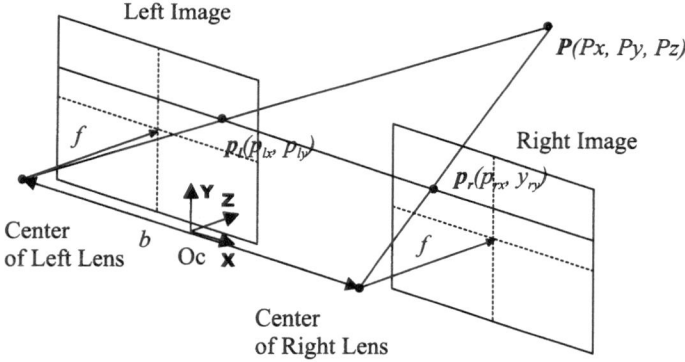

Figure 7.4   Geometry of binocular stereoscopic projection.

In this situation, let us assume that a marker whose 3-D position is $P(P_x, P_y, P_z)$ is projected onto the left and right images with 2-D positions of $p_l(p_{lx}, p_{ly})$ and $p_r(p_{rx}, p_{ry})$, respectively. Then the following equation stands:

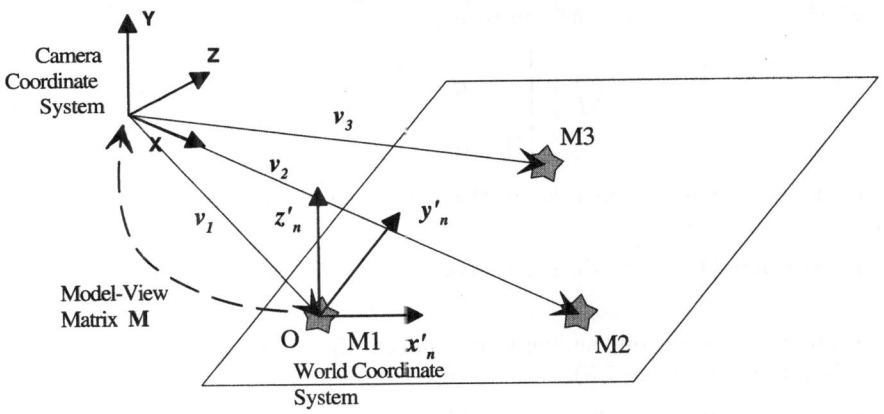

Figure 7.5   The relation between world and camera coordinates.

$$P_x = \frac{b(p_{lx} + p_{rx})}{2(p_{lx} - p_{rx})},$$

$$P_y = \frac{b(p_{ly} + p_{ry})}{2(p_{lx} - p_{rx})}, \tag{7.1}$$

$$P_z = \frac{fb}{p_{lx} - p_{rx}},$$

where, f is the focal length of both camera lenses,
   b is the baseline length a stereo camera system.

From Equation (7.1), when f and b are known, $P(P_x, P_y, P_z)$ is calculated from $p_l(p_{lx}, p_{ly})$ and $p_r(p_{rx}, p_{ry})$.

Calculation of model-view matrix

In order to compose the image of a real environment and the CG image of a virtual object, a transformation matrix (model-view matrix) from the world coordinate to the camera coordinate is required (Figure 7.5). Among a model-view matrix - M, a position of a point in world coordinate - w, and its position in the camera coordinate - c, the following equation stands.

$$c = Mw. \tag{7.2}$$

The matrix M can be decomposed into a 3×3 rotation matrix R and a 3×1 translation vector T. Thus, M can be expressed as follows.

$$M = \left[\begin{array}{ccc|c} & R & & T \\ \hline 0 & 0 & 0 & 1 \end{array} \right].$$

$$(7.3)$$

In the present prototype system, the world coordinate system is defined as follows:

- The origin O is set at the position of the maker labeled as No. 1 ($M1$ in Figure 7.5).

- The x-axis is set on the line that connects the markers labeled as No. 1 and No. 2 (x'_n in Figure 7.5).

- The $x - y$ plane is defined as the plane on which three markers reside.

According to the definition above, the translation component T can be given as the camera coordinate position of the maker No. 1. The rotation component R can be calculated with the following steps.

1. When the positions of makers No. 1, 2, and 3 are given as v_1, v_2, and v_3, respectively, the directional vector of each axis of the world coordinate (x_n, y_n, z_n) can be obtained as follows:

$$\begin{aligned} x_n &= v_2 - v_1, \\ y_n &= (v_3 - v_1) - \frac{x_n \cdot (v_3 - v_1)}{x_n \cdot x_n} x_n, \\ z_n &= x_n \times y_n. \end{aligned}$$

$$(7.4)$$

2. Normalize x_n, y_n, and z_n into x'_n, y'_n, and z'_n, respectively.

$$x'_n = \frac{x_n}{\|x_n\|}, \quad y'_n = \frac{y_n}{\|y_n\|}, \quad z'_n = \frac{z_n}{\|z_n\|}.$$

$$(7.5)$$

3. R can be represented by using x'_n, y'_n, and z'_n as follows.

$$R = [x'_n \quad y'_n \quad z'_n].$$

$$(7.6)$$

By using T and R, the model-view matrix M can be uniquely determined from Equation (7.3).

7.2.2   Depth Estimation of User's Hand

In this section, the depth estimation of the user's hand in the real world is described. The depth estimation is used to compose the image of the virtual objects and the image of the user's hand without occlusion conflicts. This means that when the user's hand is closer to the user than the virtual objects, the user's hand occludes virtual objects and vice versa. The hand area is simply assumed to be on a flat surface that is perpendicular to the optical axis of the camera. The following steps describe the method to estimate the depth of the hand area.

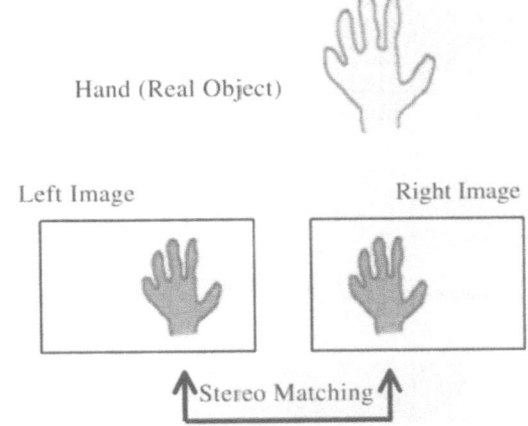

Figure 7.6   Illustration of hand depth estimation.

Step 1 A pair of stereo images are converted into the HIS color space from the RGB color space. Then skin color regions are extracted.

Step 2 Stereo matching on the skin color regions is performed and the binocular disparity d is computed. The disparity d is determined so as to maximize the overlapping hand area (Figure 7.6).

Step 3 The depth of the hand region is computed by using the following equation.

$$Z_{hand} = \frac{fb}{d},$$ (7.7)

where f is the focal length of the camera,
b is the baseline length between the stereo cameras.

7.2.3   Composition of Real Scene Image and CG Object

By using the model-view matrix, aspect ratio and other projection parameters, CG images of virtual objects are drawn. At this stage, the depth values of the hand regions and each virtual object are compared. When the hand area is closer to the user, a transparent virtual object is drawn at the 3-D position where the hand area exists. By using a hardware z buffering algorithm, the virtual objects that are farther than the transparent object are not drawn on the frame buffer. Therefore the composed image looks as if the hand is occluding the virtual objects. These rendering steps are illustrated with examples in Figure 7.7. First, only the background image is rendered on the frame buffer. Z-buffer value is also set to the farthest value through out the screen. Then, the z-buffer of the hand region is set to the depth of the hand. Note that the hand region is illustrated by a checker board pattern in Figure 7.7. Finally, the virtual objects are rendered using the estimated model-view matrix and the z-buffer.

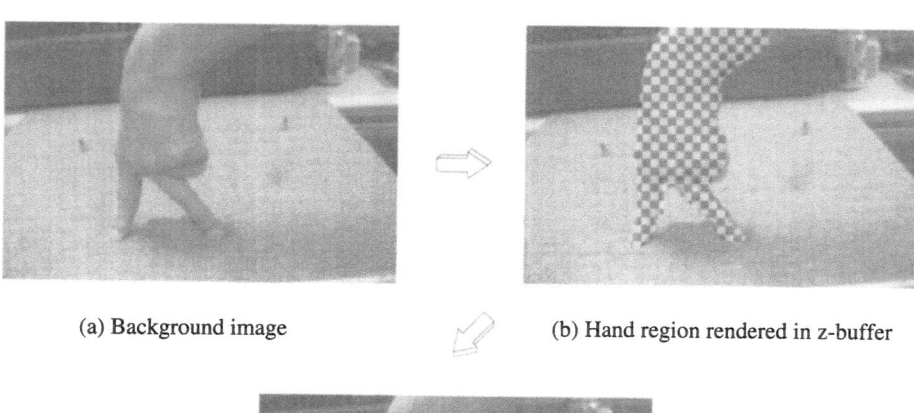

(a) Background image   (b) Hand region rendered in z-buffer

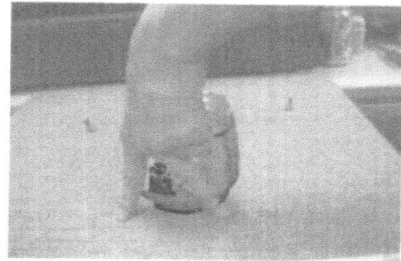

(c) CG virtual object is rendered with z-buffer comparison

Figure 7.7   Image composition using Z-buffer.

7.3   Experiments and Discussion

7.3.1   Implementation and Results

The proposed algorithms are implemented on a graphics workstation (SGI Onyx2 IR: 16CPU MIPS R10000 195MHz). Two small equivalent CCD cameras (Toshiba IK-UM42) are mounted on a HMD (Olympus Media Mask) as shown in Figure 7.8. An optical see-through function of the HMD is not used, that is, it works in closed-view mode, and a video see-through function using the proposed algorithm is employed in the following experiment. The baseline length of the stereo cameras is set to 6.5 cm. The optical axes of the cameras are set to be parallel to the gaze direction of the user (actually, the user's head direction). Images captured by a pair of HMD-mounted stereo cameras are inputted to the graphics workstation via DIVO digital video interface. Each digitized image consists of 720 × 486 pixels. The resulting composed image of the real world and virtual objects is outputted via DIVO interface and displayed on the HMD stereoscopically. The hardware configuration of the whole system is illustrated in Figure 7.9.

The sample output images of the system are shown in Figure 7.10 in two stereo-scopic viewing modes: parallel viewing and cross viewing. Figure 7.10 (a) shows the composition of a virtual object (a sake cask) in a real scene. The composition is

Figure 7.8   Appearance of stereoscopic video see-through HMD.

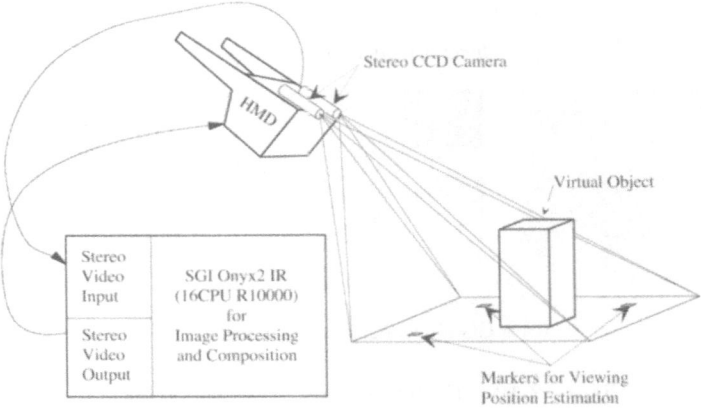

Figure 7.9   Hardware configuration of stereoscopic video see-through MR system.

achieved based on the camera parameters recovered from the three visible markers placed on the desktop. Figure 7.10 (a) exhibits that the proposed algorithm successfully estimates the camera parameters so that the composition is done without a sense of incompatibility.

Figures 7.10 (b) and (c) show the results of the image composition when there is a user's hand in the real scene. Figure 7.10 (b) is the case when the user's hand is behind the virtual object. Figure 7.10 (c) is the case when the user's hand is in front of the virtual object. The result shows that the hand properly occludes the virtual object behind it. It is proven that the stereoscopic images of both virtual and real objects reproduce a good depth sensation and the user can observe the mixed scene with considerably natural sensations.

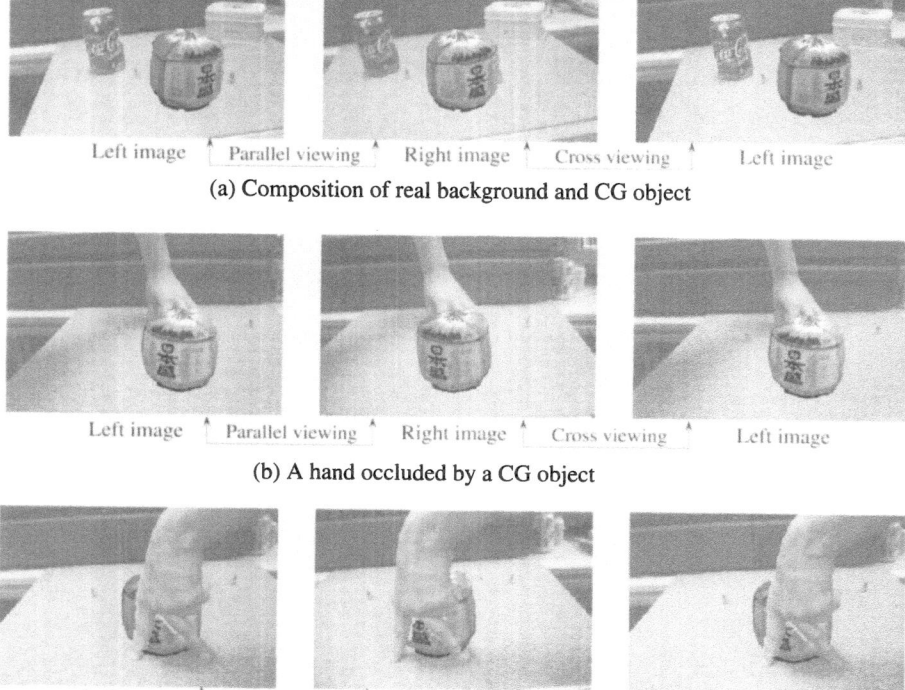

(a) Composition of real background and CG object

(b) A hand occluded by a CG object

(c) A CG object occluded by a hand

Figure 7.10   Examples of composed stereo images.

7.3.2   Discussion

The image update rate of the prototype system is fifteen frames per second, when the virtual object of a sake cask consisting of 3700 polygons with a texture is used. This means that every second frame of a video stream is merged with CG objects and 66 milliseconds (ms) are allocated for processing at each stereo image composition. The approximate processing time for each processing stage of the algorithm is as follows: 19 ms for the camera parameter estimation, 14 ms for the hand depth estimation, and 33 ms for the stereo image composition of real and CG objects, respectively. The rest of the time is used for the video input. Due to the nature of video see-through mixed reality, there is no synchronization error between the real scene and virtual objects. On the other hand, an entire scene has a time latency of about 99 ms.

In the case of only less than three markers successfully found in the image of the real environment, the system can not estimate the model-view matrix because of under constraint of equations. To relax this problem, we can set more than three markers in the real environment, so that the system should estimate the model-view matrix more stably than it is. Another approach is to employ some kind of prediction or smoothing filters in the process of estimating the model-view matrix.

Vision-based tracking essentially has no limitations in measurement range, while 3-D position sensors usually limit their measurement spaces. However, the present prototype system actually has a limitation in measurement range because the system uses a limited number of predefined markers. If we would automatically detect and track feature points other than markers in a real scene to calculate the model-view matrix, the measurement range of the system could be extended.

In the present prototype system, real objects which may occlude virtual CG objects are limited to user's hands. Resolving occlusions between CG objects and arbitrary real objects in a real environment would improve the usefulness of the system drastically. This requires the acquisition of dense depth map of the real scene [20]. Static real objects can be properly treated by the state of the art of stereo vision techniques in time-critical applications such as video see-through mixed reality using standard workstations; for example, [21] [22]. However, dynamically moving arbitrary objects are still difficult to be mixed with CG objects in real-time without special hardwares [23], maintaining correct occlusions among them. This problem should further be investigated.

7.4 Conclusion

As a pilot study on the stereo vision based mixed reality system, we have developed a prototype system that uses a pair of HMD-mounted stereo cameras to estimate the viewer's head position. The user's hand region is also detected by using color and stereo matching and the occlusion of a virtual object with the user's hand is correctly simulated in the system. The algorithm described is a combination of well-known computer vision techniques and can be easily implemented on a standard graphics workstation with a video I/O interface. The prototype system can successfully perform the image composition nearly in video rate. We have proven that the stereo vision based video see-through mixed reality system is feasible and has the possibility of building up actual applications.

In the future work, we will concentrate our attention on robust estimation of the model-view matrix in several ways; for example, (1) using more than three markers, (2) automatic detection and tracking of feature points in a real scene without predefined markers.

References

[1] P. Milgram and F. Kishino: "A taxonomy of mixed reality visual display," *IEICE Trans. on Information and Systems*, vol.E77–D, no.12, pp.1321–1329, 1994.

[2] R. T. Azuma: "A survey of augmented reality," *Presence: Teleoperators and Virtual Environments*, vol.6, no.4, pp.355–385, 1997.

[3] S. Feiner, B. MacIntyre, and D. Seligmann: "Knowledge-based augmented reality," *Comm. ACM*, vol.36, no.7, pp.52–62, 1993.

[4] P. Milgram, H. Takemura, A. Utsumi, and F. Kishino: "Augmented reality: A class of displays on the reality-virtuality continuum," *Proc. SPIE, vol.2351: Telemanipulator and Telepresence Technologies*, pp.282–292, 1994.

[5] T. Ohshima, K. Satoh, H. Yamamoto, and H. Tamura: "AR^2 Hockey: A case study of collaborative augmented reality," *Proc. IEEE Virtual Reality Annual Int. Symp. (VRAIS'98)*, pp.14–18, 1998.

[6] R. L. Holloway: "Registration error analysis for augmented reality," *Presence: Teleoperators and Virtual Environments*, vol.6, no.4, pp.413–432, 1997.

[7] K. Kiyokawa, H. Takemura, H. Iwasa, and N. Yokoya: "Collaborative immersive workspace through a shared augmented environment," *Proc. SPIE, vol.3517: Intelligent Systems in Design and Manufacturing*, pp.2–13, 1998.

[8] A. State, M. A. Livingston, W. F. Garrett, G. Hirota, M. C. Whitton, E. D. Pisano, and H. Fuchs: "Technologies for augmented reality systems: Realizing ultrasound-guided needle biopsies," *Proc. SIGGRAPH'96*, pp.439–446, 1996.

[9] K. N. Kutulakos and J. R. Vallino: "Calibration free augmented reality," *IEEE Trans. on Visualization and Computer Graphics*, vol.4, no.1, pp.1–20, 1998.

[10] Y. Nakazawa, S. Nakano, T. Komatu, and T. Saitou: "A system for composition of real moving images and CG images based on image feature points," *The J. Institute of Image Information and Television Engineers*, vol.51, no.7, pp.1086–1095, 1997 (in Japanese).

[11] U. Neumann and Y. Cho: "A self-tracking augmented reality system," *Proc. ACM Symp. on Virtual Reality Software and Technology (VRST'96)*, pp.109–115, 1996.

[12] M. Uenohara and T. Kanade: "Vision-based object registration for real-time image overlay," *Proc. IEEE Conf. on Computer Vision, Virtual Reality and Robotics in Medicine (CVRMed'95)*, pp.13–22, 1995.

[13] M. Bajura and U. Neumann: "Dynamic registration correction in video-based augmented reality systems," *Computer Graphics and Applications*, vol.15, no.5, pp.52–60, 1995.

[14] O. Faugeras: *Three-Dimensional Computer Vision, A Geometric Viewpoint*, The MIT Press, 1993.

[15] R. M. Haralick, C. N. Lee, and K. Ottenberg: "Analysis and solutions of the three point perspective pose estimation problem," *Proc. IEEE Computer Society Conf. on Computer Vision and Pattern Recognition (CVPR'91)*, pp.592–598, 1991.

[16] G. J. Klinker, K. H. Ahlers, D. E. Breen, P. Y. Chevalier, C. Crampton, D. S. Greer, D. Koller, A. Kramer, E. Rose, M. Tuceryan, and R. T. Whitaker: "Confluence of computer vision and interactive graphics for augmented reality," *Presence: Teleoperators and Virtual Environments*, vol.6, no.4, pp.433–451, 1997.

[17] J. Rolland, R. Holloway, and H. Fuchs: "A comparison of optical and video see-through head-mounted displays," *Proc. SPIE, vol.2351: Telemanipulator and Telepresence Technologies*, pp.293–307, 1994.

[18] E. K. Edwards, J. P. Rolland, and K. P. Keller: "Video see-through design for merging of real world and virtual environment," *Proc. IEEE Virtual Reality Annual Int'l Symp. (VRAIS'93)*, pp.222–233, 1993.

[19] T. Okuma, K. Kiyokawa, H. Takemura, and N. Yokoya: "An augmented reality system using a real-time vision based registration," *Proc. 14th IAPR Int'l Conf. on Pattern Recognition (ICPR'98)*, vol.II, pp.1226–1229, 1998.

[20] N. Yokoya: "Surface reconstruction directly from binocular stereo images by multiscale-multistage regularization," *Proc. 11th IAPR Int'l Conf. on Pattern Recognition (ICPR'92)*, vol.I, pp.642–646, 1992.

[21] S. Matsubara, H. Iwasa, H. Takemura, and N. Yokoya: "Fusion of virtual objects and real world images based on stereo depth estimation," *Proc. Meeting on Image Recognition and Understanding (MIRU'96)*, vol.1, pp.43–48, 1996 (in Japanese).

[22] M. M. Wloka and B. G. Anderson: "Resolving occlusion in augmented reality," *Proc. 1995 ACM Symp. on Interactive 3D Graphics*, pp.5–12, 1995.

[23] T. Kanade , A. Yoshida, K. Oda, and M. Tanaka: "Stereo machine for video-rate dense depth mapping and its new application," *Proc. IEEE Computer Society Conf. on Computer Vision and Pattern Recognition (CVPR'96)*, pp.196–202, 1996.

Chapter 8

Photometric Modeling for Mixed Reality

Katsushi Ikeuchi
Yoichi Sato
Ko Nishino
Imari Sato

The University of Tokyo, Japan

8.1 Introduction

The mixed reality technology is considered as one of the key technologies for enhancing a wide variety of applications ranging from manufacturing to entertainment. The mixed reality technology differs from the virtual reality technology in that users can feel immersed in a space which consists of not only computer-generated objects but also real objects. Thus seamless integration of the virtual and real worlds is essential for mixed reality in addition to reality of the virtual world.

Our efforts in the mixed reality research span two aspects: how to create models of virtual objects and how to integrate such virtual objects with real scenes. For model creation, we have developed two methods, the model-based rendering method and the eigen-texture method, both of which automatically create such rendering models by observing real objects. The model-based rendering method first analyzes input images of real objects, obtains reflectance parameters from this analysis, and then, using the determined reflectance parameters, generates the virtual image [1]. This method stores very compact information, namely surface shapes and reflectance parameters, and works well when an object surface follows a known reflectance model.

For other objects that do not follow simple reflectance models, we have developed the eigen-texture method [2]. This method samples appearances of a real object, pastes the images onto the surface of the 3D model of the object, and then compresses them on the 3D surface (not on the image plane). Later, the compressed data can be used for generating a virtual image of the object. The method does not assume any known reflectance model, nor does it require any detailed reflectance analysis, as was the case in model-based rendering. Using the 3D model of the object, the method can also generate shadows cast by the virtual object, something that the image-based rendering fails to do. Moreover, the method achieves a very compact storage of the object's appearances because the compression of the appearances is performed along pixels corresponding to the same physical points.

For integration of virtual object with real scenes, we have developed a method that renders virtual objects based on real illumination distribution [3]. First, a radiance distribution in the real scene is determined from two omni-directional images of the scene. Then the measured radiance distribution is used for rendering virtual objects superimposed onto the scene image. The proposed method has the ability to synthesize a convincing image even for a complex radiance distribution, a task at which other methods often fail.

8.2   Creating Models from Observation

Currently, most virtual reality systems utilize image-based rendering [4]–[6]. The image-based rendering samples a set of color images of a real object and stores them on the disk of a computer. A new image is then synthesized either by selecting an appropriate image from the stored set or by interpolating multiple images. Apple's QuickTime VR is one of the earlier successful image-based rendering methods. Image-based rendering does not assume any reflectance characteristics of objects nor does it require any detailed analysis of the reflectance characteristics of the objects; rather, the method needs only to take images of an object. The method can be applied to a wide variety of real objects. And because it is also quite simple and handy, image-based rendering is ideal for displaying an object as a stand-alone without any background for the virtual reality.

On the other hand, image-based methods have critical disadvantages on application to mixed reality. Few image-based rendering methods employ accurate 3D models of real objects. Thus, it is difficult to make cast shadows under real illuminations corresponding to the real background-image.

Unlike image-based rendering, model-based rendering assumes a reflection model of an object and determines reflectance parameters through detailed reflectance analysis. Later, the method uses those reflectance parameters to generate virtual images by considering illumination conditions of the real scene. Since the reflectance parameters are obtained at every surface point of the object, integration of synthesized images with the real background can be accomplished quite realistically, *i.e.*, the method can generate a realistic appearance of an object as well as of the shadows cast by that object onto the background.

8.2.1 Model-Based Rendering

We have developed a model-based rendering method for modeling object reflectance properties, as well as object shapes, by observing real objects [1]. First, an object surface shape is reconstructed by merging multiple range images of the object. By using the reconstructed object shape and a sequence of color images of the object, parameters of a reflection model are estimated in a robust manner. The key point of the proposed method is that, first, the diffuse and specular reflection components are separated from the color image sequence, and then, reflectance parameters of each reflection component are estimated separately. This approach enables estimation of reflectance properties of real objects whose surfaces show specularity as well as diffusely reflected lights. The recovered object shape and reflectance parameters are then used for synthesizing object images with realistic shading under arbitrary illumination conditions.

Shape modeling

The image acquisition system used in our experiments is illustrated in Figure 8.1. The object whose shape and reflectance parameters are to be recovered is mounted on the end of a robotic arm. Using the system, a sequence of range and color images of the object is obtained as the object is rotated at a fixed angle. A range image is obtained using a light-stripe range finder with a liquid crystal shutter and a color CCD video camera [7]. Using optical triangulation, 3D locations of points in the scene are computed at each image pixel. Each range-image pixel represents a 3D location of a corresponding point on an object surface. The same color video camera is used for acquiring color images. Therefore, the range image pixels and those of the color images directly correspond to one another. A single incandescent lamp is used as a point light source. In our experiments, the light source direction and the light source color are measured by calibration.

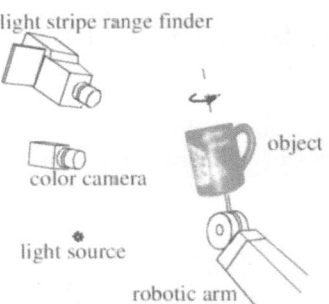

Figure 8.1 Image acquisition system.

A sequence of range images of the object is used to construct the object shape as a triangular mesh. For constructing object shapes as triangular mesh models from multiple range images, we used the volumetric method developed by Wheeler, Sato, and Ikeuchi [8]–[10]. In addition, surface orientation is necessary for estimating

reflectance parameters. Polygonal orientation computed from a triangular surface
mesh model can approximate real surface orientation only when the object surface
is relatively smooth and does not have high curvature points. For instance, using
the 3D points from the range images with a least square fitting, we compute surface
orientation at regular grid points within each triangle.

Parameter estimation

Once the object shape has been obtained, we measure the reflectance parameters
of the object surface using the obtained shape and the input color images. First,
the two fundamental reflection components (*i.e.*, the diffuse and specular reflection
components) are separated from the input color images. Then, the parameters
for the two reflection components are estimated separately. Separation of the two
reflection components enables us to obtain a reliable estimation of the specular
reflection parameters. Also, the specular reflection component (*i.e.*, highlight) in
the color images does not affect the estimated diffuse reflection parameters of the
object surface.

The Torrance-Sparrow model [11] [12] is used for representing the diffuse and
specular reflection components

$$L = K_d f_d(\theta) + K_s f_s(\psi, \alpha), \tag{8.1}$$

where $f_d(\theta) = \cos(\theta)$, $f_s(\psi, \alpha) = e^{-\alpha^2/(2\sigma^2)}/\cos\psi$. θ is the angle between the surface
orientation and the light source direction, ψ is the angle between the surface orien-
tation and the viewing direction, α is the angle between the surface normal and the
bisector of the light source direction and the viewing direction, K_d, K_s are constants
for the diffuse and specular reflection components, and σ is the standard deviation
of a facet slope.

We employ the separation algorithm, originally introduced for the case of a moving light source by Sato and Ikeuchi [13]–[15]. In order to handle the case of a moving
object as opposed to a moving light source, we align color images from different poses
onto a 3D-mesh model using known camera parameters. From this alignment, we
can determine a sequence of observed color values at each point on the object surface
during the object rotation.

Measurement at each point on the object surface provides an observation matrix,
consisting of three sequences of intensity values (R, G, and B), where M is the
number of measurements.

$$A = \begin{bmatrix} L_1^R & L_1^G & L_1^B \\ L_2^R & L_2^G & L_2^B \\ \vdots & \vdots & \vdots \\ L_M^R & L_M^G & L_M^B \end{bmatrix} \tag{8.2}$$

This $M \times 3$ observation matrix can be represented as a product of an $M \times$
2 intensity matrix, consisting of the intensity values of the diffuse and specular
reflection component with respect to the illumination/viewing directions, and of a
2×3 color matrix, consisting of the diffuse and the specular color vectors.

$$
\begin{bmatrix}
f_1(\theta_1) & f_2(\psi_1, \alpha_1) \\
f_1(\theta_2) & f_2(\psi_2, \alpha_2) \\
\vdots & \vdots \\
f_1(\theta_M) & f_2(\psi_M, \alpha_M)
\end{bmatrix}
\begin{bmatrix}
K_d^R & K_d^G & K_d^B \\
L_s^R & L_s^G & L_s^B
\end{bmatrix}
\tag{8.3}
$$

From this observation, we can conclude that the rank of the observation matrix A is 2, and the observation matrix can be decomposed into an intensity matrix and a color matrix by using the singular value decomposition. By extracting the intensity matrix, we can obtain how intensity values of the diffuse and specular components vary depending on the illumination/viewing directions.

Using the angles of the illumination/viewing directions computed from the rotation angle, the diffuse reflection parameters are estimated by fitting the first term of the Torrance-Sparrow reflection model, a cosine function, to the separated diffuse reflection component, the first column of the intensity matrix. The diffuse reflection parameters are estimated at regular grid points within each triangle.

The specular reflection parameters are computed in the same way as the diffuse reflection parameters. However, since the specular reflection component is usually observed only from a limited range of viewing directions, we have to interpolate, over the entire surface, the specular reflection parameters obtained at a fixed number of points selected using several evaluation measures [1].

Image synthesis

Using the reconstructed object shape, the surface normals, the diffuse reflection parameters, the specular reflection parameters and the reflection model, we can synthesize color object images under arbitrary illumination/viewing conditions. Figure 8.2 shows three synthesized images as well as one frame of the input color image sequence. One of those synthesized images was generated using the same illuminating/viewing condition as the input color image. It can be seen that the synthesized image closely resembles the corresponding real image. In particular, highlights, which generally are a very important cue of surface material, appear on the side and the handle of the mug naturally in the synthesized images.

Summary

This section describes an algorithm for model-based rendering. The object surface shape is reconstructed by merging multiple range images of the object. By using the reconstructed object shape and multiple color images of the object, parameters of the Torrance-Sparrow reflection model are estimated. For estimating reflectance parameters of the object robustly, our method is based on separation of the diffuse and specular reflection components from a color image sequence. Using separated reflection components, reflection model parameters for each of the two components are estimated separately. Our experiments have demonstrated that our model-based rendering method can be effectively used for synthesizing realistic object images under arbitrary illumination and viewing conditions.

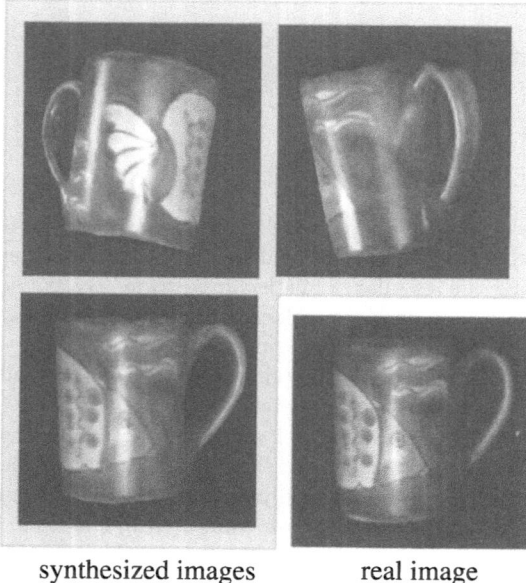

synthesized images   real image

Figure 8.2   Input and synthesized images.

8.2.2   Eigen-Texture Method

The previous section describes a model-based rendering method. This method can generate realistic appearances of an object from reflectance parameters estimated from real objects. Unfortunately, however, some classes of object surfaces reveal complicated reflectance models, and the model-based rendering method cannot be applied to those classes of objects.

For those object surface classes, we propose a new rendering method, which we refer to as the eigen-texture method [2] [16] [17]. Figure 8.3 displays an overview of the proposed method. The eigen-texture method creates a 3D model of an object from a sequence of range images. The method aligns and projects color images of the object onto the 3D surface of the object model. Then, the method compresses those images in the coordinate system defined on the 3D-model surface. This compression is accomplished using the eigenspace method [18]. The synthesis process is achieved using the inverse transformation of the eigenspace method. Shadows cast by the object are generated by using the 3D shape of the object. A virtual image under a complicated illumination condition is generated by summation of component virtual images sampled under single illuminations thanks to the linearity of image brightness.

Appearance compression based on 3D shape model

The eigen-texture method samples a sequence of color and range images. Once these two sequences are input to the system, a 3D triangular mesh model of the object is

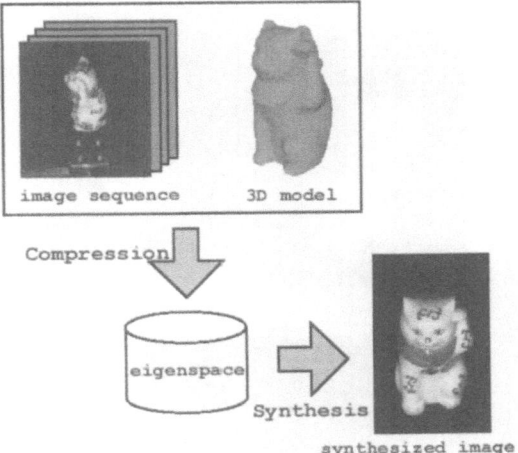

Figure 8.3   Eigen-texture method.

created from the range images. Each color image is then aligned with the 3D model, as was the case in the model-based rendering method. Each color image is divided into small areas that correspond to triangle patches on the 3D model. Each triangle patch is normalized to have the same shape and size as that of the others. Color images on the small areas are warped on a normalized triangular patch. We refers to this normalized triangular patch as a cell and to its color image as a cell image. A sequence of cell images from the same cell is collected as shown in Figure 8.4. Here this sequence depicts appearance variations on the same physical patch of the object surface under various viewing/illumination conditions.

These cell images are compressed using the eigenspace method [18]. Note that the compression is done in a sequence of cell images, whose appearance change is caused only by the change of shading, e.g., brightness and specularity of the triangular patch. Thus, high compression ratio can be expected with the eigenspace method. On the other hand, current image-based rendering methods usually compress images in the image coordinate systems or in other coordinate systems which are defined based on the image coordinate system. Each pixel in the sequence corresponds to different physical points in the image sequence. Thus, the variance carries both those from the imaging geometries and those from the physical locations.

Eigenspace compression on cell images can be achieved by the following steps. First, each cell image is converted into a $1 \times 3N$ vector X_m by arranging color values for each color band RGB in a raster scan manner. Here, m is the pose number of the real object, M is the total number of poses, and N is the number of pixels in each cell image.

$$X_m = \left[x_{m,1}^R \cdots x_{m,N}^R x_{m,1}^G \cdots x_{m,N}^G x_{m,1}^B \cdots x_{m,N}^B \right] \tag{8.4}$$

The sequence of cell images can be represented as a $M \times 3N$ matrix.

The average of all color values in the cell image set is subtracted from each

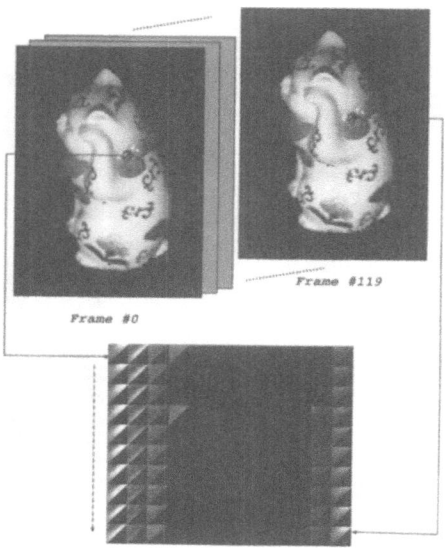

Figure 8.4   Cell images.

element of the matrix. This ensures that the eigenvector with the largest eigenvalue represents the dimension in eigenspace in which the variance of images is maximum in the correlation sense. The resulting matrix is represented as $X = [X_1 X_2 \cdots X_M]^T$.

With this resulting $M \times 3N$ matrix, we define a $3N \times 3N$ matrix Q, and determine eigenvectors e_i and the corresponding eigenvalues λ_i of Q by solving the eigenstructure decomposition problem: $Q = X^T X$ and $\lambda_i e_i = Q e_i$.

At this point, the eigenspace of Q is a high dimensional space, *i.e.*, $3N$ dimensions. Although $3N$ dimensions are necessary to represent each cell image in an exact manner, a small subset of them is sufficient to describe the principal characteristics as well as to reconstruct each cell image with adequate accuracy. Accordingly, we extract $k(k \ll 3N)$ eigenvectors which represent the original eigenspace adequately; by this process, we can substantially compress the image set. The k eigenvectors can be chosen by sorting the eigenvectors by the size of the corresponding eigenvalues, and then computing the eigenratio:

$$\sum_{i=1}^{k} \lambda_i / \sum_{i=1}^{3N} \lambda_i \geq T, \tag{8.5}$$

where $T \leq 1$.

Using the k eigenvectors $\{e_i | i = 1, 2, \ldots, k\}$, each cell image can be projected onto the eigenspace composed by matrix Q by projecting each matrix X. And the projection of each cell image can be described as a $M \times k$ matrix $G : G = XV$ where $V = [e_1, e_2, \ldots, e_k]$.

To put it concisely, the input color image sequence is converted to a set of cell image sequences, and each sequence of cell images is stored as the matrix V, which is the subset of eigenvectors of Q, and the matrix G, which is the projection onto the eigenspace. Each sequence of cell images corresponding to $M \times 3N$ matrix X is compressed to $3N \times k$ matrix V and $M \times k$ matrix G, so that the compression ratio becomes: $k(M + 3N)/3MN$.

Each synthesized cell image can be computed by the following equation. Here, R is a $M \times 3N$ matrix which corresponds to the synthesized sequence of cell images, and each column of matrix R corresponds to the appearance of each triangular mesh (cell) of the object. Accordingly, a virtual object image of one particular pose (pose number) can be synthesized by aligning each corresponding cell appearance of every R to the 3D model:

$$R_m = \sum_{i=1}^{k} g_{m,i} e_i^T. \tag{8.6}$$

Implementation

We have implemented the eigen-texture method, and have applied it to a real object. The same experimental setup in the model-based rendering experiment is used to sample sequence of color and range images of a real object. Figure 8.5 shows the images synthesized by using 8 dimensional eigenspace. Details of the images are reconstructed accurately, and the synthesized images are indistinguishable from the input images. The compression ratio of the input image sequence was 7.9%. This ratio is quite high when compared with that of other image compression methods, and this experimental result demonstrates that the eigen-texture method is effective in terms of compression. This is because the eigen-texture method compresses the sequence of cell images that come from the same physical area on the object surface. The variances in the cell images are only caused by the relative directional change of the light source and the position of the viewer. Cell images are thus highly correlative.

The eigen-texture method cannot create images of an object under unfamiliar illumination conditions. Thus, we have to sample all possible illumination conditions. This is impossible. Fortunately, however, the irradiance at one point on the surface of the object from the whole illumination environment is a linear combination of the irradiance due to each light source composing the illumination environment. For this reason, we first decompose the real illumination environment into base point light sources and then sample the color images of the object under each base point light source separately. Those color images are compressed for each point light source by using the eigen-texture method. Then, arbitrary images of the object can be obtained as a linear combination of images synthesized for those point light sources.

As an example, we took three color images under 3 different point light sources, lighting them separately. Under each point light source, we took 40 color images and synthesized 120 virtual object images for each light source with the threshold 0.999 in eigenratio with interpolation in eigenspace. Then we synthesized the virtual object image sequence with all lights on by taking a linear combination of each point light source virtual object image. Figure 8.6 shows the result of this linear combination compared with a real image with all three lights turned on.

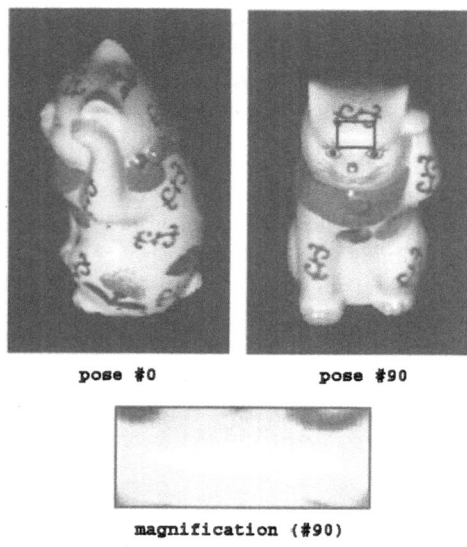

Figure 8.5   Synthesized images (8 dimensions).

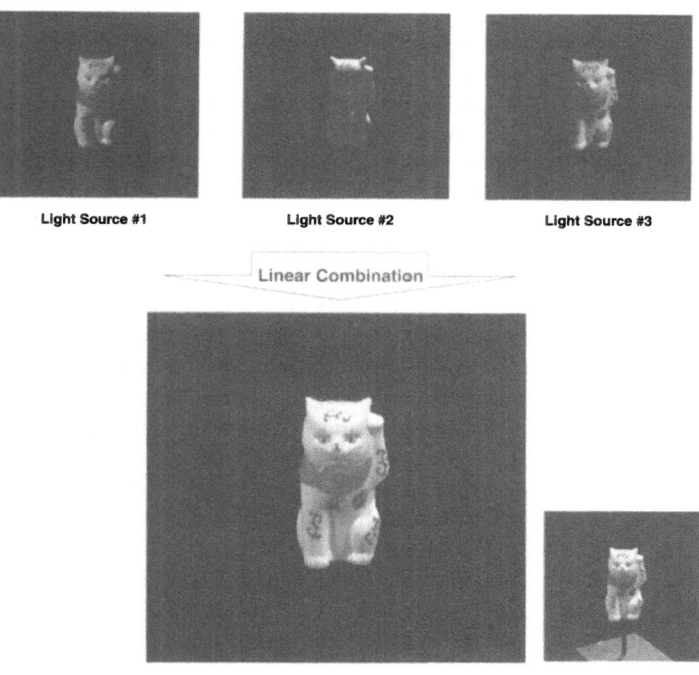

Figure 8.6   Linear combination of light sources.

Summary

This section describes the eigen-texture method for synthesizing virtual images of an object from a sequence of range and color images. Experimental results of the method proves its effectiveness, in particular, its high compression ratio. Using the model-based rendering method and the eigen-texture method, we can create appearances of a virtual object. The next section considers how to integrate these appearances with real scenes for mixed reality.

8.3 Integrating Virtual Objects with a Real Scene

For superimposing virtual objects onto a real scene appropriately [19] [20], the following three aspects have to be taken into account: geometry, illumination, and time. More specifically, the virtual object has to be located at a desired location in the real scene, and the object must appear at the correct location in the image (**consistency of geometry**). Also, shading of the virtual object has to match that of other objects in the scene, and the virtual object must cast a correct shadow, *i.e.*, a shadow whose characteristics are consistent with those of the shadows in the real scene (**consistency of illumination**). Lastly, motions of the virtual object and the real objects have to be coordinated (**consistency of time**).

In this section, we are mainly concerned with illumination consistency. Illumination of the real scene is directly measured and used for rendering virtual objects [3]. Our method measures a radiance distribution from all directions by a pair of omnidirectional images taken by a CCD camera with a fisheye lens. If we use only one omni-directional image, we can determine only the radiance distribution seen from the particular point where the omni-directional image was captured. To overcome this limitation, our method uses an omni-directional stereo algorithm for measuring a radiance distribution of the real scene as a 3D spatial distribution (3D triangular mesh). Once the radiance distribution is obtained as a triangular mesh, the measured radiance distribution is used for rendering a virtual object and for generating shadows cast by the virtual object onto the real scene wherever the virtual object is placed in the scene.

8.3.1 Constructing a Radiance Map Using Omni-Directional Stereo

A CCD camera with a fisheye lens is placed at two known locations in the scene to capture two omni-directional images from different locations. An acquisition system for the omni-directional images is illustrated in Figure 8.7. From a pair of omnidirectional images, an illumination distribution of the scene is estimated.

We first extract feature points with high contrast in the two omni-directional images by using the feature extraction algorithm proposed by Tomasi and Kanade [21]. In the algorithm, an image pixel with high gradient values in two orthogonal directions, e.g., a corner point, is extracted as a feature point.

Most of incoming light energy in a real scene comes from direct light sources such as a fluorescent lamp and a window to the outside, while the rest of the incoming light energy comes from indirect illumination such as reflection from a wall. Thus it is important to know the accurate locations of direct light sources to represent an illumination distribution of a real scene. Fortunately, direct light sources usually appear as significantly bright points in an omni-directional image. Therefore, it should be relatively easy to identify direct light sources in the image.

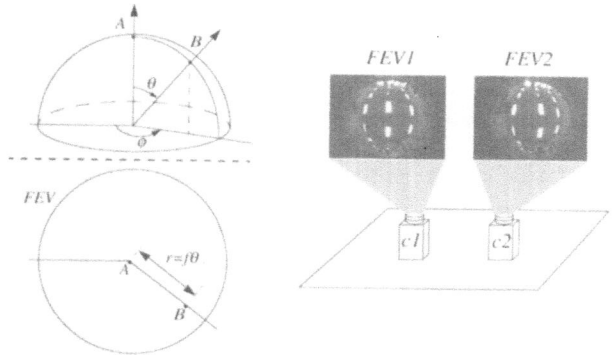

Figure 8.7   Acquisition of omni-directional images.

After feature points are extracted, 3D coordinates of points in the real scene corresponding to the extracted feature points are determined by using a stereo algorithm. This is done by obtaining an intersection point of a pair of lines-of-sight given by a pair of feature points. The direction of line-of-sight is determined from the image coordinate of the point corresponding to the light-of-sight using a known geometry of the fisheye lens.

In order to include indirect light sources such as walls and ceilings, we set up a 3D triangular mesh of the entire scene. First, we construct a 2D triangular mesh by applying 2D Delaunay triangulation to the feature points in the first omni-directional image. That determines the connectivity of a 3D triangular mesh whose vertices are the 3D points corresponding to the feature points in the image. Then, using the connectivity, a 3D triangular mesh is created from the 3D feature points.

The obtained triangular mesh approximates an entire shape of the real scene, e.g., the ceiling and walls of a room, which act as direct or indirect light sources. After the shape of the real scene is obtained as a triangular mesh, the radiance of the scene is estimated by using the image brightness of the omni-directional images. This radiance value is warped on the corresponding triangular mesh. The 3D textured mesh is referred to as a radiance map. An example of an obtained radiance map is shown in Figure 8.8(d).

8.3.2   Generating Soft Shadow

For superimposing virtual objects onto an image of a real scene, the ray casting algorithm is used. In this section, we assume that a virtual object always exists between the camera projection center and real objects in the scene and that the virtual object is located on the table, of which the position is known.

For each pixel in the input image of the real scene, a ray extending from the camera projection center through the pixel is generated by using the transformation between the world coordinate system and the image coordinate system. If the ray intersects a virtual object, we consider that the pixel corresponds to a point on the virtual object surface. Then we compute a color to be observed at the surface point under the measured radiance distribution of the real scene with the Torrance-Sparrow reflectance model. The computed color is stored in the pixel as the surface color of the virtual object at the pixel.

If a ray through an image pixel does not intersect with a virtual object, the color of a real object surface corresponding to the image pixel needs to be modified so that a shadow cast by the virtual object is created on the real object surface.

First, we obtain a 3D coordinate of a point on a real object surface where a ray through an image pixel intersects the real object. Then, a total irradiance E_1 at the surface point is computed from the measured illumination distribution by assuming that a virtual object does not occlude incoming light at the surface point.

$$E_{1,c} = \int_{-\pi}^{\pi} \int_0^{\frac{\pi}{2}} L_c(\theta_i, \phi_i) cos\theta_i \sin\theta_i d\theta_i d\phi_i \qquad c = R, G, B \qquad (8.7)$$

where $L(\theta_i, \phi_i)$ $(i = 1, 2, .., n)$ are the estimated illumination radiance.

Then, a partial irradiance E_2 at the surface point given by non-occluded light sources is computed as

$$E_{2,c} = \int_{-\pi}^{\pi} \int_0^{\frac{\pi}{2}} L_c(\theta_i, \phi_i) S(\theta_i, \phi_i) cos\theta_i \sin\theta_i d\theta_i d\phi_i \qquad c = R, G, B \qquad (8.8)$$

where $S(\theta_i, \phi_i)$ are occlusion coefficients; $S_i = 0$ if $L(\theta_i, \phi_i)$ is occluded by the object; Otherwise $S_i = 1$.

The ratio of the total radiance E_2 to the total radiance E_1 is multiplied to the observed color at the intersection between the ray and the plane on which the cast shadow is generated.

$$I_c' = I_c \frac{E_{2,c}}{E_{1,c}} \qquad c = R, G, B \qquad (8.9)$$

8.3.3 Experimental Results

An image was taken of a tabletop and miscellaneous objects on the tabletop in our laboratory. From the same camera position, another image of the tabletop with a calibration board was taken. The input image and the calibration image are shown in Figure 8.8(a) and (b).

First, regularly spaced dots on the calibration board were extracted in the calibration image to determine their 2D image coordinates. From pairs of the 2D image coordinates and the 3D world coordinates that were given a priori, the transformation between the world coordinate system and the image coordinate system was estimated by using the camera calibration algorithm [22]. Two omni-directional images of the scene, e.g., the ceiling of the laboratory in this experiment, were taken. Figure 8.8(c) shows the two omni-directional images. From this pair of images, the radiance map shown in Figure 8.8 is obtained using the omni-stereo algorithm. Then

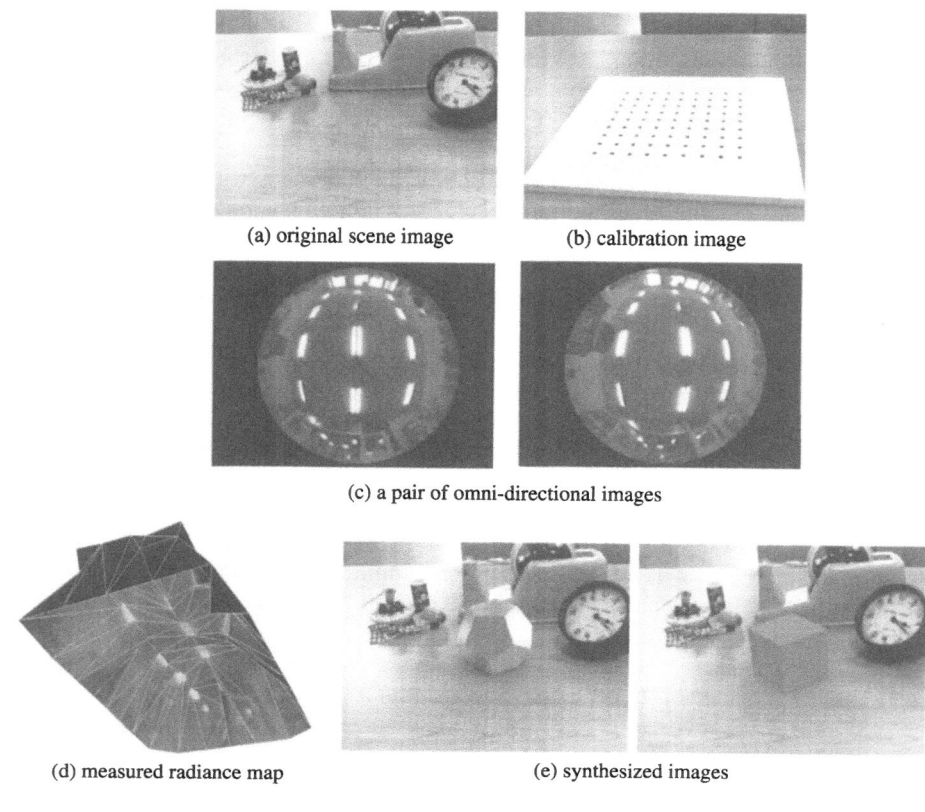

(a) original scene image   (b) calibration image

(c) a pair of omni-directional images

(d) measured radiance map   (e) synthesized images

Figure 8.8   Results.

the radiance map was used for rendering virtual objects in the real scene. Figure 8.8 (e) shows the resulting image that contains a virtual object with a soft shadow on the tabletop.

Figure 8.9 shows another example of integration of a virtual image with a real background.

8.4   Conclusions

We have explored the automatic generation of photorealistic object models from observation. For those objects with known reflectance models, we proposed a model-based rendering method that obtains reflectance parameters from sequences of range and color images. For other classes of objects, we have developed the eigen-textured method that stores and compresses appearances of objects on their 3D mesh models. Our experiments have shown that both of our rendering methods can be effectively used for synthesizing realistic object images. We have also developed a new method of superimposing virtual objects onto images of real scenes by taking into account

Figure 8.9   Another example of a mixed reality image.

the radiance distribution of the scenes. To demonstrate the effectiveness of the proposed method, we have successfully tested our method by using real images.

Acknowledgments

This research has been supported in part by Ministry of Education, Japanese Government, under grant number 09NP1401(shin-pro) and 09450164(kakenhi).

References

[1] Y. Sato, M. Wheeler, and K. Ikeuchi: "Object shape and reflectance modeling from observation," *Proc. SIGGRAPH'97*, pp.379–387, 1997.

[2] K. Nishino, Y. Sato, I. Sato, and K. Ikeuchi: "Eigen-texture method - Appearance compression based on 3D model -," *Information Processing Society of Japan Symposium Series*, vol.98, no.10, pp.I19–I26, July 1998 (in Japanese).

[3] I. Sato, Y. Sato, and K. Ikeuchi: "Acquiring a radiance distribution to superimpose virtual objects onto a real scene," *IEEE Trans. on Visualization and Computer Graphics*, 1999 (to appear).

[4] M. Bajura, H. Fuchs, and R. Ohbuchi: "Merging virtual objects with the real world," *Proc. SIGGRAPH'92*, pp.203–210, 1992.

[5] S. Gortler, R. Grzeszczuk, R. Szeliski, and M. F. Cohen: "The Lumigraph," *Proc. SIGGRAPH'96,* pp.43–54, 1996.

[6] M. Levoy and P. Hanrahan: "Light field rendering," *Proc. SIGGRAPH'96,* pp.31–42, 1996.

[7] K. Sato, H. Yamamoto, and S. Inokuchi: "Range imaging system utilizing nematic liquid crystal mask," *Proc. 1st Int'l Conf. on Computer Vision*, pp.657–661, 1987.

[8] M. Wheeler, Y. Sato, and K. Ikeuchi: "Consensus surfaces for modeling 3D objects from multiple range images," *Proc. 6th Int'l Conf. on Computer Vision,* pp.917–924, 1998.

[9] B. Curless and M. Levoy: "A volumetric method for building complex models from range images," *Proc. SIGGRAPH'96,* pp.303–312, 1996.

[10] H. Hoppe, T. DeRose, T. Duchamp, J. McDonald, and W. Stuetzle: "Surface reconstruction from unorganized points," *Proc. SIGGRAPH'92,* pp.71–78, 1992.

[11] K. Torrance and E. Sparrow: "Theory for off-specular reflection from roughened surface," *J. Opt. Soc. Am.,* vol.57, pp.1105–1114, 1967.

[12] S. Nayar, K. Ikeuchi, and T. Kanade: "Surface reflection," *IEEE Trans. on Pattern Analysis and Machine Intelligence,* vol.13, no.7, pp.611–634, 1991.

[13] Y. Sato and K. Ikeuchi: "Temporal-color space analysis of reflection," *J. Opt. Soc. Am. A,* vol.11, no.11, pp.2990–3002, November 1994.

[14] K. Ikeuchi and K. Sato: "Determining reflectance properties of an object using range and brightness images," *IEEE Trans. on Pattern Analysis and Machine Intelligence,* vol.13, no.11, pp.1139–1153, 1991.

[15] G. K. Klinker, S. Shafer, and T. Kanade: "The measurement of highlights in color images," *Int'l J. Computer Vision,* vol.2, pp.7–32, 1988.

[16] K. Pulli, M. Cohen, T. Duchamp, H. Hoppe, L. Shapiro, and W. Stuetzle: "View-based rendering," *Proc. 8th Eurographics Workshop on Rendering,* June 1997.

[17] Z. Zhang: "Modeling geometric structure and illumination variation of a scene from real images," *Proc. 6th Int'l Conf. on Computer Vision,* pp.1041–1046, 1998.

[18] H. Murase and S. K. Nayar: "Visual learning and recognition of 3-D objects from appearance," *Int'l J. Computer Vision,* vol.14, pp.5–24, 1995.

[19] R. T. Azuma: "A survey of augmented reality," *Presence,* vol.6, no.4, pp.355–385, August 1997.

[20] A. State, G. Hirota, D. Chen, W. Garrett, and M. Livingston: "Superior augmented-reality registration by integrating landmark tracking and magnetic tracking," *Proc. SIGGRAPH'96,* pp.429–438, August 1996.

[21] C. Tomasi and T. Kanade: "Shape and motion from image streams under orthography: a factorization method," *Int'l J. Computer Vision,* vol.9, no.2, pp.137–154, 1992.

[22] R. Tsai: "A versatile camera calibration technique for high accuracy machine vision metrology using off-the-shelf TV cameras and lenses," *IEEE Trans. on Robotics and Automation,* vol.3, no.4, pp.323–344, 1987.

[23] S. Chen: "Quicktime VR," *Proc. SIGGRAPH'95*, pp.29–38, 1995.

[24] A. Fournier, A. Gunawan and C. Romanzin: "Common illumination between real and computer generated scenes," *Proc. Graphics Interface '93*, pp.254–262, 1993.

[25] G. Kay and T. Caelli: "Inverting an illumination model from range and intensity maps," *CVGIP: IU*, vol.59, pp.183–201, 1994.

Chapter 9

The Ray-Based Approach to Augmented Spatial Communication and Mixed Reality

Takeshi Naemura
Hiroshi Harashima
The University of Tokyo, Japan

9.1 Introduction

Most of the previous works on three-dimensional (3-D) image communication have been concentrated on how to compress the huge amount of data, required for the stereoscopic visual effect [1]. Some additional requirements, however, are recently identified in order to enhance the flexibility of visual communication. The display-independent feature, which the authors proposed in [2], is one of the promising approaches to outgrow the framework of conventional image communication and to open up a novel field of technology, which should be called the "spatial communication". Not only image data captured by a camera, but also spatial data (such as stereo pairs, multi-view images, integral photographs, holograms, volumetric data, and geometric models of objects), will be utilized seamlessly in the next-generation visual communication system. For this purpose, it is important to establish a method of representing several kinds of spatial data in the same way, which is neutral for any input and output system (See Figure 9.1).

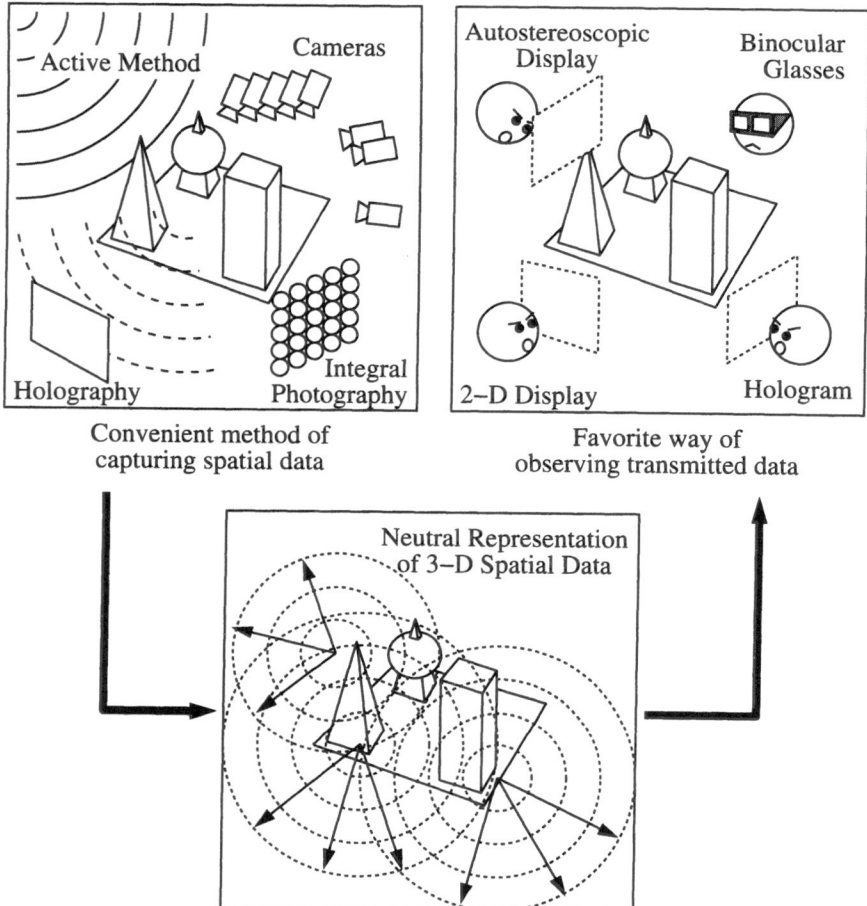

Figure 9.1 The integrated 3-D visual communication system.

As the neutral representation, the authors have proposed and investigated a method, in which a light ray is considered as a primitive element of visual cues. Theoretically, we can decompose any visual data into a set of light ray data, and synthesize any visual cue appropriate for each display system from the set of light ray data. As a practical matter, however, a method of interpolating light ray data is essential. On the other hand, the neutral representation of spatial data, which has been developed in the field of visual communication, has a close connection with the mixed reality technology, since it aims to treat any visual data in the same way. The combination of physical and virtual objects will augment the spatial communication, and it may be one of the goals of mixed reality technology.

This chapter reviews the results of the studies on the neutral representation of visual cues (especially on the light ray interpolation), and shows the applicability of the method to the mixed reality technology.

9.2   Integrated 3-D Visual Communication System

For the 3-D image communication, the development of various supporting technologies (such as 3-D capturing, 3-D representation, 3-D compression, 3-D transmission, 3-D display and 3-D handling) is required. As for the 3-D displays, several promising technologies (such as binocular glasses, head-mounted displays, lenticular methods, parallax barrier methods, focused-beam arrays, holographic methods and eye-position tracking methods) have made rapid progress. Consequently, most of the works on 3-D image processing have been specialized for the corresponding display technologies. The 3-D image communication itself, however, has not been discussed at a comprehensive level.

Considering such a situation, the authors proposed the concept of "Integrated 3-D Visual Communication System" [2] illustrated in Figure 9.1. The key feature of the concept is the display-independent representation of visual cues; all the differences between input (capturing) and output (displaying) methods are absorbed by the intermediate representation, which is neutral for any visual data. So it allows us to select our favorite ways of observing the transmitted spatial data, regardless of the input methods. Furthermore, we can capture the spatial data by our convenient methods, regardless of the observer's display system. This function will promote the progress of 3-D image communication systems, even before the 3-D display technology reaches maturity.

The authors have investigated following two practical approaches : (a) structure-based approach [3] [4] and (b) ray-based approach [5]–[7].

(a) Structure-based approach   It is straightforward that the structure-based approach is a practical way of representing scene objects. For instance, VRML (Virtual Reality Modeling Language) [8] is one of the standardized methods of representing geometric models in the virtual world. This approach is well-suited for handling, as well as displaying, objects in the virtual world.

(b) Ray-based approach   Since the visual sensation is excited by rays of light incident on the retina, each of these rays can be seen as a primitive element of visual cues, and any kind of visual data can be decomposed into a set of light ray data. Conversely, arbitrary display-dependent visual data can be synthesized from the set of light ray data. This approach is useful for displaying objects in the physical world as well as in the virtual world.

In order to apply the structure-based method (a) to the physical world, we must construct geometric models of every object in the world. It is, however, not easy to measure the accurate shapes, surface textures and reflection models of the physical objects. To solve this problem, a great deal of improvement in the field of structure recovery from images will be required. What is essential for the visual communication, however, is not the structural properties of scene objects, but the light rays emitted from them. The merit of the ray-based approach (b) is that we can avoid the problems of structure recovery. From this point of view, the following sections will clarify the basic concept and applications of the ray-based approach (b).

9.3   Ray-Based Representation of Visual Cues

The ray-based approach was introduced to the field of 3-D image communication including holographic displays in [5] after the proposal of structure-based representation of light rays [9]. It has a close connection with the concept of image-based rendering [10]–[13], which attracts a great deal of attention in the field of computer graphics. Both of them aim to avoid difficulties in constructing geometric models of objects in the physical world. Since an image can be regarded as a set of light rays, the ray-based approach involves the image-based one.

9.3.1   5-D Data Space and Its Sub-Spaces

A set of light rays passing through any point in a 3-D space is called a pencil, which is distinguished by its position (X, Y, Z). Each of the light rays which constitute a pencil can be identified by its directions (θ, ϕ), representing its longitude and latitude angle of propagation. Therefore, by providing a five-dimensional data space $XYZ\theta\phi$, we can store the light ray data (power or wavelength) $f(X, Y, Z, \theta, \phi)$ at any point in any direction, separately. This data space is called the "Plenoptic Function" [14] or the "Ray-Space" [5].

An environment map [15], which records the incident light arriving from all directions at a point (X_a, Y_a, Z_a), can be represented as follows :

$$f'(\theta, \phi) = f(X, Y, Z, \theta, \phi)|_{(X,Y,Z)=(X_a,Y_a,Z_a)}. \tag{9.1}$$

This means that an environment map corresponds to a 2-D sub-space of the 5-D data space. We can quickly synthesize any outward looking view of the environment from $f'(\theta, \phi)$ [16]. From this point of view, $f(X, Y, Z, \theta, \phi)$ can be regarded as a set of environment maps at any location in a 3-D space.

In the same way, multi-view images, which are captured by cameras aligned on a line $Y = Z = 0$, can be stored in a 3-D sub-space of the 5-D data space as follows :

$$f'(X, \theta, \phi) = f(X, Y, Z, \theta, \phi)|_{(Y,Z)=(0,0)}. \tag{9.2}$$

Furthermore, multi-view images, which are captured by cameras arranged on a plane $Z = 0$ can be stored in a 4-D sub-space as follows :

$$f'(X, Y, \theta, \phi) = f(X, Y, Z, \theta, \phi)|_{Z=0}. \tag{9.3}$$

A hologram, located on the plane $Z = 0$, also corresponds to the 4-D data space $f'(X, Y, \theta, \phi)$.

In the 4-D sub-space, the world is divided by the plane $Z = 0$ into two zones; the visual zone $(Z \geq 0)$ and the captured zone $(Z < 0)$. By assuming that rays go straight on without any variation in the directions of propagation, we can even synthesize virtual views at any viewpoint (X_a, Y_a, Z_a) within the visual zone (see Figure 9.2). The reduction of a 5-D data space to that of four or less dimensions is always accompanied with a loss of information. This results in the restriction of the region, inside or outside of which observers can walk around. In other words, we can effectively suppress the redundancies in a 5-D data space by utilizing a 4-D data

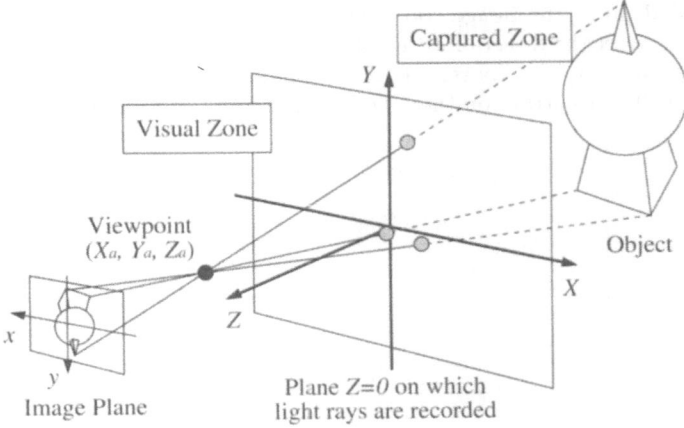

Figure 9.2   Synthesizing virtual views at any viewpoint within the visual zone.

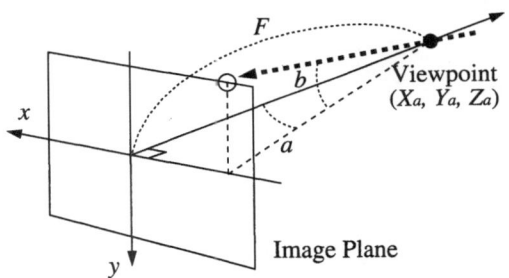

Figure 9.3   Image coordinates (x, y) and directions of light rays (θ, ϕ).

space, as long as the observers are staying within the visual zone. In many cases, a 4-D data space is enough to represent 3-D images, though it is sure that there are some applications requiring a 5-D data space. Thus, a 4-D data space can be regarded as an efficient representation of spatial data. From this point of view, the authors have investigated the 4-D data space representation of visual cues. More general cases, in which light rays are recorded on a parametric surface $S(P, Q)$ including spherical and cylindrical surfaces, are discussed in [7]. In the followings, the most simple method, where the surface $S(P, Q)$ is defined as $(X, Y, Z) = (P, Q, 0)$, is reviewed.

9.3.2   Simple Example of 4-D Data Space

Figure 9.3 illustrates an example of the relationship between image coordinates (x, y) and directions of light rays $(\theta, \phi) = (a, b)$. From this figure, the following equations are derived.

$$x = -F \tan\theta, \quad y = -F\frac{\tan\phi}{\cos\theta} \tag{9.4}$$

where (x, y) denotes the image coordinates, and F focal length. These equations allow us to convert the direction (θ, ϕ) of a light ray into its recorded position (x, y). This means that we can convert $f'(X, Y, \theta, \phi)$ into $f'(X, Y, x, y)$. The following discussions will concentrate on the 4-D data space in the form of $f'(X, Y, x, y)$.

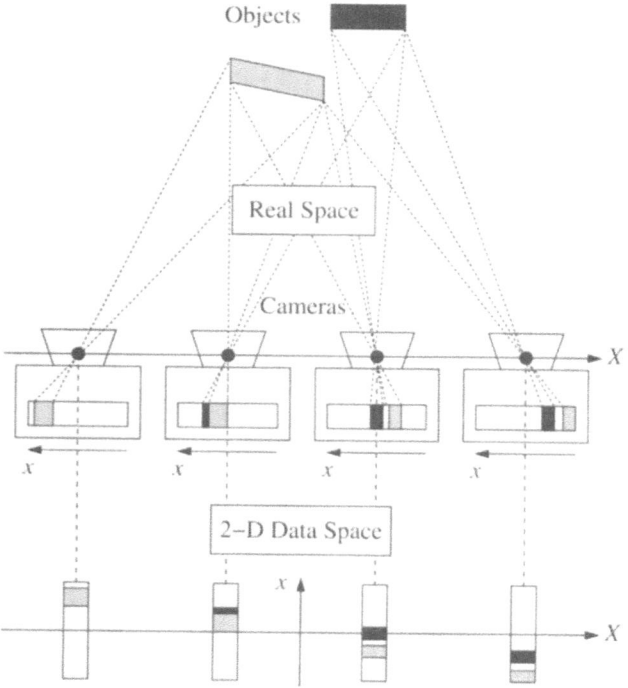

Figure 9.4   Real space to data space.

Figure 9.4 illustrates a top view of a camera array system, in which four cameras are aligned horizontally. X coordinates denote the positions of the cameras on a horizontal line in a 3-D space, and x, the positions of pixels on a horizontal line in each image. In the same way, we can consider the case where the cameras are aligned vertically. Y coordinate denote the positions on a vertical line in the 3-D space and y, the positions of pixels on a vertical line in each image.

Let the Xx-plane denote the 2-D data space, whose axes are X and x as illustrated in Figure 9.4. In the same way, we can define the Yy-plane. The Xx- and Yy-planes are called EPI (Epipolar Plane Image) [17]. The followings are the important characteristics of EPIs.

- All the pixels, corresponding to a point in the space, are aligned on a straight line in an EPI.

- The inclination of the straight line depends on the distance between the camera array and the point in the space.

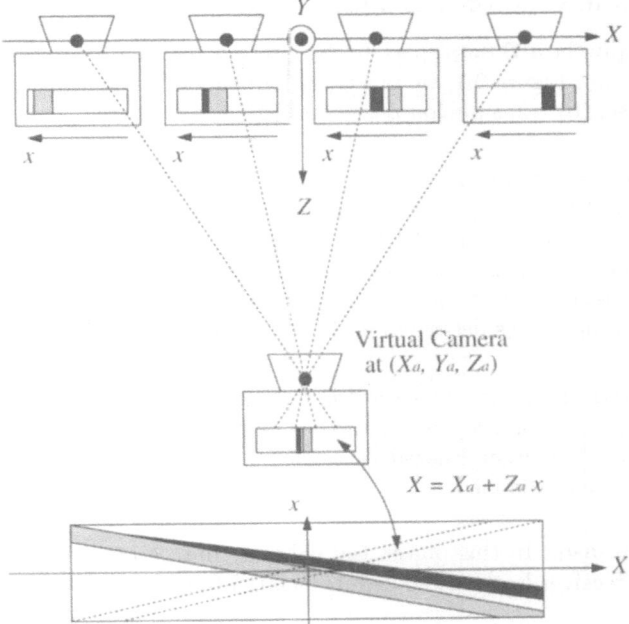

Figure 9.5   How to synthesize a virtual camera's view.

When large number of cameras are densely aligned, continuous structure will appear in the Xx- and Yy-planes as shown in the bottom of Figure9.5. The straight lines, corresponding to the point (X_a, Y_a, Z_a) in the space, can be represented as follows :

$$X = X_a + Z_c x, \quad Y = Y_a + Z_a y. \tag{9.5}$$

From Eq.(9.5), we can explain the followings.

- In the Xx plane, a line $X = ax + b$ $(a < 0)$ corresponds to an object point in the captured zone $Z < 0$. The process of structure recovery can be regarded as the process of decomposing the Xx plane into a set of lines $X = ax + b$, $(a < 0)$.

- In the Xx plane, a line $X = ax + b$ $(a \geq 0)$ corresponds to a virtual viewpoint in the visual zone $Z \geq 0$. By gathering the light ray data on the line, we can synthesize a virtual view corresponding to the viewpoint. Especially, a line $X = b(const.)$ corresponds to a horizontal line in the image captured on the plane $Z = 0$.

The Yy-plane has the same characteristics as the Xx plane. Consequently, by considering a 4-D data space $f'(X, Y, x, y)$, we can synthesize an image $I_a(x, y)$, whose viewpoint is located at (X_a, Y_a, Z_a) in the 3-D space, as follows :

$$I_a(x, y) = f'(X, Y, x, y)|_{(X,Y)=(X_a, Y_a)+Z_a(x,y)}. \tag{9.6}$$

9.3.3   Framework of 4-D Data Space System

In order to apply the 4-D data space to the intermediate representation of any visual data illustrated in Figure 9.1, we must consider the following three principal parts; (a) input phase, (b) transmission phase and (c) output phase.

(a) Input phase   In this phase, input visual data is converted into light rays and mapped onto a 4-D data space. We can also synthesize light rays from geometric models in the virtual world by the ray-tracing technique. For 2-D image input, the camera position (X_a, Y_a, Z_a) and the relationship between image coordinates (x, y) and directions (θ, ϕ) of rays are required to be given or estimated. In many cases, interpolation of light rays described in **9.3.4** will be required in this phase.

(b) Transmission phase   Since a 4-D data space still contains a huge amount of data, an efficient compression method is strongly required in this phase. Vector quantization [12], fractal-based [18], DCT-based [19], and disparity-based [20] methods have been examined.

(c) Output phase   In this phase, the light ray data appropriate for the display system at the receiver is extracted or interpolated from the transmitted data space.

9.3.4   Interpolation of Light Rays

It is not easy to arrange cameras densely enough to produce a continuous 4-D data space. Consequently, a sparsely sampled data space as shown in the bottom of Figure 9.4 will be obtained from a camera array system. It is important to fill up the data space by interpolating cracks between the sampled data.

This process can be performed in both input phase (a) and output phase (c). In order to optimize the quality of interpolation, very high computational power will be required. In this case, we may perform the interpolation process before the transmission phase (b), because the process of compression will degrade the quality of interpolation. After the dense data space is transmitted, we can synthesize virtual views at the receiver without any interpolation process. For the purpose of real time applications, however, we may transmit the data space with many cracks first, and then perform a real time interpolation at the receiver, at the cost of the quality of virtual views. The authors have proposed and compared some of the interpolation methods introduced below.

- Nearest neighbor method; the most simple approach for real time applications

- Plane approximation method; utilizing simple structure model for real time applications

- Fractal method; sophisticated interpolation without any structure model

- Local structure model method; combination of several structure models

They are briefly reviewed in the following.

Nearest neighbor method

One of the easiest ways of this process is the nearest neighbor method. Let K denote the number of cameras. We can represent the positions of cameras as (X_k, Y_k) $(k = 0, \ldots K-1)$. Let $f'_R(X_k, Y_k, x, y)$ denote the data space sampled by the cameras. The aim of the interpolation is to synthesize a densely sampled data space $f'(X, Y, x, y)$ from a sparsely sampled $f'_R(X_k, Y_k, x, y)$.

The nearest neighbor method is formulated as follows :

$$f'(X, Y, x, y) = f'_R([X], [Y], x, y) \tag{9.7}$$

where

$$[X] = X_{k_o}, \quad [Y] = Y_{k_o} \tag{9.8}$$

$$(X - X_{k_o})^2 + (Y - Y_{k_o})^2 = \min_k \left\{ (X - X_k)^2 + (Y - Y_k)^2 \right\}. \tag{9.9}$$

Figure 9.6(a) shows an example of the results of the method. We can see discontinuities in the interpolated data space.

Plane approximation method

In order to suppress the discontinuities, the following method can be formulated.

$$f'(X, Y, x, y) = f'_R([X], [Y], \langle x \rangle, \langle y \rangle) \tag{9.10}$$

where,

$$\langle x \rangle = x - \frac{X - [X]}{Z_i}, \quad \langle y \rangle = y - \frac{Y - [Y]}{Z_i} \tag{9.11}$$

Figure9.6(b) shows an example of the result of the method. In this method, all the scene objects are assumed to be placed on a plane $Z = Z_i$ $(Z_i < 0)$. From Eq.(9.11), we can regard the nearest neighbor method as a special case of this method, in which the objects are assumed to be placed on $Z_i = -\infty$. We can see that the discontinuities are suppressed but still remain.

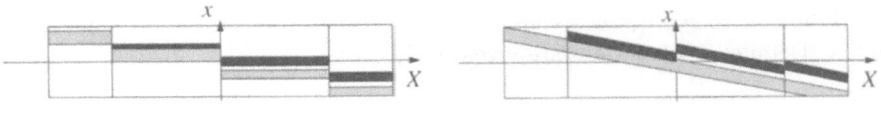

(a) Nearest neighbor method   (b) Plane approximation method

Figure 9.6   Interpolation of a sparsely sampled data space.

Fractal method

In order to blur the discontinuities, the linear interpolation [12] or fractal interpolation [18] is applicable. Especially, the fractal-based method can achieve both of the interpolation and the data compression of light rays. Figure 9.7 shows results of the fractal interpolation.

This approach will not require any structural property of scene objects, but just exploit the self-similarity of a 4-D data space. It requires, however, very high computational power to detect the self-similarity.

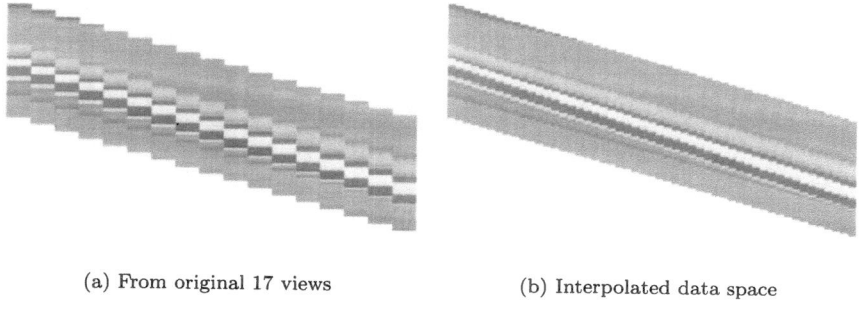

(a) From original 17 views   (b) Interpolated data space

Figure 9.7   Results of fractal interpolation.

Local structure model method

For the purpose of light ray interpolation, structural properties of scene objects can be utilized. As previously stated, it is not easy to recover the accurate structure from images. Structure models, however, are helpful for interpolating light rays, although not accurate [13].

It is also effective to utilize several structure models of the same object selectively [7]. Consider the case where we have several views of an object. For each view, a corresponding structure model, which is optimized for the view, can be estimated and synthesized. We call this model "local structure model". In most cases, the local models are incomplete. It is, however, very useful within a certain zone, in which errors in estimating structural properties from images are not conspicuous. We can take advantage of the invisibility of the errors to interpolate light rays, though it is indeed difficult to synthesize a complete global model from several incomplete local models. Detailed explanations are given in [7].

9.4   Applications of Ray-Based Approach

The authors have developed several applications of the ray-based representation of visual cues. A real time application, a high quality application, and a hybrid application of ray-based and structure-based approaches are reviewed below.

9.4.1 Real Time Interaction with 3-D Scenes in Motion Using Plane Approximation Method

First, a real time method of rendering arbitrary views of 3-D scenes in motion is outlined [21]. In order to obtain the data space $f'(X, Y, x, y)$ in real time, multiple views of a 3-D scene must be captured by a computer simultaneously. Figure 9.8 shows the configuration of our system. We have sixteen cameras and five quad

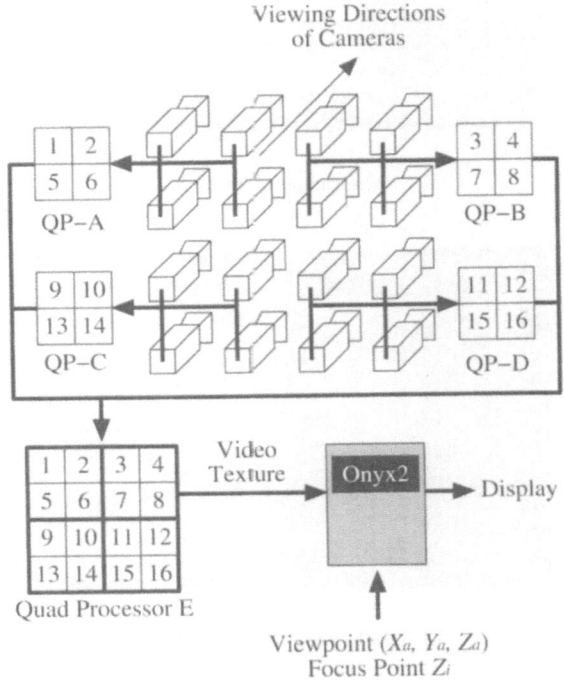

Figure 9.8   System configuration.

processors (QP). Four video sequences are combined into one sequence by a quad processor. Four quad processors (QP-A, B, C, D) combine sixteen video sequences into four sequences. Then, the combined sequences are integrated into one sequence by the fifth quad processor (QP-E). Thus, sixteen video sequences are captured by a computer simultaneously.

The input video sequence can be regarded as the sparsely sampled data space $f'_R(X_k, Y_k, x, y)$. The plane approximation method, described in **9.3.4**, is applied here. The computer renders virtual views $I_a(x, y)$ according to the position of observer's eye (X_a, Y_a, Z_a) and the interpolation parameter (focus point) Z_i, by utilizing the video-texture technology, which can warp and stitch video input in real time.

It is, however, not practical to align the directions of optical axes of all cameras, precisely. This is because the directions are slightly different from each other, even if

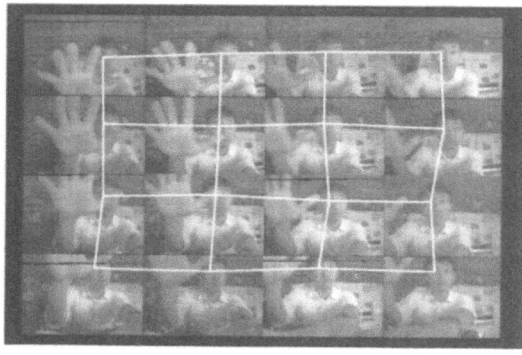

(a) Original input data

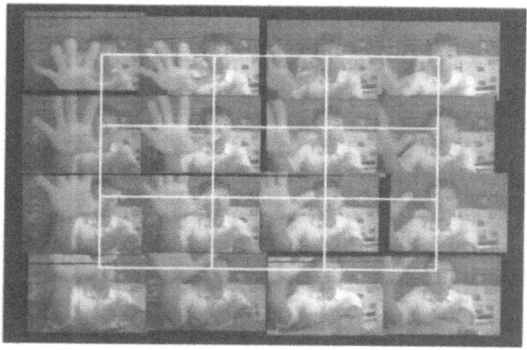

(b) Result of real time adjustment

Figure 9.9   Software approach to suppress the differences between cameras.

the cameras are the same products. Considering the cost performance, the authors concluded that a software approach to suppress the differences is more efficient than constructing an accurate hardware system. Figure 9.9(a) shows an example of the combined image. White lines illustrate how the corresponding points of the sixteen views appear irregularly on the combined image. In order to arrange the corresponding points regularly, we can translate each view as shown in Figure9.9(b). This adjustment can be applied to every frame in real time by the video-texture technology.

Figure 9.10 shows the results of real time video synthesis. The virtual view is synthesized from the set of real views in real time. We can see how the real views are warped and stitched in the virtual views. Since we realize a real time application at the cost of image quality, we can see discontinuities between the stitched views. In this experiment, the interpolation parameter Z_i is set to suppress the discontinuities

(a) Combined input views

(b) Synthesized virtual view

Figure 9.10   Result of real-time video-based rendering.

on the region of a man. There are two walls in front of sculptures and the man. We can see how the occlusion occurs between the walls and the sculptures. This visual effect is very three-dimensional, because we cannot reproduce such effect by scaling a two-dimensional image.

9.4.2   Sophisticated Interpolation for Static 3-D Scenes Using Local Structure Model Method

Secondly, a sophisticated method, which can improve the quality of interpolated light rays at the cost of high computational power, is outlined [7]. The multi-view image database, presented by University of Tsukuba, is used as input images (See Figure 9.11). For each image, a disparity map between the image and its four neighbors is estimated. Figure 9.12 shows the estimated results. The disparity values are coded

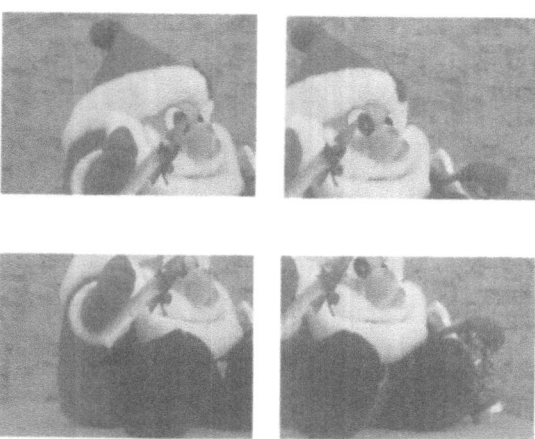

Figure 9.11   Parts of original multi-view images (9 × 9 views).

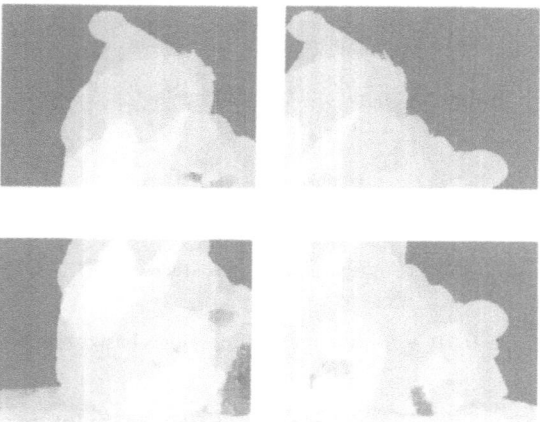

Figure 9.12   Local structure models corresponding to Figure 9.11.

by intensity; larger disparity is indicated with brighter intensity. We can synthesize local structure models from the estimated disparity maps and camera parameters. In this sense, Figure 9.12 shows the shape of each local models. We can see that the shapes of local models are different from each other. By stitching and warping the original views according to their corresponding local models, we can interpolate light rays to fill up a 4-D data space [7].

(a) Plane approximation method

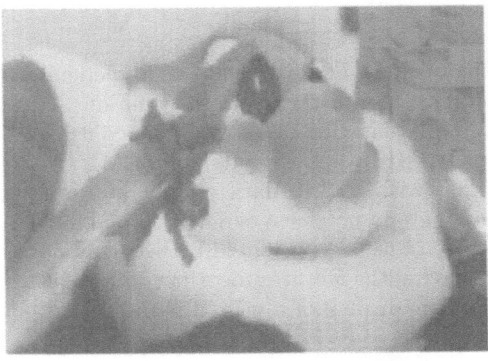

(b) Local structure method

Figure 9.13 Virtual views synthesized from the interpolated 4-D data space (see color pages).

Figure 9.13 compares a result of the plane approximation method (a) and that of the local structure model method (b). Thanks to the sophisticated interpolation technique explained before, the synthesized view (b) is so natural that we wonder whether it is a real photograph or a computer-generated image. We can see how effective the local structure model method is.

9.4.3 Integration of Virtual Scene and Physical Object

Finally, a kind of mixed reality applications, in which physical objects represented by light ray data are placed in the virtual world represented by geometric models, is outlined [22] [23]. Figure 9.14 shows an example of a cyber space called CyberMirage. Components inside a building are presented using polygonal models and dresses are

Figure 9.14   CyberMirage.

presented by the light ray data. These objects are implemented so that the observer can move viewpoints and operate them in real time on SGI Indigo2 IMPACT or equivalent machines. Representation of a virtual space using polygonal model data is based on a subset of VRML 1.0. Descriptors of VRML are expanded so that they can merge light ray data with VRML data. Light ray data are objects which can only be observed since they do not have explicit shape models. However, a scheme is built to move or rotate objects represented by light ray data in the CyberMirage system. CyberMirage having the features mentioned above is useful for representing products realistically in the display of a virtual mall.

9.5   Conclusions

First, this chapter explained the concept of "Integrated 3-D Visual Communication", aiming to realize a display-independent method of representing 3-D scenes. This concept will promote the progress of 3-D image communication, while several 3-D display technologies have not reached maturity. Secondly, a ray-based method was introduced as a fundamental approach to realize this concept. The 4-D data space representation was applied to the display-independent representation of any visual data. Some practical methods of interpolating light rays were outlined and examples of a real time application, a high quality application, and a mixed reality application were presented.

   While the ray-based method is developed in the field of visual communication, it has a wide range of application for visual data processing. Almost all the techniques developed in the field of visual data processing can be re-defined in the ray-based domain because of its neutrality. So, it is very useful for several emerging new technologies. Especially, the method is essentially well-suited for the mixed reality technology, since any visual cue, from both of the physical and virtual worlds, is treated in the same way.

References

[1] T. Naemura, M. Kaneko, and H. Harashima : "Compression and representation of 3-D images - A survey," *IEICE Trans. Inf. & Syst.*, vol.E82-D, no.3, 1999 (to appear).

[2] H. Harashima : "Three-dimensional image coding, future prospect," *Proc. Picture Coding Symp. of Japan (PCSJ'92)*, pp.9–12, 1992 (in Japanese).

[3] T. Fujii and H. Harashima : "Data compression and interpolation of multi-view image set," *IEICE Trans. Inf. & Syst.*, vol.E77-D, no.9, pp.987–995, 1994.

[4] T. Naemura, M. Kaneko, and H. Harashima : "3-D object based coding of multi-view images, *Proc. Picture Coding Symp. (PCS'96)*, pp.459–464, 1996.

[5] T. Fujii : "A basic study on the integrated 3-D visual communication," Ph.D thesis, Course of Elec. Eng., The Univ. of Tokyo, 1994 (in Japanese).

[6] T. Yanagisawa, T. Naemura, M. Kaneko, and H. Harashima : "Handling of 3-dimensional objects in ray space," *Proc. IEICE Inf. & Syst. Society Conf. of IEICE*, D-169, 1995 (in Japanese).

[7] T. Naemura, M. Kaneko, and H. Harashima : "Orthographic approach to representing 3-D images and interpolating light rays for 3-D image communication and virtual environment," *Signal Process. : Image Commun.*, vol.14, pp.21–37, 1998.

[8] http://www.web3d.org

[9] T. Fujii and H. Harashima : "Coding of an autostereoscopic 3-D image sequence," *Proc. SPIE Visual Commun. and Image Process.(VCIP'94)*, vol.2308, pp.930–941, 1994.

[10] A. Katayama, K. Tanaka, T. Oshino, and H.Tamura : "A viewpoint independent stereoscopic display using interpolating of multi-viewpoint images," *Proc. SPIE Stereoscopic displays and virtualreality systems II*, vol.2409, pp.11–20, 1995.

[11] L. McMillan and G. Bishop : "Plenoptic modeling : An image-based rendering system," *Proc. SIGGRAPH'95*, pp.39–46, 1995.

[12] M. Levoy and P. Hanrahan : "Light field rendering," *Proc. SIGGRAPH'96*, pp.31–42, 1996.

[13] S. Gortler, R. Grzeszczuk, R. Szeliski and, M. Cohen, "The Lumigraph," *Proc. SIGGRAPH'96*, pp.43–54, 1996.

[14] E. Adelson and J. Bergen : "The plenoptic function and the elements of early vision," in (M. Landy and J. Mcvshon, eds.) *Computer Models of Visual Processing*, Chapter 1, MIT Press, 1991.

[15] N. Greene : "Environment mapping and other applications of world projections," *Computer Graphics and Applications*, vol.6, no.11, pp.21–29, 1986.

[16] S. E. Chen : "QuickTimeVR - an image-based approach to virtual environment navigation -," *Proc. SIGGRAPH'95*, pp.29–38, 1995.

[17] R. C. Bolles, H. H. Baker, and D. H. Marimont : "Epipolar-plane image analysis : an approach to determining structure from motion," *Int'l J. Computer Vision*, vol.1, pp.7–55, 1987.

[18] T. Naemura and H. Harashima : "Fractal coding of a multi-view 3-D image," *Proc. IEEE Int'l Conf. on Image Process. (ICIP'94)*, vol.III, pp.107–111, 1994.

[19] T. Takano, T. Naemura, M. Kaneko, and H. Harashima : "3-D space coding based on light ray data - Local expansion of compressed light ray data -," *J. Inst. of Image Inform. and Television Eng.*, vol.52, no.9, pp.1321–1327, 1998 (in Japanese).

[20] T. Fujii, T. Kimoto, and M. Tanimoto : "Data compression of 3-D spatial information based on ray-space coding" , *J. Inst. of Image Inform. and Television Eng.*, vol.52, no.3, pp.356–363, 1998 (in Japanese).

[21] T. Naemura and H. Harashima : "Real-time video-based rendering for augmented spatial communication," *Proc. SPIE VCIP'99*, 1999 (to appear).

[22] http://www.x-zone.canon.co.jp/CyberMirage/index-e.html, http://www.hc.t.u-tokyo.ac.jp/3D/CyberMirage/index-e.html

[23] S. Uchiyama, A. Katayama, A. Kumagai, H. Tamura, T. Naemura, M. Kaneko, and H. Harashima : "Collaborative CyberMirage : A shared cyberspace with mixed reality," *Proc. Virtual System and MultiMedia (VSMM'97)*, pp.9–18, 1997.

Chapter 10

Building a Virtual World
from the Real World

Michitaka Hirose
Tomohiro Tanikawa
The University of Tokyo, Japan

Takaaki Endo
Mixed Reality Systems Laboratory, Japan

10.1 Introduction

Generally speaking, conventional Computer Graphics (CG) consisting of 3-D geometrical polygon models is used to synthesize a large-scale virtual environment such
as a virtual city walk-through [1] [2]. However, in this case, it is difficult to generate
a photorealistic landscape because the preparation of a huge amount of polygons is
very labor-intensive and tedious. In addition, since the CG image is too geometrical and looks artificial, it might be impossible to represent the delicate nuances of
natural landscapes in principle.

  If just an outlook of the landscape is important, a geometrical 3-D model may
not be necessary. We may consider a methodology that generates a 3-D virtual world
directly from real images without using an explicit 3-D model. This methodology
is called "Image-Based Rendering (IBR)," and has become popular because it can
generate a highly photorealistic virtual world. In other words, IBR can enrich the
virtual world by importing information from the real world.

"Aspen Movie Map" was the pioneering work of IBR technology [3]. This system consists of a computer-controlled laser disc, which records the images along the town (Aspen) streets. The user can "walk" along the street where images are captured and can also select the street he/she wants to proceed along at the crossing. However, in this environment, the user can only look at the images from the original viewpoint of the camera. "Virtual Dome" [4] and "QuickTime VR" [5] can be used to generate a virtual world where we can look around, but can never walk around in these environments.

If we generalize the above three systems, they are based on the idea of recording images with the position information of the camera, rearranging them in a virtual world, and generating an interactive photorealistic virtual world. A large image database is required for these systems. In other words, the principle of "rote memorization" is used here, which will become possible only when highly integrated memory devices become available more cheaply.

Along with this principle or philosophy, we have been developing a prototype system, which can record and reproduce a broad area of a real city in a computer [6]. This paper describes the newest system in which a user can walk through and look around by combining the panorama-based method, morphing method, and 3-D layer method.

10.2   Data Capturing System

To record images with the position information of the camera automatically and effectively, we developed a mobile data capturing system equipped with video cameras and position data sensors. An outlook of the system is shown in Figure 10.1.

10.2.1   Overview

The data capturing system is equipped with various pieces of equipment as shown in Table 10.1. The system configuration is as shown in Figure 10.2.

Eight video cameras, a GPS and a gyroscope are mounted on the automobile in order to capture the panoramic images with position and direction information [6]. Each video camera captures images at 30 frames per second. When the automobile moves at 30 kilometers per hour, the resultant sampling rate of images is 3.4 frames per meter.

All of the images and position information is indexed by time codes so that the two information streams can be integrated afterwards.

Table 10.1   List of instruments used in the system.

Instrument	Manufacturer	Type of product
Automobile	Toyota	Noah Field Tourer
Video Camera	Sony	DCR-VX1000
PC	Custom made	CPU Pentium Pro 200MHz
GPS	Trimble	4400
Angle sensor	Data Tech	GU-3020
Terrestrial Magnetism Sensor	TOKIN	TMC-2000

Figure 10.1 Data capturing system.

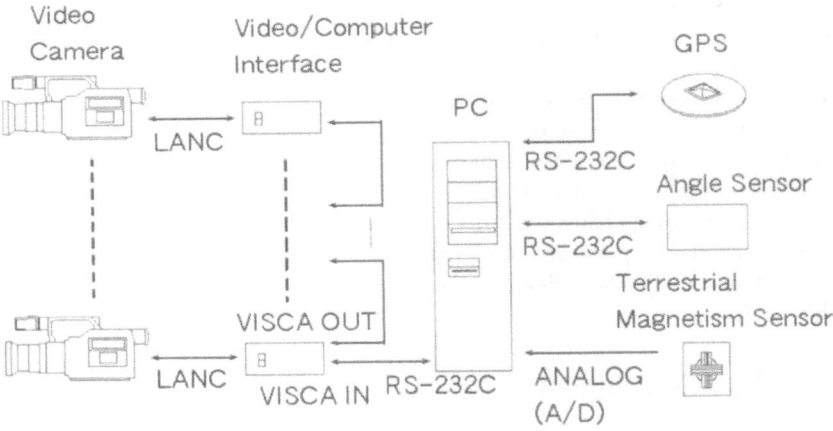

Figure 10.2 Image recording and position sensing system.

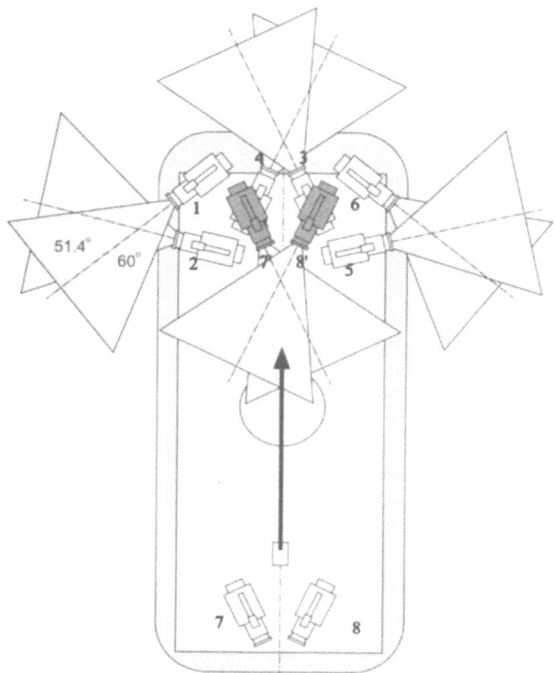

Figure 10.3   Arrangement of 8 cameras.

10.2.2   Configuration of Cameras

The horizontal viewing angle of each camera is 60 degrees. The geometrical arrangement of 8 cameras is as shown in Figure 10.3. A pair of cameras is placed symmetrically and the difference between their viewing directions is 51.4 degrees. This camera configuration supports a total horizontal view field of 360°.

The lens centers of the 8 cameras are not identical. In particular, cameras 7 and 8 are positioned backwards in order to avoid capturing the automobile itself within their view field. To shorten the differences of the lens centers between cameras 7 and 8 and the rest, the image sequences captured by cameras 7 and 8 are delayed for several frames by shifting the time code depending on the vehicle speed. After this condensation, the positions of cameras 7 and 8 can be regarded as cameras 7' and 8' in Figure 10.3. The influence of these differences of lens centers is evaluated in Table 10.2.

10.2.3   Position Sensors

To measure the viewpoint information of the cameras, a GPS sensor, a terrestrial magnetism sensor and a 3-axis angle sensor are equipped on the automobile. The GPS sensor is used to measure the automobile's location. In this system, we used

Table 10.2 Differences between lens centers of two paired cameras (1: calculated assuming that the automobile speed is 20km/h).

Two paired cameras	Difference between cameras centers (cm)
Camera 1 and 2, 5 and 6	14.4
Camera 2 and 3, 4 and 5	49.0
Camera 3 and 4, 7 and 8	12.3
Camera 6 and 7, 8 and 1	29.7 - 35.8 [1]

Kinematics GPS because its resolution is less than 3cm if it captures signals from 5 satellites. This resolution is considered sufficient for image data indexing. One problem is that real-time position calculation is impossible for a Kinematics GPS, and postprocessing of the sampled signal is required. Another problem is that if it cannot capture signals from 5 satellites, its resolution worsens, and an other compensation method is required. The signal-sampling rate of the GPS is 5Hz.

A 3-axis angle sensor consisting of 3 oscillating gyros and 3 acceleration sensors is used to measure the orientation. The angle measuring error is 0.5 degree for the pitch and roll angles, and 0.9 degree for the yaw angle. Since this angle sensor has "drift", a terrestrial magnetism sensor is also used to compensate the drift of the sensor output. Its instrumental error is 2 degrees. The sampling rate of this orientation sensor is 60Hz.

Of course, this measuring error should be further minimized. One idea which has not yet been implemented is the estimation of the position/orientation from the captured images. If camera parameters such as field of view are known, we should be able to calculate the camera position from the captured images.

10.3 Image Reproduction Systems

In order to construct a virtual world from the images with position and orientation information taken by the mobile data capturing system mentioned above, we developed several image reproduction systems.

The easiest implementation will be the construction of panoramic images. By stitching images taken at the same position, the panoramic image can be synthesized. When the user looks around, the corresponding part of the panoramic image should be displayed. By preparing many panoramic images as a sequence, panoramic "Movie Maps" can be constructed. When the user moves, the appropriate panoramic image is selected by using the position data as a key. Although this system is very simple, the user can walk around interactively and look in any direction.

However, since the simple panorama-based method uses a kind of "rote memorization," it requires a huge amount of disk space to extend a motion range. So, it is important to reduce the size of the image database, in other words, to reduce the sampling rate of images. There is also another problem in that the system, by just switching panoramic images, cannot provide an image at a position other than the prerecorded camera positions. To overcome these problems, we tried two interpolation methods, the "morphing method" and the "3-D layer method".

As mentioned before, by using the morphing method or the 3-D layer method, we can interpolate images between already existing images. Since this technique has an effect of smoothening the change of images for a given image data set, it is possible to reduce the image data amount and increase the user's movable area under a given memory capacity. A combination of the simple panorama-based method and the interpolation methods also gives design freedom to use the panorama-based method for a street walk-through where the user walks in one direction, and use the interpolation methods for an open space where the user will move freely around the area.

In this section, we first introduce the panorama-based method and then introduce the morphing method and the 3-D layer method in detail.

10.3.1 Panorama-Based Method

A panoramic image is synthesized by projecting each image captured by the 8 cameras onto a cylindrical surface and then aligning them horizontally after compensating for distortion (Figure 10.4) [6]. For looking around, this combined image data set is sufficient. However, for walking through, we have to prepare many data sets along the user's path, and switch the data set according to the user's position as shown in Figure 10.5). This method is a simple extension of the Virtual Dome [4].

Figure 10.4   Examples of synthesized panoramic images.

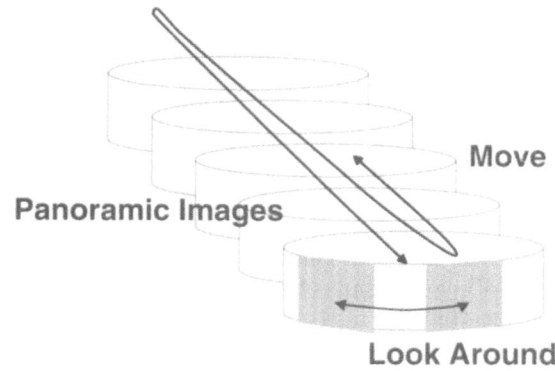

Figure 10.5   The principle of the panorama-based method.

One of the technological issues is how to quickly access image data in the database. A kind of caching system should be designed. We used the following approaches in order to make it possible to render images directly read from the hard disk in real-time.

Utilizing RAID Disk   In order to speed up the disk access at a hardware level, we employed a RAID disk, which can transfer data from the disk to the main memory at 100 Mbps.

Shrinking the Size of the Images   As a software consideration, we used images halved in width and height. The size of the halved panoramic image is 2,136 x 243 pixels (approx. 1.4 MB).

Utilizing Multithreading Technology   Both reading and rendering processes work in parallel in order to reduce the CPU idle time at the time of disk access. Image data near the current position should be loaded on the main memory before it is actually displayed by the rendering process. Therefore, our system reads both the previous and next panoramic image in the memory so that the system can render previously read images irrespective of whether the observer moves forward or backward. So, at least about 4.2 Mbytes of the main memory area should be reserved for the image cache. Since our ONYX2 Graphics Workstation has 256 Mbytes main memory, several panoramic image sets can be loaded in the image cache.

By employing these approaches, we obtained of approximately 10 frames per second.

10.3.2   Morphing Method

In the panorama-based method, a user's movement is strongly restricted, and huge amounts of data storage will be required to increase the freedom of user's motion. If we use an interpolation algorithm, we will be able to generate infinite image data from the finite image data.

After several trials, we realized the importance of optical flow when changing images. If we employ an insufficient morphing algorithm, disorder of the optical flow will occur, and the naturalness of the displayed image will be significantly damaged. Consequently, we developed a distortion-free interpolation algorithm based on the View Morphing method [7] as follows.

1. Obtain interpolation parameters from the coordinate positions of the viewpoint and the cameras of the three images I_1, I_2, and I_1 (Figure 10.6).

2. Rotate I_1 and I_2 so that the optical axes of I_1 and I_2 are parallel and the axes and the baseline are perpendicular.

3. Linearly interpolate the corresponding coordinate points in the rotated images of I_1 and I_2 using the ratio of $s : 1 - s$, and obtain interpolated image I_4.

4. Rotate I_3 and I_4 so that the optical axes of I_3 and I_4 are parallel and the axes and the baseline are perpendicular.

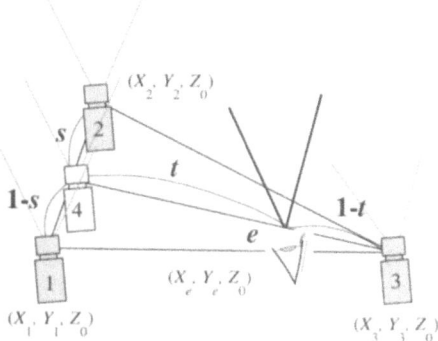

Figure 10.6   Interpolation parameters.

5. Linearly interpolate the corresponding coordinate points in the rotated images of I_3 and I_4 using the ratio of $t : 1 - t$, and obtain the image from the viewpoint I_e.

6. Rotate I_e so that its optical axis coincides with the viewing direction of the user.

Note that it is impossible to rotate images so that the optical axes of the images are parallel and the axes and the baseline are perpendicular if the center of one camera is within a viewing area of another camera as shown in Figure 10.7. In this case, we cannot compensate the distortion in the method described above. However, we can generate a distortion-free image by rotating two images so that the optical axes of the two images are identical.

Figure 10.7   Singular views.

Figure 10.8 shows a result of the morphing method applied to the images captured at Marunouchi, Tokyo. We will discuss this experiment in detail later. When implementing this kind of walk-through system, we have to be careful about which original data to use for interpolation. After the selection of images I_1, I_2, and I_3, images I_2, I_3, and I_4 should be selected (Figure 10.9).

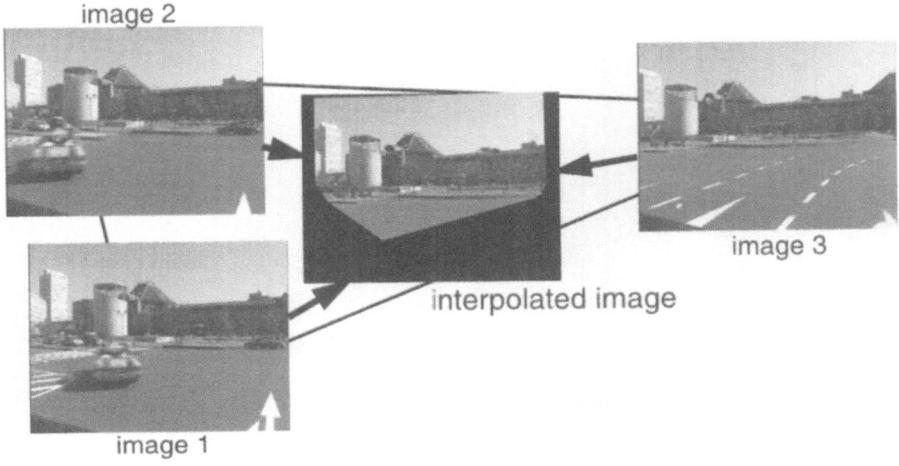

image 2

interpolated image

image 3

image 1

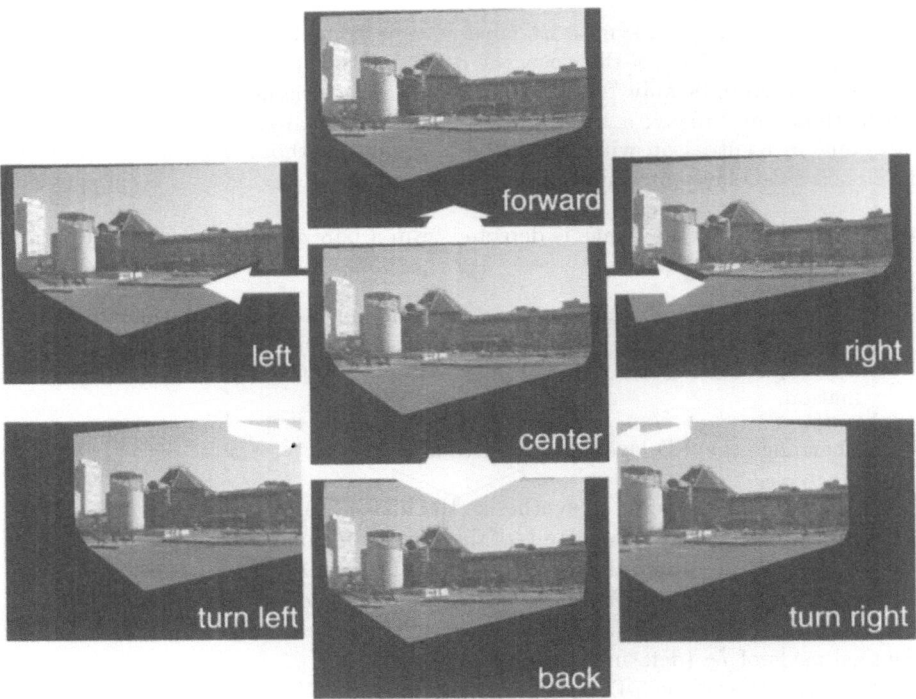

forward

left

right

center

turn left

turn right

back

Figure 10.8   A result of the morphing method (see color pages).

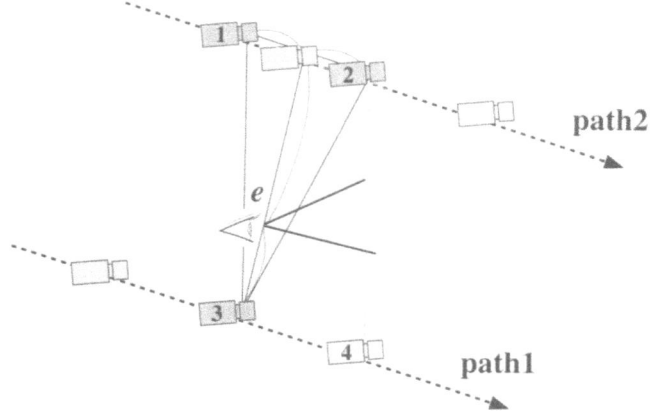

Figure 10.9   Switching three original images.

10.3.3   3-D Layer Presumption Method

Another methodology to synthesize novel images from the prerecorded images is to use simple 3-D models. In this case, the 3-D models can be "simple" because novel viewpoints need to be only "slightly" shifted from the original viewpoints. However, this method may still give a wider possible interpolation area from the original image viewpoint in comparison with the above morphing method. For this purpose, we developed a 3-D layer presumption method as follows.

1. Optical flow of the image is detected from a pair of images with position and orientation information.

2. Calculate depth values for each pixel of the image by using optical flow.

3. Decompose the image into several 3-D layers by referring to this depth information.

4. Rearrange the 3-D layers in the virtual space.

Then, we should be able to synthesize the image from novel viewpoints.

Figure 10.10 illustrates the details of the decomposition process. If we can assume a location p and an orientation α of a layer L of image I_A, by deforming I_A we can synthesize the image of L from the viewpoint of image I_B. So, by deforming I_A according to the various p and α values set, we can derive "3-D layers" which are identical parts of I_C (deformed I_A) and I_B.

By rearranging these 3-D layers, we can synthesize the novel images from novel viewpoints. Figure 10.11 shows a result of the 3-D layer presumption method applied to the images captured at Marunouchi, Tokyo.

Note that this methodology is only possible when the requested viewpoint is "slightly" shifted from the original viewpoint, because the derived 3-D models are

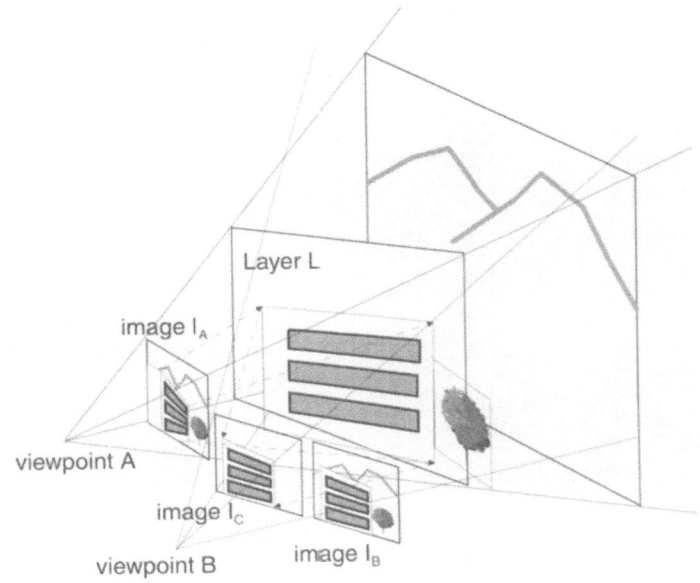

Figure 10.10   Algorithm of 3-D layer presumption method.

not complete. Therefore, we have to switch 3-D layers to walk-through the wide area, as already introduced in Section 10.3.2. Namely, according to the user's movement, reference images have to be reloaded from the image database as shown in Section 10.3.1, and based on the novel reference images, new 3-D layers can be regenerated.

This method is based on a very simple principle, but it provides photorealistic images. Of cause, several problems still remain: for example, this method cannot avoid an occlusion problem. However, by increasing the reference viewpoints, these problems can be overcome and the quality of synthesized images will be improved. It will be a trade-off between image quality and memory capacity.

10.4   Image-Based Walk-Through System

As a demonstration of our principle, we implemented a walk-through system where the synthesized images of a city street are projected on three large screens and a user can walk-through the virtual space using a joystick (Figure 10.12). In this system, we integrated the three methods introduced in the previous section.

A force feedback joystick (Impulse Engine) is used for this prototype. The information of the joystick is transmitted to a PC and then sent to ONXY2 through the network. The ONYX2 workstation reads the information of the joystick from the shared memory and then renders images according to the information by reading them from the RAID disk onto the main memory. Since this joystick has force feedback function, we plan to add some guidance feedback for navigation.

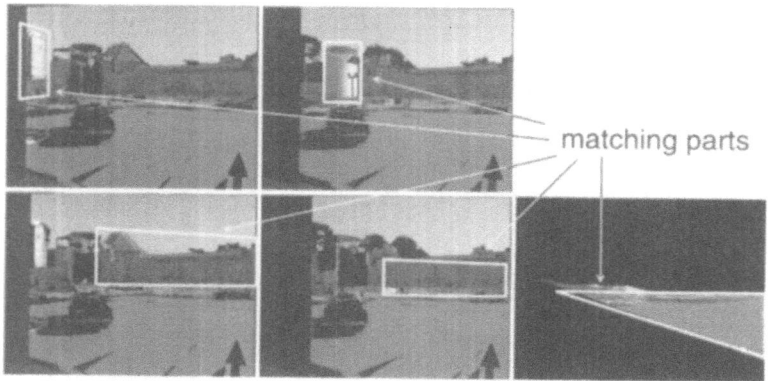

Arrangement

Figure 10.11   A result of the 3-D layer method.

Figure 10.12   Image-based walk-through system.

The 3-D images are rendered in a single image. These images are separated through three digital scan converters into the images for each screen and then sent to projectors. The projectors cast these separated images onto corresponding screens so that the observer can observe scenes of 180 degrees and can move and look around using a joystick (Figure 10.12).

The user can move left and right by inclining the joystick left and right while pressing a button, back and forth by inclining the joystick back and forth. When the user inclines the stick to the left and right, the scene is switched as though he/she is actually looking around him/her. Another button is used to display a map in which an arrow shows the current position and direction of the user.

In the current system, three methods: (the panorama-based method, the morphing method and the 3-D layer presumption method) are implemented separately. When the user comes to a special position in the virtual environment, he/she can switch and experience the different methods by pushing a button. Note that the panoramic images are read directly from the disk but the images for the morphing method and the 3-D layer presumption method are read into the main memory beforehand.

This system allows the user to move around an approximate area of 800 x 200 square meters in Marunouchi, Tokyo. The number of panoramic images used in this system was about 5000 and the total size was 7 GB.

10.5   Conclusion and Discussion

In this paper, we described a new approach to build a virtual space in which a user can walk through and look around interactively. Through the development of a prototype system, it is proved that it is possible to generate a wide-range, photorealistic virtual world by combining the panorama-based method and two interpolation methods, the morphing method and the 3-D layer method.

The limitation of these methods is that only a still environment can be constructed. In order to treat a moving environment, we need advancement in these methods. Precisely speaking, time is not constant, but is dependent on the position because we cannot capture all images simultaneously. We have to "scan" the real world. Consequently, sometimes strange phenomena, such as the disappearance of an object out of the view field, can occur. To avoid this occurrence, we need to process original images in advance. (For example, the elimination of moving objects might be needed.)

The methodology introduced in this paper was originally developed as a method to transfer a huge amount of information existing in the real world into the virtual world. However, in future, many more applications will be considered. For example, as original images, we may use CG images generated by a sophisticated rendering method such as ray tracing which cannot be used in real-time world generation at present.

Basically, there will be two extreme methodologies, one of which is used to generate the virtual world from images and its viewpoint (we may call this data-intensive methodology), and in the other this is done using 3-D models and rendering software (we may call this algorithm-intensive methodology). Between the two extreme methodologies shown in Figure 10.13, we may consider various methodologies.

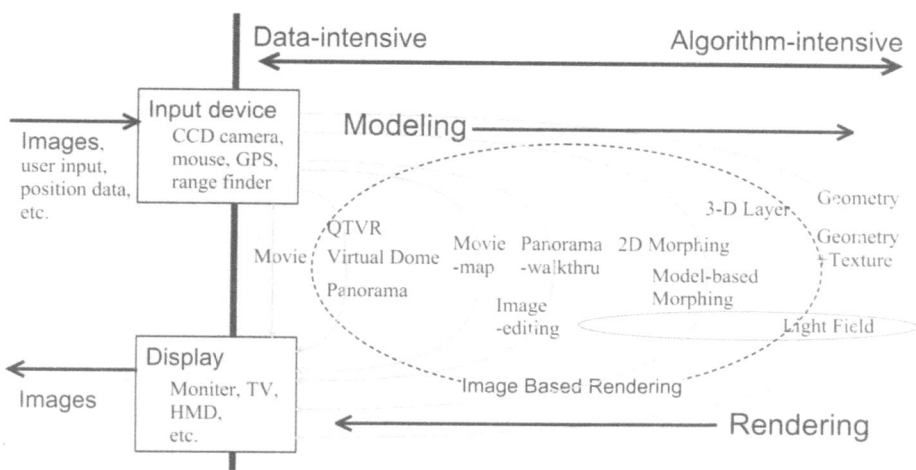

Figure 10.13   Image-based modeling and rendering.

We have to understand that by the use of a purely idealized methodology such as the "simple model and simple rendering" principle, we will not be able to model the complex real world. This problem is extremely important in the field of mixed reality technology. We have to learn from the failure of artificial intelligence in the '80s. The real world is extremely complicated to express using a simple theory.

References

[1] ModelCity™Philadelphia: http://www.bentley.com/modelcity/

[2] CyberCity Berlin: http://www.echtzeit.de/products/CC.html

[3] A. Lippman: "Movie Maps: An application of the optical videodisc to computer graphics," *Proc. SIGGRAPH'80*, pp.32–43, 1980.

[4] M. Hirose, K. Yokoyama and S. Sato: "Transmission of realistic sensation: Development of a virtual dome," *Proc. VRAIS'93*, pp.125–131, 1993.

[5] S. E. Chen: "QuickTime VR - An image-based approach to virtual environment navigation," *Proc. SIGGRAPH'95*, pp.29–38, 1995.

[6] M. Hirose, S. Watanabe, and T. Endo: "Generation of wide-range virtual spaces using photographic images," *Proc. VRAIS'98*, pp.234–241, 1998.

[7] S. M. Seitz and C. R. Dyer: "View morphing," *Proc. SIGGRAPH'96*, pp.21–30, 1996.

Part III

Multi-Sensory Augmentation

One of the overall goals of virtual reality (VR) systems is to create every aspect of the virtual environment such that an observer feels present within it, through all the human senses. In spite of this, most VR systems have been designed to appeal only to observer's sense of vision. Because humans receive 70–80% of external stimuli through their eyes, it may thus seem natural to concentrate on visual information in VR, and it is perhaps as a consequence of this that the amount of research on auditory or tactile/haptic interfaces is comparatively low in VR. This same tendency is even more apparent in augmented reality (AR) and mixed reality (MR) research where, for example, the notion of visual see-through displays easily comes to mind, but auditory or tactile mixtures are less common.

With this in mind, Jack M. Loomis and his group have been at the forefront of research on auditory distance, perception in real and virtual environments. Chapter 11 begins with the informational basis for auditory localization of direction and distance, followed by a review of some of the research on auditory distance perception in real environments. Then, after noting a few of the issues pertaining to virtual sound in virtual and mixed environments, the chapter concludes with a description of the group's experience in using virtual sound as part of a navigation system for blind travelers.

Hiroo Iwata has been studying the analogous concept of haptic displays and force feedback using haptic displays. His ideas extend to haptic augmented reality, which he calls *Feel-through*. Chapter 12 proposes a system in which an observer can feel synthetic forces via a force display superimposed on real forces from real objects, while experiencing visual AR by wearing a see-through HMD.

Whereas Iwata's integration of force displays with visual displays re-emphasizes the importance of the tactile senses in human-computer interaction, Hiroshi Ishii's approach to this field takes the concept one step further. According to him, it is very important to use graspable and tactile feelings for both input to and output from computers. Ishii's Tangible Bits, presented in Chapter 13, is noteworthy as a novel concept indicating potential new directions for human-computer interaction research at the intersection between humans and cyberspace.

Chapter 11

Auditory Distance Perception in Real, Virtual, and Mixed Environments

Jack M. Loomis
University of California, Santa Barbara, U.S.A.

Roberta L. Klatzky
Carnegie Mellon University, U.S.A.

Reginald G. Golledge
University of California, Santa Barbara, U.S.A.

11.1 Introduction

Human localization of a sound source is three-dimensional. The egocentric location of a source is specified in terms of the two parameters indicating its direction, azimuth (lateral direction with respect to the facing direction of the head) and elevation (direction with respect to the ear-level plane), and the third parameter indicating its distance. Most of the hundreds of research papers on spatial hearing have dealt with directional localization of sources, with only a handful of studies addressing the perception of distance.

While some spatial tasks can be accomplished solely on the basis of direction (e.g., localizing the direction of a threat so that the head may be turned to acquire it visually), distance is often crucial (e.g., perceiving whether a threat is so close as to require an immediate response). The emerging technology of virtual acoustics is generating increased interest in auditory distance perception, for even apart from those tasks for which distance perception is essential, the phenomenological aspects of auditory distance perception cannot be undervalued. For example, in entertainment (e.g., musical performance, computer games), the impression of sounds varying in distance adds inestimably to the aesthetic impact.

11.2   Information for Directional Localization

Sound transmitted from a source to the listener's head travels along direct and indirect paths. The arriving sound is modified by the head, shoulders, and pinnae (the visible structures of the external ears). The stimulus cues for localizing sound in direction (azimuth and elevation) are well understood (e.g., [1]–[6]). The most important stimulus information can be described with reference to a spherical model that is an approximation of the head without pinnae. The aural axis is defined by the two ear canals. The angle between the aural axis and the source direction is referred to as the lateral angle [7]. Reflection, absorption, and diffraction of the incoming sound by the spherical head gives rise to the cue of interaural intensity difference (IID). If the source is off to one side of the mid-sagittal plane (the vertical plane that bisects the head through the nose), the ear on the opposite side receives a less intense signal than that on the same side. The IID cue is minimal at low frequencies but grows with frequency. For the spherical approximation, the IID cue is symmetric about the aural axis. A source at a given lateral angle (e.g., 20°) defines a cone of symmetry about the aural axis; sources located on this cone produce the same value of IID. A lateral angle of 90° corresponds to sounds in the mid-sagittal plane, within which the IID is zero at all frequencies.

The other primary cue for direction is interaural time difference (ITD). For a source off to one side, the path length that the sound must travel is greater for the ear on the opposite side than for the ear on the same side. When the sound lies near the aural axis off to one side, ITD is maximal, with a value of just under 0.7 ms for the average adult. As lateral angle increases toward 90°, the ITD decreases to zero. Under the spherical approximation, ITD, like IID, is constant for different locations having the same lateral angle.

The symmetry of IID and ITD about the aural axis means that sources equal in lateral angle and a constant distance ought to be indiscriminable by a listener whose head is stationary. While there is a tendency for a listener with stationary head to confuse sources having equal lateral angles (e.g., two sounds equally above and below the aural axis or two sounds equally in front of and behind the aural axis), such a listener can discriminate between such sources well above chance [3] [6]. This means that representing IID and ITD in terms of the spherical model is an oversimplification [8] [9].

The complete mathematical specification of how sound arriving at the ear canal is modified by the head, shoulder, and pinnae is referred to as the head-related transfer function (HRTF). It represents the complex variations in IID and ITD that depend upon sound frequency and upon source azimuth and elevation (e.g., [1] [5]

[9]). The variations above and beyond those due to the spherical model are the result of diffraction by structures not encompassed in the spherical model (e.g. the pinnae and shoulder). A listener is most accurate in localizing the azimuth and elevation of a source when he/she listens to sounds with his/her own HRTFs [4] [10].

Although front/back and up/down confusions are fairly common for a listener with stationary head, these confusions largely disappear when the stimulus lasts long enough to permit the listener to derive additional information by means of head rotations [7] [11] [12]. The analysis of directional localization under head rotations reveals that perceived sound direction can be understood in terms of the sensed change in lateral angle relative to the sensed head rotation [7]. To understand why, consider a source that is initially within the ear-level plane at 50° right azimuth. At the outset, it has a lateral angle of 40°. If the listener rotates the head counterclockwise 10° so that the lateral angle has diminished to 30°, the 10° reduction in lateral angle for a 10° clockwise rotation signifies to the listener that the source was initially in the ear-level plane and in front of the aural axis. This analysis implies that sounds should be localizable in azimuth and elevation by means of head rotations around various axes, even without the supplementary directional cues provided by the pinnae, non-spherical head, and shoulders.

11.3 Information for Distance Localization

There are a number of potential cues to egocentric distance that are available to an observer whose head is stationary. The first of these is sound level. In an anechoic environment, sound level for a source of constant intensity falls off by 6 dB for each doubling of distance [13]. Sound level serves as an absolute distance cue if the observer has independent knowledge about source intensity; otherwise, sound level is informative only about the changing distance of a source [14]. Experiments by Mershon and King [15] and Zahorik [16], among others, have shown that sound level does act as a relative distance cue in influencing the observer's judgment of distance over multiple presentations of the same source. A cue that serves to specify absolute distance, even on the first presentation of a stimulus, is termed "reverberation"; it is the ratio of direct-to-reverberant energy in an echoic environment, with lower ratios signifying more distant sources [15]–[22]. Another cue, which probably serves to indicate only relative distance, is the change in the spectral content of a sound due to selective absorption of high frequencies with passage through the air [23] [24].

When an observer translates through space, additional information about source distance becomes available under the assumption that the source is stationary. One of these is absolute motion parallax, the changing direction of any source that is initially off to one side [25]. The other is acoustic tau [26]–[30], a computed variable associated with the increasing sound level of any source that is being approached. Tau specifies the time to collision with the source, whether or not it is stationary; if the source is stationary, its distance is given by the product of acoustic tau and the observer's travel velocity. Speigle and Loomis [25] found that, in comparison with a condition in which only sound level and reverberation were available, also making motion parallax and acoustic tau available only slightly improved their observers' judgments of source distance. Ashmead et al. [26] found a larger effect of acoustic tau but in a study where sound level was deliberately rendered unreliable.

11.4   Measurement of Perceived Distance Using Perceptually Directed Action

Zahorik [16] has recently reviewed most of the studies using verbal report as a measure of auditorially perceived distance. He also conducted experiments in an indoor environment in which he systematically manipulated the various cues to distance. His research and the studies he reviewed indicate that perceived distance increases much more slowly than the physical distance of the sound source, even when multiple cues are varied in concert, much as they do in natural environments. In most cases, the function relating perceived to physical distance was well fit by a power function with an exponent considerably less than 1.0, signifying a compressive nonlinearity intervening between physical and perceived distance.

An alternative to verbal reports of perceived distance is the use of "perceptually directed action". Here the observer indicates the perceived target location by means of some open-loop spatial behavior. In connection with visual perception, for example, one such behavior is visually directed walking. Here, the observer views a visual target at some distance and then closes the eyes and attempts to walk to it without further information about its location (also, the target is silently removed so that the observer does not collide with it). Many studies have been conducted on visually directed walking and show that, under full visual cues, perception is quite accurate for targets up to $20m$ away [31]–[39]. This is consistent with much of the research using verbal reports (see [40] for review).

Other variants of visually directed action involve triangulation. In "triangulation by pointing", the observer views a target and then walks blindly along an oblique path, while attempting to continue pointing in the direction of the previously viewed and now imaginally updated target [33]. The terminal pointing direction is used to triangulate the initially perceived target location, and, hence, its perceived distance from the viewing location. In "triangulation by walking", the observer views a target and then walks blindly along an oblique path; at some unanticipated location, the observer is instructed to turn and begin walking toward the target [33]. The heading or travel direction following the turn is used to triangulate the initially perceived target location, and, hence, its perceived distance from the viewing location. In another variant of triangulation by walking, the observer walks to the target along direct and indirect path (on different trials). For the latter, the observer walks blindly along an oblique path, turns on command, and then attempts to walk the full distance to the target [35] [41].

Only very recently have these methods been applied to the study of auditory distance perception [25] [26] [35]. All three studies employed walking to a previously heard target; the study by Loomis et al. [35] also used the method of walking to the target along direct and indirect paths. The direct walking experiments used a procedure similar to that used in the vision studies. First the observer is presented with an auditory target by way of loudspeaker. After the sound is extinguished, the observer attempts to walk to the location of the speaker without further perceptual information about its location (and the speaker is removed in the meantime to avoid collisions). For the indirect walking trials of Exp. 3 of Loomis et al. [35], the subject walked for 5m along a path oblique to the target and then turned and attempted to the rest of the way to the target.

Figure 11.1 summarizes the results of five of the experiments in the three afore-mentioned studies. Perceived distance, which corresponds to the walked distance, is plotted against source distance. Although the five functions differ considerably in terms of the y intercept, they are consistent in showing that in real outdoor environments, observers do perceive large variations in distance that are systematically related to source distance. This result is consistent with our everyday experience. At the same time, however, the experiments also confirm the general finding from research using verbal report that perceived auditory distance is compressed relative to source distance. In particular, these three studies, all done in outdoor environments, show that perceived auditory distance varies over a range that is only about half of that of the physical distance. In addition, the study by Loomis et al. [35] obtained verbal reports from observers and found that they were largely in agreement with those obtained with perceptually directed walking, the primary difference being that the verbal measures exhibited greater variability.

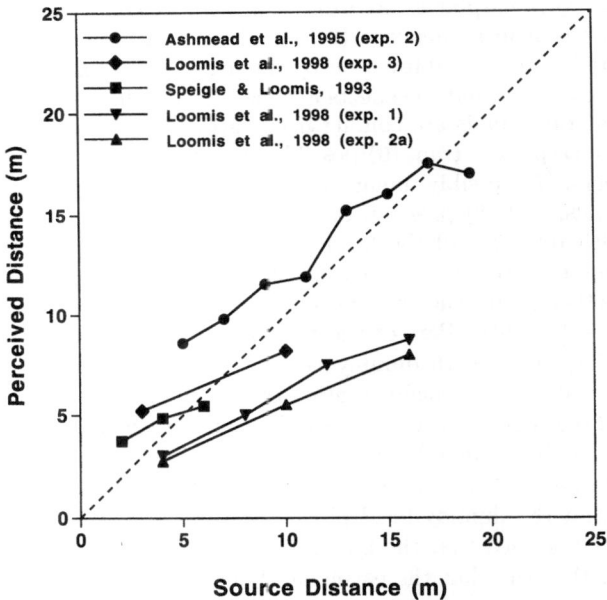

Figure 11.1   The results of five experiments using perceptually directed walking to measure the perceived egocentric distance of a sound source. Sounds were presented by loudspeaker in natural outdoor settings. After termination of the sound, observers attempted to walk to the location of the source. Perceived distance, which corresponds to walked distance, is plotted against source distance.

11.5   Externalization of Earphone-based Virtual Sound

The assumption underlying earphone-based virtual acoustics is that a person who obtains the same binaural stimulation with earphones that he/she would obtain in the presence of real sound sources ought to have the same auditory perception in the two situations. Computer implementation of a virtual acoustic display involves tracking rotations and translations of the observer's head and, for each momentary orientation and position of the head, synthesizing and delivering binaural signals that mimic those from a real source after modification by the environment and the observer's head. If synthesized properly, the virtual sounds ought to appear identical to those from the real source. Producing realistic earphone-based virtual sound, especially sound that appears to vary in perceived distance as much as it does in real environments, remains a major challenge confronting researchers in virtual acoustics.

Leaving aside the challenge of producing virtual sounds that appear meters away in distance, we note that just getting earphone-delivered sounds to appear outside the head has stymied audio engineers for years (e.g., [42]). A number of possible factors have been suggested as contributing to whether sounds are heard inside or outside the head when earphones are worn. These include (1) availability of natural reverberation in the acoustic imagery, (2) availability of binaural signals appropriate to absorption, reflection, and diffraction by the head, (3) availability of appropriate pinna cues above and beyond the effects associated with the head, (4) knowledge or lack thereof that the sounds are coming from earphones, (5) pressure on the head coming from the earphone strap, (6) possible distortion in the earphone signals delivered to the ears, (7) possibly unnatural acoustic coupling between the earphones and the ear canals, and (8) presence or absence of changes in the acoustic signals concomitant with rotations of the head [5]–[7] [43]–[49]. In particular, the belief has arisen rather recently that the externalization of virtual sound depends critically upon whether pinna cues are correctly rendered in the sounds delivered to the ears (e.g. [5] [6] [49]). Part of the motivation for this belief is the fact that binaural recordings made with dummy heads with artificial pinnae produce quite compelling externalization. Documentation of this comes from the work of Plenge [47] who showed that externalization is readily achieved when the observer listens to binaural signals coming from a dummy head (with pinnae) situated in a reverberant environment. However, Plenge neither manipulated the presence/absence of pinnae nor concluded that the dummy head pinnae were in any way critical to externalization. He merely argued that the binaural stimuli need to be "ear-adequate", by which he meant that the that the earphone signals "would be essentially equal to those occurring when external sources are perceived" ([47] p.946). More recently, Zahorik [16] has demonstrated unambiguous extracranial localization of binaural recordings made using microphones within the ear canals of people. The evidence is accumulating, however, that pinna cues are relatively unimportant for externalization and that the primary determinants are reverberation and the presence of IID and ITD associated with reflection, absorption, and diffraction by the head (with or without pinnae). As we have reported elsewhere (e.g. [50]), externalization

Figure 11.2   A simple listening device that demonstrates perceptual externalization of sounds presented through earphones. The device consists of in-ear earphones, a sound-attenuating hearing protector worn over the earphones, microphones mounted on top of the hearing protector earcups, and a stereo amplifier driving the earphones according to the microphone signals. Even though a person wearing this device hears the world indirectly with altered HRTFs, compelling externalization is experienced.

is readily achieved with a simple electronic device by which the listener hears the surrounding environment. The device (Figure 11.2) consists of in-ear earphones, a sound-attenuating hearing protector worn over the earphones, microphones mounted on top of the hearing protector earcups, and a stereo amplifier driving the earphones in concert with the microphone signals. A person wearing this device hears the world indirectly without his/her own pinnae and thus with HRTFs that are quite different from normal. Even though the person listens with drastically altered "pinnae", externalization is complete, with sounds appearing to come from meters away when the source is indeed quite distant. We have demonstrated this device to well over 100 people and in only 3 or 4 cases was intracranial localization reported. These informal results indicate that pinna cues are quite unimportant for externalization. Consistent with this, Durlach et al. [43] cite two informal studies involving binaural recordings with real observers and with dummy heads, in which pinna cues were inconsequential for externalization. Also, Loomis et al. [46] reported informal results indicating externalization with a virtual acoustic display implementation that lacked detailed modeling of pinna cues. Despite this accumulation of evidence, what is clearly needed is a formal investigation in which a variety of factors (e.g., IID and ITD cues associated with a spherical head, pinna cues, reverberation, cognitive set) are manipulated to establish once and for all the extent to which these factors contribute to externalization.

The more important result of the above formal and informal research is the conclusion that earphone listening does not preclude realistic sound appearing to come from considerable distances. In particular, it shows that many of the potential factors cited above, such as knowledge that one is listening with earphones, play at best a minor role in determining whether sound appears inside or outside the head. Thus, there is reason to be optimistic that computer-synthesized virtual sound produced with earphones will someday attain the same level of realism that has been demonstrated in these formal and informal studies.

11.6   Virtual Sound in Mixed Environments

One of the advantages of using audition instead of vision in mixed environments is the relative ease of implementation. With vision, the usual implementation involves optical or digital mixing of the virtual and real imagery. Optical mixing precludes occlusion and results in a ghost-like appearance of one or the other environment. Digital mixing affords naturally appearing occlusion of either the real or virtual imagery but at the expense of range estimation in the real environment and complex rendering software. For many applications of virtual sound, it is sufficient merely to produce the acoustic equivalent of optical mixing whereby the acoustic virtual imagery is superimposed on ambient sound from the real environment. This can be accomplished using open-ear earphones that minimally attenuate or distort the ambient sound. However, auditory masking still presents a problem, for virtual sounds can mask ambient ones and vice-versa. This is especially of concern for display of virtual information to blind individuals who are moving around in the environment, for any disturbance of the high-frequency information needed for obstacle avoidance can be very detrimental.

11.7   An Application of Mixed Auditory Environments: The Personal Guidance System

For over a decade now, we have been making progress on the development of a navigation system for the visually impaired that we call the Personal Guidance System [50]–[53]. From the beginning [54], our concept has been of a navigation system that leads a blind traveler along specified routes by means of virtual auditory beacons appearing at waypoints (turn points) along the routes. In addition, the system should indicate the positions of important off-route landmarks by having their labels, spoken by speech synthesizer, appear as virtual sounds at the appropriate locations within the auditory space of the traveler, as if they were coming from loudspeakers at these locations. We have thought that these virtual landmarks will assist blind travelers in developing better mental representations of the environment than is currently the case. We believe that this is most likely to occur if the traveler perceives the virtual sounds close to the intended locations, both in terms of direction and distance. Incidentally, our system is not intended to provide the visually impaired person with detailed information about the most immediate environment (e.g. obstacles); accordingly, the blind traveler will still have to rely on the long cane, seeing-eye dog, or ultrasonic sensing devices for this information.

Our system, like others being developed (for example, see [55]), uses the Global Positioning System (GPS) to locate the traveler within the environment. With differential correction by way of radio link from a GPS reference receiver in the local vicinity, localization is accurate to around $1m$ under good conditions of reception. A portable computer carried by the traveler contains a Geographic Information System (GIS) comprising both a geocoded database of the surrounding environment and software giving the desired functionality. In particular, the software uses the differentially corrected GPS signals to locate the traveler within the digitally represented environment. A virtual acoustic display (currently the Alphatron from Crystal River Engineering) along with stereophonic earphones allows the traveler to receive spatialized auditory information about the locations of waypoints along the desired route as well as off-route features of the environment. The current implementation of our system weighs $11kg$ and is carried in a backpack worn by the user (Figure 11.3), but future versions ought to be truly wearable. For details on the hardware, see [53].

Figure 11.3   The UCSB Personal Guidance System, being worn by the first author (who is sighted). This implementation weighs 11 kg but future versions ought to be truly wearable.

We have conducted a number of informal demonstrations of the system at our test site, the UCSB campus. We have shown the capability of the system to guide an unsighted person with normal binaural hearing to some specified destination using a sequence of virtual beacons; under conditions of good satellite availability, the differential GPS subsystem functions well enough to keep the traveler within sidewalks varying from 3 to $5m$ in width.

The formal research we have done consists of two experiments. Both experiments compared virtual sound with conventional synthesized speech as means of displaying spatial information. The first experiment ([53]) was concerned with the relative effectiveness of virtual auditory beacons and synthesized speech in guiding a traveler along a route. On a given trial the subject was guided by the computer along a $71m$ path of 9 straight segments. In one condition, waypoints along the path were signaled by auditory beacons created using virtual sound. In other conditions, the subject was given guidance information relative to the desired path by way of synthesized speech (e.g., "left 30 degrees"). Ten blind individuals performed in the experiment in all conditions. Virtual sound proved more effective than synthesized speech, both in terms of two performance measures (travel time and total walked distance) and in terms of preference ratings [53].

The second experiment was concerned with the learning of spatial layout. In the training phase, the subject was guided five times around a square using virtual beacons located at the four vertices [56]. Along each leg, the subject received information about the location of each of three off-route landmarks, identified by simple names spoken by a speech synthesizer. In the virtual sound condition, subjects heard the name as spatialized speech from the virtual display. In the other condition, the subjects heard non-spatialized speech giving the approximate relative bearing to each landmark in terms of a clockface (e.g. "gate, 3 o'clock"). Spatial learning was assessed using tactual sketch maps as well as direction estimates obtained during a sixth traverse of the square. The data from the 9 blind subjects who performed in both conditions showed no difference between the two conditions [56].

The two experiments together indicate that spatialized sound from a virtual acoustic display shows promise as part of the user interface of a blind navigation system. Performance was at least as good as with the other display modes investigated, and spatialized speech has the additional advantage of consuming less time than conventional speech, for the latter must include the spatial information as part of the utterance. On the other hand, there are currently two problems associated with the use of virtual sound. First, present earphone designs attenuate or distort some of the environmental information that is important to visually impaired travelers. This problem might be eliminated, however, by using small transducers mounted a few centimeters from the ears. Second, as mentioned earlier, producing realistic virtual sound that appears to come from moderate to large distances has been very difficult. However, given that there is no fundamental difficulty with earphone display per se (see earlier discussion), it is probably just a matter of time before more effective algorithms for virtual sound are developed.

Acknowledgments

Both our basic research on auditory distance perception and our applied research and development relating to the navigation system have been supported by Grant 9740 from the National Eye Institute. The authors thank Pavel Zahorik for his helpful comments on an earlier version of this article.

References

[1] J. Blauert: *Spatial Hearing: The Psychophysics of Human Sound Localization*, Cambridge, MA, MIT Press, 1983.

[2] R. Gilkey and T. R. Anderson: *Binaural and Spatial Hearing in Real and Virtual Environments*, Hillsdale, NJ, Lawrence Erlbaum Associates, 1997.

[3] S. R. Oldfield and S. P. A. Parker: "Acuity of sound localisation: a topography of auditory space. I. Normal hearing conditions," *Perception*, vol.13, pp.581–600, 1984.

[4] S. R. Oldfield and S. P. A. Parker: "Acuity of sound localisation: a topography of auditory space. II. Pinna cues absent," *Perception*, vol.13, pp.601–617, 1984.

[5] F. L. Wightman and D. J. Kistler: "Headphone simulation of free-field listening. I: Stimulus synthesis," *J. Acoustical Society of America*, vol.85, pp.858–867, 1989.

[6] F. L. Wightman and D. J. Kistler: "Headphone simulation of free-field listening. II: Psychophysical validation," *J. Acoustical Society of America*, vol.85, pp.868–878, 1989.

[7] H. Wallach: "The role of head movements and vestibular and visual cues in sound localization," *J. Experimental Psychology*, vol.27, pp.339–368, 1940.

[8] J. C. Middlebrooks and D. M. Green: "Directional dependence of interaural envelope delays," *J. Acoustical Society of America*, vol.87, pp.2149–2162, 1990.

[9] J. C. Middlebrooks, J. C. Makous and D. M. Green: "Directional sensitivity of sound-pressure levels in the human ear canals," *J. Acoustical Society of America*, vol.86, pp.89–108, 1989.

[10] E. M. Wenzel, M. Arruda, D. J. Kistler, and F. L. Wightman: "Localization using nonindividualized head-related transfer functions," *J. Acoustical Society of America*, vol.94, 111-123, 1993.

[11] W. R. Thurlow and P. S. Runge: "Effects of induced head movements on localization of direction of sound sources," *J. Acoustical Society of America*, vol.42, pp.480–488, 1967.

[12] W. R. Thurlow, J. W. Mangels, and P. S. Runge: "Head movements during sound localization," *J. Acoustical Society of America*, vol.42, pp.489–493, 1967.

[13] P. D. Coleman: "An analysis of cues to auditory depth perception in free space," *Psychological Bulletin*, vol.60, pp.302–315, 1963.

[14] D. H. Mershon: "Phenomenal geometry and the measurement of perceived auditory distance," in (R. H. Gilkey and T. R. Anderson, eds.) *Binaural and spatial hearing in real and virtual environments*, Hillsdale, NJ, Lawrence Erlbaum Associates, pp.257–274, 1997.

[15] D. H. Mershon and L. E. King: "Intensity and reverberation as factors in the auditory perception of egocentric distance," *Perception & Psychophysics*, vol.18, pp.409–415, 1975.

[16] P. A. Zahorik: "Experiments in auditory distance perception," Unpublished doctoral dissertation, Department of Psychology, University of Wisconsin, Madison, Wisconsin, 1998.

[17] R. A. Butler, E. T. Levy, and W. D. Neff: "Apparent distance of sounds recorded in echoic and anechoic chambers," *J. Experimental Psychology: Human Perception and Performance*, vol.6, pp.745–750, 1980.

[18] D. H. Mershon, W. L. Ballenger, A. D. Little, P. L. McMurtry, and J. L. Buchanan: "Effects of room reflectance and background noise on perceived auditory distance," *Perception*, vol.18, pp.403–416, 1989.

[19] D. H. Mershon and J. N. Bowers: "Absolute and relative cues for the auditory perception of egocentric distance," *Perception*, vol.8, pp.311–322, 1979.

[20] C. W. Sheeline: "An investigation of the effects of direct and reverberant signal interaction on auditory distance perception," Doctoral Thesis, Stanford University, 1983.

[21] D. R. Begault: "Perceptual effects of synthetic reverberation on three-dimensional audio systems," *J. Audio Engineering Society*, vol.40, pp.895–904, 1992.

[22] S. H. Nielsen: "Auditory distance perception in different rooms," *J. Audio Engineering Society*, vol.41, pp.755–770, 1993.

[23] A. D. Little, D. H. Mershon, and P. H. Cox: "Spectral content as a cue to perceived auditory distance," *Perception*, vol.21, pp.405–416, 1991.

[24] P. D. Coleman: "Dual role of frequency spectrum in determination of auditory distance," *J. Acoustical Society of America*, vol.44, pp.631–632, 1968.

[25] J. Speigle and J. M. Loomis: "Auditory distance perception by translating observers," *Proc. the IEEE Symposium on Research Frontiers in Virtual Reality*, San Jose, CA, Oct. 25-26, Washington, DC: IEEE, pp.92–99, 1993.

[26] D. H. Ashmead, L. D. DeFord, and A. Northington: "Contribution of listeners' approaching motion to auditory distance perception," *J. Experimental Psychology: Human Perception and Performance*, vol.21, pp.239-256, 1995.

[27] D. N. Lee: "Getting around with light or sound," in (R. Warren and A. H. Wertheim, eds.) *Perception and control of self-motion*, Hillsdale, NJ, Lawrence Erlbaum Associates, pp.487–505, 1990.

[28] R. Guski: "Acoustic tau: An easy analogue to visual tau?" *Ecological Psychology*, vol.4, pp.189–197, 1992.

[29] B. K. Shaw, R. S. McGowan and M. T. Turvey: "An acoustic variable specifying time-to-contact," *Ecological Psychology*, vol.3, pp.253–261, 1991.

[30] W. Schiff and R. Oldak: "Accuracy of judging time to arrival: Effects of modality, trajectory, and gender," *J. Experimental Psychology: Human Perception and Performance*, vol.16, pp.303–316, 1990.

[31] D. Elliott: "The influence of walking speed and prior practice on locomotor distance estimation," *J. Motor Behavior*, vol.19, pp.476–485, 1987.

[32] D. Elliott, R. Jones, and S. Gray: a"Short-term memory for spatial location in goal-directed locomotion," *Bull. Psychonomic Society*, vol.8, pp.158–160, 1990.

[33] S. S. Fukusima, J. M. Loomis, and J. A. Da Silva: "Visual perception of egocentric distance as assessed by triangulation," *J. Experimental Psychology: Human Perception and Psychophysics*, vol.23, pp.86–100, 1997.

[34] J. M. Loomis, J. A. Da Silva, N. Fujita, and S. S. Fukusima: "Visual space perception and visually directed action," *J. Experimental Psychology: Human Perception and Performance*, vol.18, pp.906–921, 1992.

[35] J. M. Loomis, R. L. Klatzky, J. W. Philbeck and R. G. Golledge: "Assessing auditory distance perception using perceptually directed action," *Perception & Psychophysics*, vol.60, pp.966–980, 1998.

[36] J. W. Philbeck and J. M. Loomis: "Comparison of two indicators of visually perceived egocentric distance under full-cue and reduced-cue conditions," *J. Experimental Psychology: Human Perception and Performance*, vol.23, pp.72–85, 1997.

[37] J. J. Rieser, D. H. Ashmead, C. R. Talor, and G. A. Youngquist: "Visual perception and the guidance of locomotion without vision to previously seen targets," *Perception*, vol.19, pp.675–689, 1990.

[38] J. A. Thomson: "How do we use visual information to control locomotion?" *Trends in Neuroscience*, vol.3, pp.247–250, 1980.

[39] J. A. Thomson: "Is continuous visual monitoring necessary in visually guided locomotion?" *J. Experimental Psychology: Human Perception and Performance*, vol.9, pp.427–443, 1983.

[40] J. A. Da Silva: "Scales for perceived egocentric distance in a large open field: Comparison of three psychophysical methods," *American J. Psychology*, vol.98, pp.119–144, 1985.

[41] J. W. Philbeck, J. M. Loomis, and A. C. Beall: "Visually perceived location is an invariant in the control of action," *Perception & Psychophysics*, vol.59, pp.601–612, 1997.

[42] M. V. Thomas: "Improving the stereo headphone sound image," *J. Audio Engineering Society*, vol.25, pp.474–478, 1997.

[43] N. I. Durlach, A. Rigopoulos, X. D. Pang, W. S. Woods, A. Kulkarni, H. S. Colburn and E. M. Wenzel: "On the externalization of auditory images," *Presence: Teleoperators and Virtual Environments*, vol.1, pp.251–257, 1992.

[44] W. E. Kock: "Binaural localization and masking," *J. Acoustical Society of America*, vol.22, pp.801–804, 1950.

[45] W. Koenig: "Subjective effects in binaural masking," *J. Acoustical Society of America*, vol.22, pp.61–62, 1950.

[46] J. M. Loomis, C. Hebert, and J. G. Cicinelli: "Active localization of virtual sounds," *J. Acoustical Society of America*, vol.88, pp.1757–1764, 1990.

[47] G. Plenge: "On the differences between localization and lateralization," *J. Acoustical Society of America*, vol.56, pp.944–951, 1974.

[48] F. E. Toole: "In-head localization of acoustic images," *J. Acoustical Society of America*, vol.48, pp.943–949, 1969.

[49] E. M. Wenzel: "Localization in virtual acoustic displays," *Presence: Teleoperators and Virtual Environments*, vol.1, pp.80–107, 1992.

[50] J. M. Loomis, R. G. Golledge, R. L. Klatzky, J. M. Speigle, and J. Tietz: "Personal guidance system for the visually impaired," *Proc. the ACM/Siggraph Conf. on Assistive Technologies*, Marina Del Rey, CA, Oct. 31–Nov. 1, pp.85–91, 1994.

[51] R. G. Golledge, J. M. Loomis, R. L. Klatzky, A. Flury, and X. -L. Yang: "Designing a personal guidance system to aid navigation without sight: Progress on the GIS component," *Int'l J. Geographical Information Systems*, vol.5, pp.373–395, 1991.

[52] R. G. Golledge, R. L. Klatzky, J. M. Loomis, J. Speigle, and J. Tietz: "A geographic information system for a GPS/GIS based personal guidance system," *Int'l J. Geographical Information Science*, in press.

[53] J. M. Loomis, R. G. Golledge, and R. L. Klatzky: "Navigation system for the blind: Auditory display modes and guidance," *Presence: Teleoperators and Virtual Environments*, vol.7, pp.193–203, 1998.

[54] J. M. Loomis: "Digital map and navigation system for the visually impaired," Unpublished manuscript, Department of Psychology, University of California, Santa Barbara, 1985.

[55] H. Makino, I. Ishii, and M. Nakashizuka: "Development of navigation system for the blind using GPS and mobile phone connection," *Proc. 18th Annual Meeting of IEEE EMBS*, Amsterdam, The Netherlands, October 31–November 3, 1996.

[56] J. M. Loomis, R. G. Golledge, and R. L. Klatzky: "GPS-based navigation systems for the visually impaired," in (W. Barfield and T. Caudell, eds.) *Augmented Reality and Wearable Computers*, Hillsdale, NJ, Lawrence Erlbaum Associates (in press).

Chapter 12

Feel-through: Augmented Reality with Force Feedback

Hiroo Iwata
University of Tsukuba, Japan

12.1 Introduction

It has been often discussed that force feedback enhances virtual environments. Current force displays are commercially available and those are combined in various applications. However, force displays have not been integrated in augmented reality (AR). A see-through HMD superimposes visual image of virtual objects on the real world. How can force sensation are superimposed on the real world? This problem is a major topic of this paper.

A user of AR systems can easily feel real objects. The user can see the real world by a see-thorough HMD and he/she can touch real objects. For example, State et al. developed a video-see-through AR system, which is applied to the medical procedure known as ultrasound-guided needle biopsy of the breast [1]. The operator of the system feels real force from the needle inserted to the breast. The Boeing Company has an AR system to guide a technician in building a wire harness for an electrical system [2]. The user of the system feels real wire harness by the hand. In case the user manipulates virtual objects superimposed in real scene, such as AR2 Hockey [3], synthetic force feedback is definitely required. Former research activities of AR have never introduced force displays.

Force feedback has been often used in desktop virtual environments. We developed a desktop force display in 1989 [4]. Current commercial force displays, such as PHANToM, are designed to be used for desktop. In those cases, visual image of virtual space is provided by a conventional monitor. The user feels force from a virtual tool displayed in the synthetic imagery.

The major objective of this paper is to propose methods of superimposing synthetic force in AR environments. We have been working on this problem for several years [5] [6]. This paper presents recent activities of our research on AR with force feedback. Section 12.2 discusses basic idea of Feel-through: adding force feedback to AR environment. Section 12.3 describes design issues in force display for Feel-through. Section 12.4 presents system configuration of a prototype including see-through HMD, tracker, and computer systems. Effects of force feedback are quantitatively tested. Experiments and its results are presented in Section 12.5. Section 12.6 introduces a wearable force display for mobile AR with force feedback. We conclude with a discussion of application areas and future work of Feel-through in Section 12.7.

12.2   Basic Idea of Feel-through

Methods of adding synthetic force to AR environments can be classified according to distance between real/virtual objects and a user. If objects are located within the range of the user's arm, the user can directly touch real objects as well as virtual objects. In this case, the system must provide real force and synthetic force to the user. Figure 12.1 illustrates this situation. If objects are located far from the user, he/she cannot touch them directly. However, the system can virtually extend the user's hand and provide synthetic force. We introduce "beam spotting" metaphor for extension of the user's hand. Figure 12.2 shows basic idea of virtual beam. The system displays virtual beam by see-through HMD and the user feels force at the spot of the beam. In this case, shapes of real objects should be modeled in the computer. If objects are located at a large distance from the user, effect of binocular parallax is reduced. Force feedback is expected to enhance recognition performance of the user.

We must consider relative position of real and virtual objects. If virtual objects are located inside of real objects, the user cannot touch virtual objects directly even if those are placed nearby the user. Virtual beam can be used in this case. Figure 12.3 illustrates virtual beam that penetrates a real object. In this case, length of the beam is fixed, so that the user feels as if he/she were handling a stick. The user feels reaction force at the end of the virtual stick.

Thus, methods of adding synthetic force can be classified into above mentioned three categories.

12.3   Force Display for Feel-through

Force display is a mechanical device that generates reaction force from virtual objects. Research activities in force display are rapidly growing recently, although the technology is still in a state of trial-and-error. There are three approaches to implement force display: tool handling type force display, exoskeleton type force

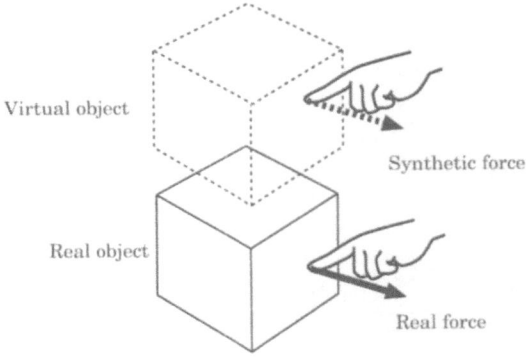

Figure 12.1   Feel-through nearby the user.

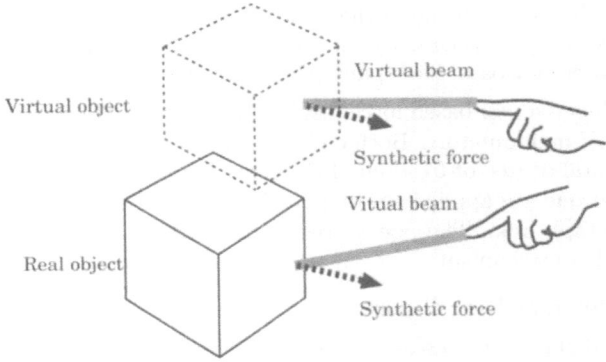

Figure 12.2   Feel-through in a large space.

display and object oriented type force display. We have developed prototypes in each category.

1. Tool handling type force display

   Tool handling type force display is easiest way to realize force feedback. Configuration of this type is similar to joystick. Virtual world technology usually employs glove-like tactile input devices. Users feel troublesome when they put or off these devices. If the glove is equipped with force feedback device, the problem is much severe. This disadvantage obstructs practical use of force display. Tool handling type force display is free from fitting it to user's hand. It cannot generate force between the fingers but it has practical advantages.

   We developed a 6 DOF (degree-of-freedom) force display that has ball grip [7]. The device is called "HapticMaster" and is commercialized by Nissho Electronics Co. The HapticMaster is a high-performance force feedback device for desktop use. This device employs parallel mechanism in which a top triangular platform and a base triangular platform are connected by three sets of

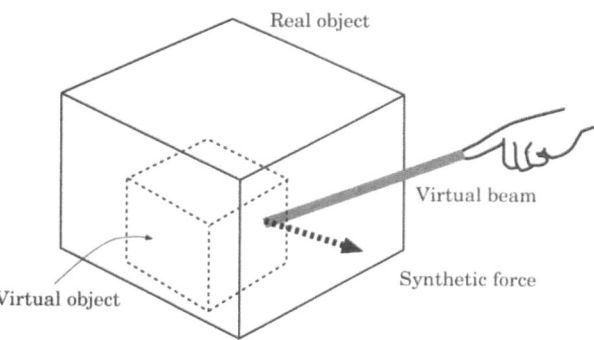

Figure 12.3   Feel-through inside a real object.

pantographs. The top end of the pantograph is connected with a vertex of the top platform by a spherical joint. This compact hardware has the ability to carry a large payload. Each pantograph has three DC motors.

We developed a pen-based force display in 1993 [8]. The force display employs two 3 DOF manipulators. Both end of the pen are connected to these manipulators. Total degree-of-freedom of the force display is 6. Three DOF force and 3 DOF torque are applied at the pen. Each 3 DOF manipulator is composed of pantograph link. The pen is free from the weight of the actuators due to this parallel mechanism.

2. Exoskeleton type force display

In the field of robotics research, master manipulators are used in teleoperation. Most master manipulators, however, have large hardware with high cost, which restricts their application areas. In 1989, we developed a compact master manipulator as a desktop force display [4]. The core element of the device is 6 DOF parallel manipulator, in which three sets of pantograph link mechanisms are employed. Three actuators are set coaxially with the first joint of the thumb, forefinger and middle finger of the operator.

3. Object oriented type force display

Object oriented type force display is a radical idea of design of force display. The device moves and deforms to present shapes of virtual object. A User of the device can contact with virtual object by its surface. It allows natural interaction compared to exoskeleton and tool handling type. However, it is fairly difficult to implement. Further more, its ability to simulate virtual objects is limited. Because of these characteristics, object oriented type is effective for specific applications. We focused on 3D shape modeling as an application of our object oriented type force display. We have developed a prototype named Haptic Screen [9]. The device employs an elastic surface made of rubber. An array of actuators is set under the elastic surface. The surface deforms by the actuators. Each actuator has force sensors. Hardness of the surface is variable by these actuators and sensors. Deformation of virtual object occurs according to force applied by the user. We demonstrated it at Enhanced Realities venue at SIGGRAPH'98.

Among those configurations of force displays, only pen-based force display is suitable for Feel-through system. The user can touch real objects at the end of the pen, so that he/she can feel real objects as well as virtual objects. Exoskeleton obstacles the users to touch real objects. HapticScreen cannot be merged in real world.

Figure 12.4 illustrates a 3 DOF pen-based force display that we used for the Feel-through system. Each joint is equipped with a potentiometer and DC motor. The maximum generated force at the grip is 500gf. Figure 12.5 shows overall view of the user.

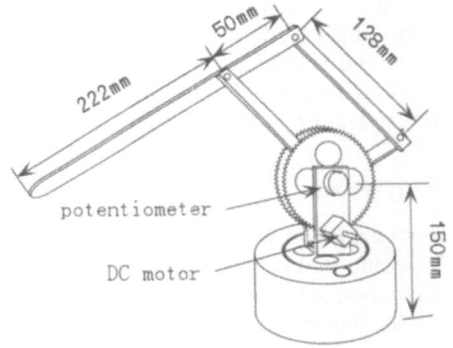

Figure 12.4   Pen-based force display.

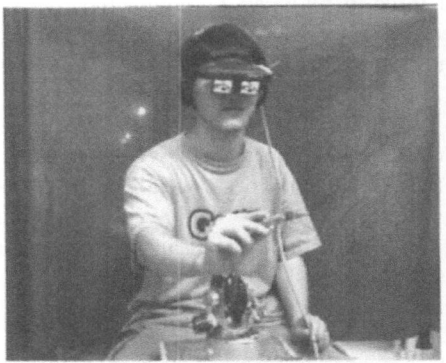

Figure 12.5   Overall view of the user.

12.4   System Design

12.4.1   Graphic Computer and Display

Real-time image of the virtual space is generated by a PC. We use Gateway 2000
with RIVA128 graphics engine. Windows NT 4.0 is chosen for OpenGL acceleration.
The CPU of the PC is Pentium II 300MHz, which manages model of virtual space.
The image on the CRT of the PC is converted to NTSC standard video signal, and
sent to the HMD. We set two windows on the CRT and each image is taken by a
video camera. We use i-glasses! HMD. The HMD has optical see-through capability
and it can presents stereoscopic image merged with real scene. The liquid crystal
display has 18,000 pixels. The effective field of view is 30 degrees.

12.4.2   Motion Tracker

A scene of the virtual space is generated corresponding with the results of motion tracking of the head. The motion of the head is measured by a Polhemus
FASTRACK. A receiver of the FASTRACK is mounted on the HMD. The device
measures 6 degree-of-freedom motion. Sampling rate of is 60Hz. The Polhemus
FASTRACK is connected to the PC via 38,400 bps RS232C.

12.4.3   Control Software

Software of the Feel-through system is implemented on our in-house software tool
named LHX [10]. LHX is composed of seven modules: device driver of force display,
haptic renderer, model manager, primitive manager, autonomy engine, visual display
manager and communication interface. Dividing into these modules, force displays
and virtual environments are easily reconfigured.
   Functions of those modules are:

   1. Device driver

      Device driver manages sensor input and actuator output for force display.
      Various types of force display can be connected to LHX by changing device
      driver. We developed device driver of above mentioned force displays so that
      they can be connected to LHX.

   2. Haptic renderer

      Currently rendering means generation of visual image. However, force sensa-
      tion also needs rendering. Hardness, weight and viscosity of virtual objects
      are generated by haptic rendering. We have developed a software package for
      haptic rendering. Haptic render of LHX has three categories according to three
      types of force displays.

   3. Model manager

      Model of virtual objects is implemented in model manager module of LHX.
      Shapes and attributes of virtual objects are defined in this module. Users of
      LHX program the methods for interaction between virtual objects and opera-
      tors.

4. Primitive manager

   Primitives of virtual object are stored in the primitive manager. Primitives include cube, sphere, cylinder, free-form surface and 3D voxels. Haptic icons for user interface are also included. This module supervises ID code of each primitive. Users of LHX interactively generate or erase primitives. Working primitives are placed in shared memory.

5. Autonomy engine

   Autonomy engine determines behavior of virtual objects. Physical laws for the virtual world are contained in this module. Gravity, elasticity, and viscosity are currently implemented. Collision between primitives is detected in real time. This module defines "time" of virtual environment. Time of virtual environment increases independently from the user. This function enables autonomous growth of virtual objects.

6. Communication interface

   LHX has network interface by which multiple force displays are connected each other. Multiple users can simultaneously interact in the same virtual environment. This function enables easy construction of groupware program. LHX supports TCP/IP so that the system can use existing Internet.

7. Visual display manager

   Visual display manager generates graphic image of virtual environment. This module translates the haptic model to OpenGL format. HMDs, stereo shutter glasses and multiple screens are supported as visual displays.

LHX is currently implemented in SGI workstation and Windows NT workstation. LHX is composed of two processes: visual feedback process and force feedback process. Visual feedback process runs visual display manager, primitive manager and autonomy engine. Force feedback process runs other modules. Shared memory is used for communication between these processes. Required update rate of force feedback is much higher than that of visual feedback. Image can be seen continuously at update rate of 10Hz. On the other hand, force feedback requires 40Hz at least. In LHX force feedback process has higher priority than visual feedback process. LHX enables high update rate of force display in complex virtual environment. Update rate of visual image of Feel-through system is 20 to 30 Hz, and update rate of force feedback is 100Hz.

12.5   Evaluation of Prototypes

12.5.1   Desktop Feel-through

Desktop Feel-through is the first method of adding force feedback to AR system: real and virtual objects are located within the range of the user's hand. A real box and a virtual box are placed on a desk and the pen-based force display is set on the desk. Figure 12.6 shows an example of imagery seen by the user.

Figure 12.6   Scene of desktop Feel-through.

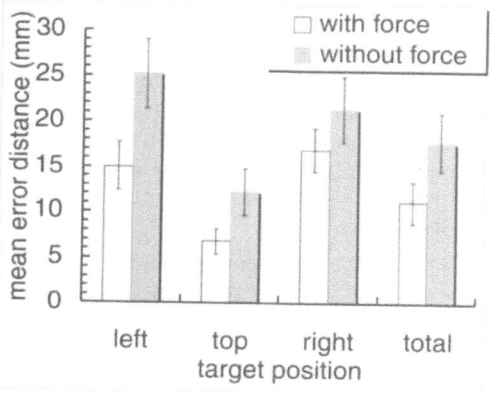

Figure 12.7   Mean error distances.

We set docking tasks for evaluation of usability of the system. The virtual box is placed on the desk at first. The subject is asked to pick the box up using the force display. A button is put on the pen and the user can pick virtual object by pressing the button. Then the subject is asked to put the virtual box on the real box. The user feels force when the virtual box hits the real box. After the task is over, the subject is asked to put the virtual box at the left side and the right side of the real box. We recorded trajectories of the virtual box and evaluate distance between the real box and the virtual box.

We compared two conditions:

mode 1) visual + force feedback

mode 2) visual feedback only

Six university students participated in the experiment. Figure 12.7 shows mean error distances. Error bars represent standard deviations. The result shows that mean error distances of visual/force feedback mode are significantly smaller than those of visual mode. The subjects reported that force feedback helps final adjustment of the position. They also reported that top positioning is relatively easier than side positioning. This is caused by occlusion problem of merged image.

12.5.2   Feel-through in a Large Space

When objects are located at a large distance from the user, we use the virtual beam in order to extend the user's hand. We set an uneven virtual wall for an evaluation test. The wall is placed at 5m from the user. The user feels reaction force from the surface using the virtual beam. Figure 12.8 illustrates the top view of the test space and Figure 12.9 shows imagery seen from the user. In Figure 12.9, the virtual beam is registrated at the top of the pen-based force display.

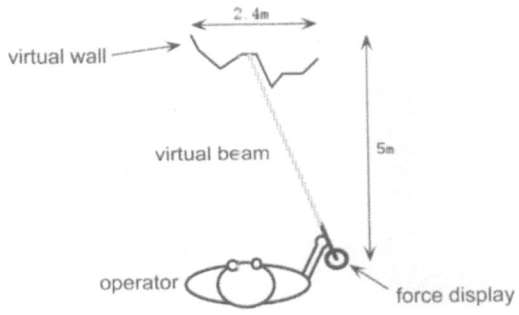

Figure 12.8   Configuration of test space.

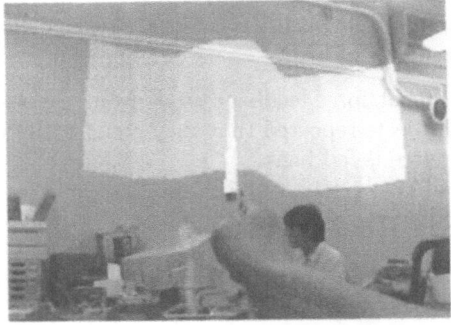

Figure 12.9   Scene of merged virtual wall.

Subjects of this experiment feel the wall freely and after that they are asked to draw the shape of the wall. The figure drawn by the subjects are digitized by scanner and width of the wall is normalized. We compared shapes of the drawn wall and model of the virtual wall. We set two conditions as same as the former experiment: visual/force feedback mode and visual feedback mode. Seven university students participated in the experiment. Figure 12.10 shows mean shapes of drawn walls. The result indicates the shape of visual/force feedback mode is closer to the original

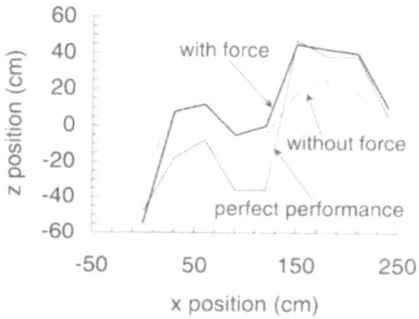

Figure 12.10   Mean shape of drawn walls.

Figure 12.11   Scene of Feel-through inside a real object.

wall. Displacement of visual/force feedback mode from the original wall averaged in 72% of visual mode. Subjects reported that they could easily recognize angle of the edge of the wall when force is fed back.

12.5.3   Feel Inside Real Objects

If virtual objects are located inside of real objects, the user cannot touch virtual objects directly even if those are placed nearby the user. We use the virtual beam in order to feel inside real objects. In this case, length of the beam is fixed, so that the user feels as if he/she were handling a stick. The user feels reaction force at the end of the virtual stick. Three virtual boxes are placed inside a real box for an evaluation test. Figure 12.11 shows imagery seen from the user.

We set three different stiffness of the virtual boxes. Stiffness of each virtual box is randomly selected. Subjects of this experiment are asked to report difference of stiffness of three virtual boxes. Each subject reports 20 sets of randomly selected stiffness. Seven university students participated in the experiment. The subjects of this experiment can easily distinguish difference of stiffness. Correct answer ratio of the subjects averaged in 93%.

12.6   Wearable Force Display and Mobile Augmented Reality

Development of AR displays that work anywhere, in any environment is one of the ultimate goals of AR research. Such AR displays require wearable devices. Virtual world technology used to employ wearable devices such as HMD or DataGlove. Users of glove-like tactile input devices, such as DataGlove, feel troublesome when they put or off these devices. They often need calibrations. If the glove is equipped with force feedback device, the problem is much severe. This disadvantage obstructs practical use of Exoskeleton type force display. Therefor, tool handling type force displays are often used in desktop VE.

There is functional difference between wearable force display and desktop force display. Wearable force displays are not grounded, so that they cannot present gravitational force or resistance against a wall. On the other hand, wearable force displays can be used in anywhere in a large space. We therefor developed a new force display that takes advantage of wearable force display and tool handling type force display [11]. The force display employs 3 DOF joystick attached to the user's arm by straps. The device is named Wearable Master. Table 12.1 presents characteristics of Wearable Master compared to other typical force displays.

Table 12.1   Configuration of force displays.

	exoskeleton	tool handling type
grounded	SARCOS Master, EXOS Arm Master	Haptic Master, PHANToM
non-grounded	Rutgers Master, Cyber Grasp	Wearable Master

Mechanical configuration of Wearable Master is illustrated in Figure 12.12 (top view) and Figure 12.13(side view). Figure 12.14 shows overall view of the device. Each joint is equipped with DC motor and potentiometer. The maximum generated force at the grip is 200gf. Overall weight of the device is 500g. A Polhemus sensor is mounted at the grip for tracking of the user's hand.Wearable Master generates force between an arm and fingertips. Fingertips are much sensitive than an arm, so that the user feels as if force were fed back from virtual objects. The device can be applied to Feel-through using the virtual beam as mentioned at Section 12.2.

12.7   Conclusions

We proposed methods of adding force feedback to AR environments. The prototype system of Feel-through was developed. Results of the evaluation experiments showed that the system effectively worked and force feedback improved performance of the tasks.

Future work of the research will be finding serious application of the Fell-through technology. Application areas of Feel-through will be classified into three categories according to three methods of adding force feedback to AR environment. Desktop

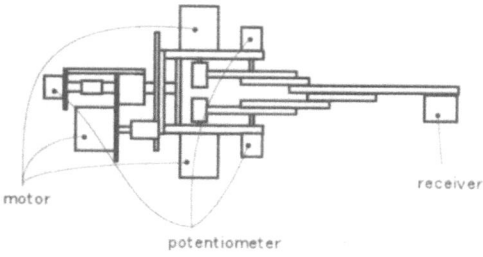

Figure 12.12   Wearable Master (top view).

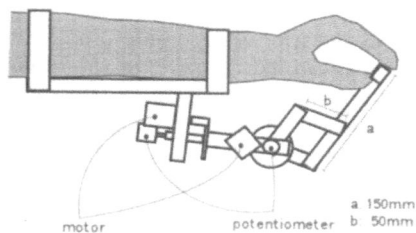

Figure 12.13   Wearable Master (side view).

Feel-through, where real and virtual objects are located nearby the user, can be applied to industrial design. A designer can merge physical mockup and virtual mockup. Feel-through in a large space using the virtual beam will be effective in architectural design. Virtual buildings merged in real ground help design work as well as assessment. Feel-through inside real objects can be effective in medical applications. A doctor sees volume graphics superimposed on a patient and feels rigidity of internal organs. Required performance of Feel-through system will be clarified through these applications.

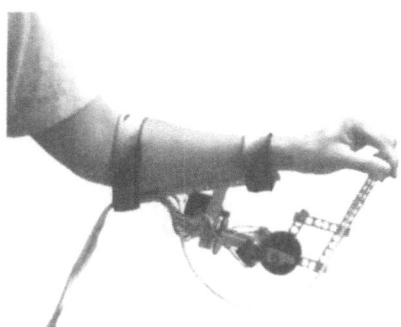

Figure 12.14   Overall view of Wearable Master.

References

[1] A. State et al.: "Technologies for augmented reality systems: Realizing ultrasound-guided needle biopsies," *Proc. SIGGRAPH'96*, pp. 439-446, 1996.

[2] D. Sims: "New realities in aircraft design and manufacture," *Computer Graphics and Applications*, p.91, 1994.

[3] T. Ohshima et al.: "AR^2 Hockey: A case study of collaborative augmented reality," *Proc. VRAIS'98*, pp.268–275, 1998.

[4] H. Iwata: "Artificial reality with force-feedback: Development of desktop virtual space with compact master manipulator," *Computer Graphics*, vol.24, no.4, 1990.

[5] H. Iwata and Y. Asada: "Augmented reality with force feedback," *Proc. the Virtual Reality Society of Japan Annual Conference*, vol.1, 1996 (in Japanese).

[6] H. Iwata and Y. Asada: " 'Feel-through' in augmented reality," *Proc. the Virtual Reality Society of Japan Annual Conference*, vol.2, 1997 (in Japanese).

[7] H. Iwata: "Desktop force display," *SIGGRAPH'94 Visual Proceedings*, 1994.

[8] H. Iwata: "Pen-based haptic virtual environment," *Proc. VRAIS'93*, 1993.

[9] H. Iwata: "Haptic screen," *SIGGRAPH'98 Conference Abstracts and Applications*, p.117, 1998.

[10] H. Iwata, H. Yano, and W. Hashimoto: "LHX : An integrated software tool for haptic interface," *Computers and Graphics*, vol.21, no.4, 1997.

[11] H. Iwata and H. Hakagawa: "Wearable force feedback Joystick," *Human Interface N&R.*, vol.13, no.2, 1998 (in Japanese).

Chapter 13

Tangible Bits: Coupling Physicality and Virtuality Through Tangible User Interfaces

Hiroshi Ishii
MIT Media Laboratory, U.S.A.

13.1 Bits and Atoms: GUI, VR, AR, MR, UbiComp

We live between two realms: our physical environment and cyberspace. Despite our dual citizenship, the absence of seamless couplings between these parallel existences leaves a great divide between the worlds of bits and atoms. At the present, we are torn between these parallel but disjoint spaces. We are almost constantly "wired" so that we can be here (physical space) and there (cyberspace) simultaneously. Our windows between these worlds are a myriad of rectangular screens, where streams of bits leak out of cyberspace as photon beams.

The interactions between people and cyberspace, however, are now largely confined to traditional GUI (Graphical User Interface)-based boxes sitting upon our desks, laps, and palms (Figure 13.1). The interactions with these GUIs are isolated from the ordinary physical environment within which we live and interact.

As humans, we have developed rich skills and work practices for processing information through haptic interactions with physical objects (e.g., scribbling messages on Post-It™notes, and spatially manipulating them on desks and walls) as well as physical world   digital world

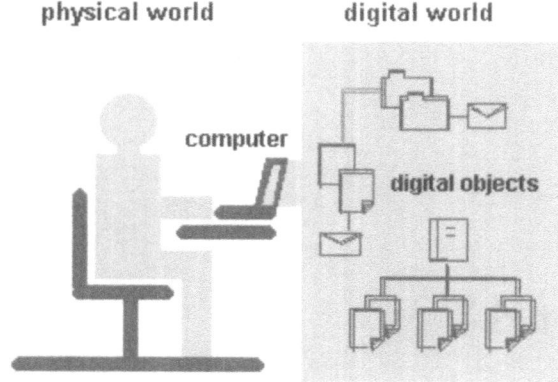

Figure 13.1   Graphical User Interface(GUI).

peripheral senses (e.g., being aware of changes in weather through ambient light). However, most of these practices are neglected in current HCI design because of poor diversity of input/output media, and exaggerated bias towards graphical output at the expense of input from the real world [1].

13.1.1   Graphical User Interface (GUI)

In 1981, the Xerox Star workstation set the stage for the first generation of GUI [2], establishing a "desktop metaphor" which simulates a physical desktop on a bit-mapped screen. The Star was the first commercial system that demonstrated the power of the mouse, windows, icons, property sheets, and modeless interactions. The Star also set several important HCI design principles, such as "seeing and pointing" vs. "remembering and typing," and "what you see is what you get." The Apple Macintosh brought this new style of HCI to public attention in 1984, creating a new stream in the personal computer industry [3]. Now, the GUI is widespread, largely through the pervasiveness of Microsoft Windows.

13.1.2   Virtual Reality (VR) and Augmented Reality (AR)

Virtual Reality (VR) systems immerse a user inside completely synthetic digital environments (Figure 13.2). These synthetic worlds may simulate the visual properties of existing real-world environment or completely imaginary spaces, sometimes accompanied by audible and haptic displays. While using such systems, the user generally cannot see or interact with the real world around him/her.

In contrast, Augmented Reality (AR) allows the user to see the real world, with virtual objects graphically superimposed upon or mixed with the real world (Figure 13.3). DigitalDesk [4] is one pioneering AR work which demonstrated a way to merge physical and digital documents by using video projection of computer display onto a real desk with physical documents.

Azuma gave the following definition of AR [5]:

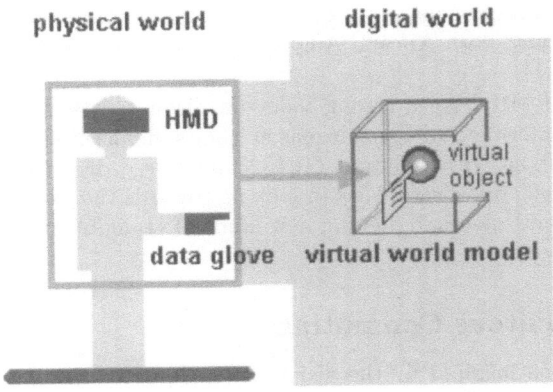

Figure 13.2   Virtual Reality(VR).

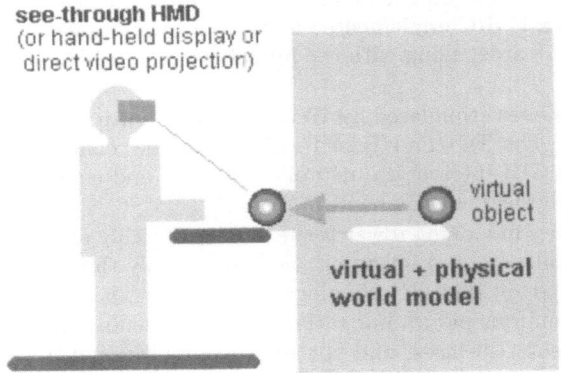

Figure 13.3   Augmented Reality(AR).

AR are systems that have the following three characteristics:

1) *Combines real and virtual*
2) *Interactive in real time*
3) *Registered in 3-D*

The most common AR approach is the visual overlay of digital information upon real-world imagery with see-through head-mounted display devices (HMD) [6] or hand-held display devices [7] [8], with some pioneering projects exploring direct video projection [4].

13.1.3   Mixed Reality

Mixed Reality (MR) was defined by Paul Milgram as the environment in which real world and virtual world objects are presented together within a single display, anywhere between a continuum of real ←→ virtual environments [9]. According to Milgram's taxonomy, both AR and Augmented Virtuality (AV) represent subsets of MR.

Both AR and MR have a strong focus on display, rather than the input and manipulation of information. For this reason, they suffer a similar imbalance between input and output as with VR and GUI. While these display-centric approaches clearly have a place, our interest lies in looking towards the richly afforded world of physical objects and spaces, designing new kinds of "tangible media" augmented by digital technology.

13.1.4   Ubiquitous Computing

Our work has been inspired by the vision of "Ubiquitous Computing" [10] as well as AR research. Mark Weiser proposed a vision of Ubiquitous Computing in which access to computational services is delivered through a number of different devices deployed into niche physically-contextualized tasks and contexts. In addition to this ubiquity, he stressed that the delivery of computation should be "transparent." His team at Xerox PARC implemented a variety of computational devices including Tabs, Pads, and Boards, along with the infrastructure enabling these devices to talk with each other.

Our work has been stimulated by Weiser's vision, but it is also marked by important differences. The Tab/Pad/Board vision is largely characterized by exporting a display-centric GUI interaction metaphor to large and small computer terminals situated in the physical environment.

Our interest lies in looking towards the bounty of richly afforded physical devices of the last few millennia and inventing ways to re-apply these elements of "tangible media" augmented by digital technology. Thus, our vision is not about making "computers" ubiquitous per se, but rather about awakening richly-afforded physical objects, instruments, surfaces, and spaces to computational mediation, borrowing perhaps more from the physical forms of the pre-computer age than the present.

13.1.5   Tangible Bits

Our approach in Tangible Bits (Tangible User Interfaces) is distinguished by a strong focus on graspable physical objects as inputs, interfaces, and physical representations of digital information rather than by considering purely visual augmentations [11]. The combination of ambient media with graspable interfaces [12] is also a unique point of Tangible Bits (Figure 13.4).

The aim of our research is to show concrete ways to move beyond the current dominant model of GUI, based upon computers with a flat rectangular display, windows, pointer, and keyboard. To make computing not only ubiquitous but also a more human medium, we seek to establish a new type of HCI that we call "Tangible User Interface" (TUI).

In a broader context, we see the locus of computation now shifting from the desktop in two major directions: i) onto our skins/bodies, and ii) into the physical environments we inhabit. The transition to our bodies is represented by recent activities in the new field of "wearable computers" [13]. We are focusing on the

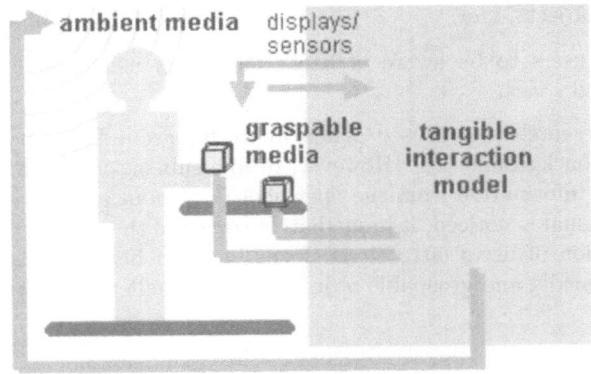

Figure 13.4   Tangible User Interface(TUI).

second path: integration of computational augmentations into the physical environment. Our intention is to take advantage of natural physical affordances [14] to achieve a heightened legibility and seamlessness of interaction between people and information.

13.2   Tangible Bits: Key Concepts

"Tangible Bits" is an attempt to bridge the gap between cyberspace and the physical environment by making digital information (bits) tangible. Our key concepts are:

1) *Interactive Surfaces*: Transformation of physical surfaces throughout architectural space (e.g., walls, desktops, ceilings, doors, windows) into active interfaces between the physical and virtual worlds;

2) *Coupling of Bits and Atoms*: Seamless coupling of graspable physical objects with associated digital information and processes; and

3) *Ambient Media*: Use of ambient media such as sound, light, airflow, and water movement as background interfaces with cyberspace at the periphery of human perception.

ClearBoard [15] can be seen as an early example of *Interactive Surfaces*, while Bricks [12] made early progress in *Coupling of Bits and Atoms*. Both Dunne and Raby [16] and Jeremijenko [17] made early design progress in relating to *Ambient Media*, inspiring our work in this area.

Ultimately, we are seeking ways to turn each state of physical matter - not only solids, but also liquids and gases - into "interfaces" between people and digital information.

We are exploring ways of both improving the quality and broadening the bandwidth of interaction between people and digital information by:

- allowing users to "grasp & manipulate" foreground bits by coupling bits with physical objects, and

- enabling users to be aware of background bits using ambient media in an augmented space.

Current HCI research has focused primarily on foreground activity, in the process neglecting the background [18]. However, people subconsciously are constantly receiving various information from the "periphery" without attending to it explicitly. If anything unusual is noticed, it immediately comes to the center of attention. This smooth transition of users' attentional focus between background and foreground using ambient media and graspable objects is a key challenge of Tangible Bits (Figure 13.5).

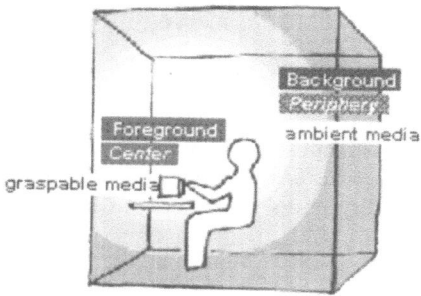

Figure 13.5   Framework of Tangible Bits: Center and periphery of User's attention within physical space.

13.3   Tangible Interface Designs

"Tangible User Interfaces" emphasize both "hands-on" foreground interactions, and background perception of ambient media at the periphery of our senses. This paper will introduce three design examples of Tangible User Interfaces:

1) **Illuminating Light**: an example exploring the use of physical objects as a means to manipulate bits in the center of users' attention (foreground),

2) **InTouch**: an application of Tangible Interface to haptic interpersonal communication, and

3) **Water Lamp and Pinwheels**: examples of ambient display media, which focus on the periphery of human perception (background).

13.4   Illuminating Light

Illuminating Light is a workbench for rapid prototyping of laser-based optical and holographic layouts. Users of this experimental tool move physical representations of various optical elements about a workspace. The system tracks these components, and projects back onto the workspace surface the simulated propagation of laser light through the evolving layout. This application is built atop the Luminous Room infrastructure; an aggregate of interlinked, computer-controlled projector-camera units called I/O Bulbs [19].

13.4.1 The I/O Bulb and the Luminous Room

The project's primary goal is the transformation of architectural space to transform each surface into an information-display-and-interaction structure. The approach we propose requires the conceptual generalization of the familiar lightbulb into the *I/O Bulb*. If an ordinary incandescent bulb is actually a low-resolution digital projector - specifically, 1x1 pixel(s) - then we must increase this resolution, so that the lightbulb is capable of projecting images into the space around it, while at the same time incorporating a tiny video camera that looks out at the world around the bulb. The resulting structure, called an *I/O Bulb*, is capable of simultaneous optical input and output (Figure 13.6). The work described here makes use of a prototype *I/O Bulb* constructed with commercially available projectors and cameras.

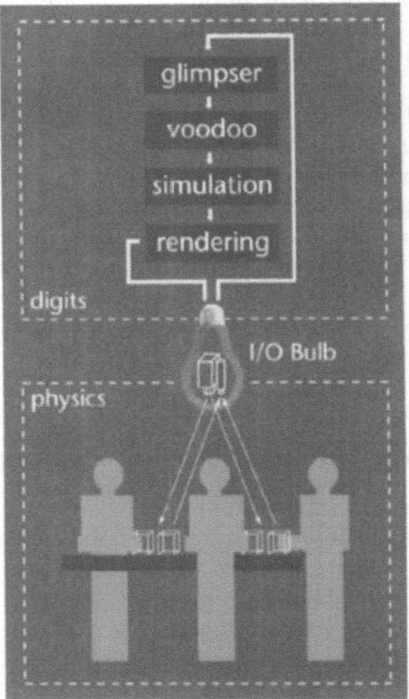

Figure 13.6 I/O Bulb and Luminous Room.

The notion of a Luminous Room infrastructure involves extrapolating from just one to a collection of many I/O Bulbs computationally interlinked and distributed throughout an architectural space. The resulting aggregate of two-way optical nodes addresses every portion of a room and achieves our original goal of Luminous Room.

13.4.2   Scenario

A stylized plastic model of a laser is placed on an ordinary table; immediately a luminous beam, projected from above onto the table's surface, appears to shoot forward from the laser model's aperture. As the model is moved about the table the beam tracks along with it, always originating from same point on the laser's front surface. A beamsplitter model placed in the path of the beam splits it in two, half-continuing forward and half-reflecting off the splitter's surface. A model mirror bounces an intercepted beam at the incidence-equals-reflection angle. Lens models spread a single beam into a fan of beams; recording-film models absorb beams incident upon them; and 'recording subjects' - a small automotive model, for example - scatter incoming beams, insuring that some are redirected to arrive at any recording film that may be nearby.

During such manipulations, the various 'inert' optics models behave very much as their real counterparts would, directing and modifying the light that passes through them; and these physically accurate 'beams' of light are wholly simulated and projected down in careful registration with the 'optics'. Simple metrics floats in appropriate positions and indicates the angle at each beam bounce location and the distance between each pair of successively traversed optics. As work progresses, a continuously updated display at the far end of the table shows the layout's relative optical path-lengths. Once a viable recording setup has been achieved a rendered simulation emerges, showing how the object would appear in an optical reconstruction of a real, analogously recorded hologram (Figure 13.7, Figure 13.8).

The scenario described above is Illuminating Light, the first working application of the I/O Bulb and Luminous Room infrastructures. It is built atop a pipeline comprising *glimpser*, a low-level machine vision system, and voodoo, a toolkit for constructing layout-based interactive simulations. Its half-physical, half-projective interaction style, an approach termed *Luminous-Tangible*, also extends Tangible Interface [11].

13.4.3   Application Domain: Optics & Holography

High-quality optical elements are simultaneously expensive and notoriously susceptible to damage. The breadboarding tables on which experiments are constructed and prototypes are built are a scarce resource. At the same time, the precision required of laser-based optical systems frequently results in setup and refinement times that greatly exceed the time spent actually running the experiment. All of this suggests that a well-designed 'simulated optics workbench' could be a valuable tool. Such a workbench should permit the optical engineer to manipulate an accurate simulation of an evolving layout and its operation, and then rapidly reproduce the setup on the real table to perform the end experiment.

In holography - a higher level 'goal' whose specific domain knowledge we chose to include in the system - a fine-grained photographic plate must be exposed simultaneously to an 'object beam', comprising light scattered from laser illumination of the

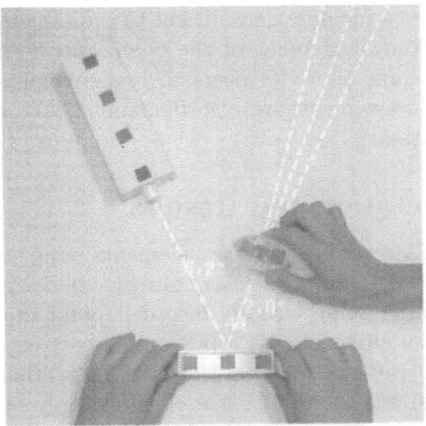

Figure 13.7   The Illuminating Light system in use.

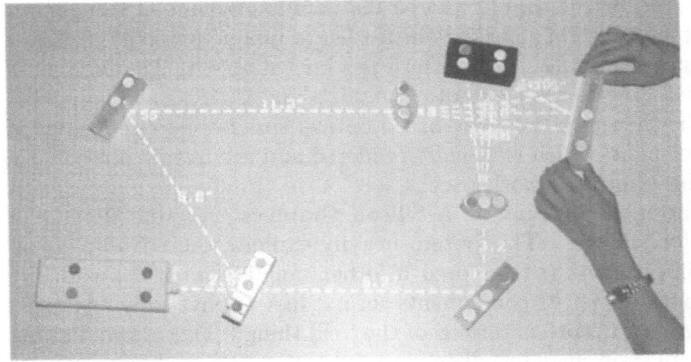

Figure 13.8   Completed Holographic layouts.

object being recorded, and to a 'reference beam': uniform, unmodulated laser light. These two beams must not only originate from the same laser but also have equal lengths. Additional geometric requirements are imposed according to physical and aesthetic demands. Thus the principal challenge of designing a working holographic layout is the simultaneous satisfaction of various geometric requisites.

13.4.4 System Design Requirements

Our intent was to build a prototyping tool for holographic recording setups whose input and output were arranged to emulate the real thing - not just visually, but haptically and spatially as well - in order to both foster and exploit geometric-optical understanding skills. Although the optical elements and the laser beams - the former moved volitionally by human users, the latter computationally generated and projectively inserted into the real space - would be implementationally decoupled, the application would convincingly cause them to appear causally coupled.

We required, in acknowledgement of the comparative efficiency of two-handed work [12], that the system allow its users to maneuver as many objects concurrently as necessary or convenient; we also demanded that our application permit collaborative manipulation.

13.4.5   Implementation and Results

Experimenters manipulate six basic optical elements required for the proper execution of hologram-recording setups: a laser, mirrors, beamsplitters, lenses, a 'holo-object' (the physical thing being visually recorded), and the holographic film plate itself. Each element has affixed to its top a unique pattern of small colored dots. Optical input from the workspace is passed from an overhead I/O Bulb to a very modest 'raw vision' system: the glimpser program simply identifies colored dots in its visual field. Built as a client-server facility, glimpser accepts commands over a network connection and, frame by frame, reports back to the originating client a collection of located dots. An application-independent geometric parsing toolkit called voodoo interprets this simple colored-dot-location output of the glimpser program, analyzing each batch of color dots into a list of unique patterns registered with it by the application it serves - one pattern per known object. The instantaneous spatial configuration (the recognized optical models and their recovered positions and orientations) is then used by a ray-based optical simulator to determine the resultant path of laser light, which is visually rendered and accurately projected via the same I/O Bulb back into the workspace.

Illuminating Light runs on a Silicon Graphics O2 (R5000) at a framerate of twenty to forty Hertz. The system heavily exploits the advantages of control via graspable implements (as explored in other tangible interface work), but its additional strength is that its components act not just as physically instantiated abstractions but as direct representations of the 'real thing'. This allows Illuminating Light to provide constant visual feedback in a form that is already intrinsic to the simulation's real-world counterpart n and so the 'virtual' part of the application does not seem distracting or glaringly distinct from its 'real' part.

13.5   InTouch

Touch is often recognized as a fundamental aspect of interpersonal communication. Touch can instantly indicate the nature of a relationship; it is sincere, immediate, and compelling. Yet while many traditional technologies allow communication through sound or image, none are designed for expression through touch. inTouch is a system for haptic interpersonal communication based on the concept of "Synchronized Distributed Physical Objects." We will describe the iterative design of inTouch as an application of Tangible User Interface to remote interpersonal communication [20].

13.5.1   inTouch Design

inTouch provides a system for haptic interpersonal communication across distance. Of these other works. As seen in the concept sketch of Figure 13.9, inTouch consists of two hand-sized objects each with three cylindrical rollers embedded within a base. Employing the Synchronized Distributed Physical Objects concept, the rollers on each base are haptically coupled such that each one feels like it is physically linked to its counterpart on the other base. To achieve the tight coupling necessary to allow simultaneous manipulation, inTouch employs bilateral force-feedback technology, with position sensors to monitor the states of the rollers and high precision motors to synchronize those states. Two people separated by distance can then passively feel the other person's manipulation of the rollers, cooperatively move the "shared" rollers, or fight over the state of the rollers, providing a means for expression through touch.

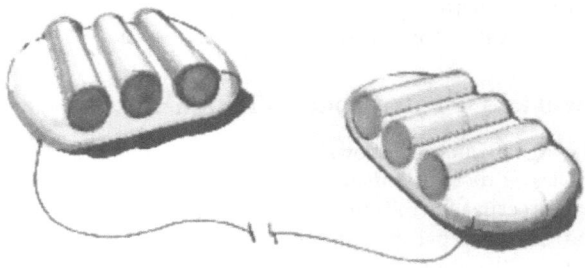

Figure 13.9   inTouch concept sketch.

13.5.2   Standalone Prototype: inTouch-1

InTouch-1 was created to implement the connection between rollers, virtually, using force-feedback technology. The system architecture for inTouch-1 is shown in Figure 13.10. Hewlett Packard optical position encoders were used to monitor the physical states of the rollers (positions were read directly, other values were interpolated) and high performance Maxon DC motors were used to synchronize those states. A 200MHz Pentium PC controlled all motor/encoder units (one unit for each roller) using Immersion Corporation's Impulse Drive Board 1.0 boards and 2-Axis Card 1.0 ISA cards.

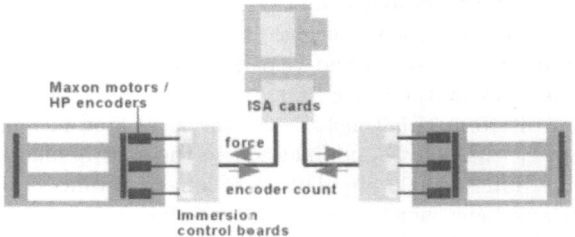

Figure 13.10   inTouch-1 system architecture(standalone prototype).

The control algorithm that ran on the host PC simulates a highly damped, stiff rotary spring between corresponding rollers. In other words, the algorithm looks at the difference in position of each pair of "connected" rollers and applies a restoring force, proportional to that difference, to bring the rollers together.

The first prototype of inTouch-1 was completed in March 1997, and has been demonstrated at sponsor meetings and at the 1997 Ars Electronica Festival, as well as tested internally. People, who knew the previous version, inTouch-0, a mechanical mockup, were surprised at how closely the interaction matched the mechanical mockup. In total, more than 500 people have tried inTouch, several of whom have made enthusiastic requests for the system to "keep in touch" with distant family and loved ones. Many people have indicated their belief that inTouch provides a means to be aware of a distant person's emotional state and sincerity, however, we have not yet formally tested this proposition.

13.5.3   Networked Prototype: inTouch-2

The next prototype, inTouch-2, allows the virtual connection of inTouch-1 to be extended over arbitrary distance, using the Internet.

The system architecture for inTouch-2 is shown in Figure 13.11. The architecture is identical to that of inTouch-1 except that the two sets of three rollers run on separate host computers, distributed over a standard network. Positions and velocities of the local rollers are passed to the remote computer using User Datagram Protocol (UDP).

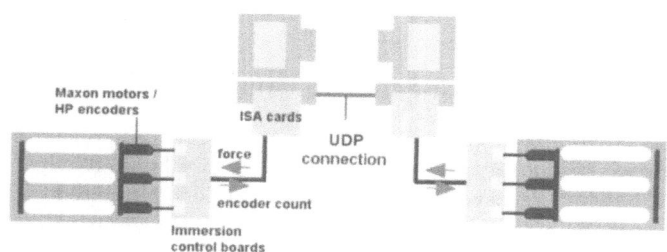

Figure 13.11   inTouch-2 system architecture(networked prototype).

The basic control algorithm for the networked design is also the same as that for inTouch-1. Each computer simply calculates the forces to impart to its three rollers given the state of each local roller (received from the local control hardware) and the most recently received position and velocity of the corresponding remote roller (passed over the network by the other PC).

We have so far distributed inTouch-2 over the local area network in our building. At this distance, with a little modification to the control algorithm, inTouch-2 behaves identically to inTouch-1. Simulations of longer distances, and consequently longer network delays, have shown promise in extending inTouch over arbitrary distances (see Appendix). Figure 13.12 shows the inTouch-2 in use.

We are now making microprocessor embedded version, inTouch-3, and are planning to use this new version to investigate the implications and appropriate appli-

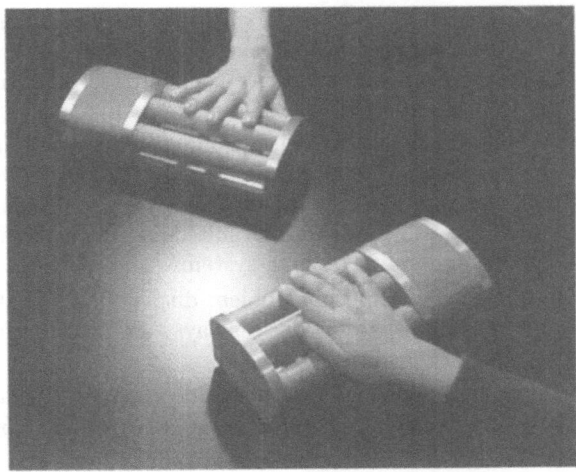

Figure 13.12   inTouch-2 in use.

cations of a haptic interpersonal communication link through experimentation and long-term user testing.

13.6   Ambient Media: Water Lamp and Pinwheels

We envision that the architectural spaces we inhabit will be an interface between humans and online digital information. We introduce ambient fixtures called *Water Lamp* and *Pinwheels:* a new approach to interfacing people with online digital information. The Water Lamp projects water ripple shadow created by a "rain of bits." The Pinwheels spin in a "bit wind." These ambient fixtures present information within an architectural space through subtle changes in light, sound, and movement, which can be processed in the background of awareness. We describe the design and implementation of the Water Lamp and the Pinwheels, and discuss design issues.

13.6.1   Ambient Media

Nature is filled with subtle, beautiful and expressive ambient displays that engage each of our senses. The sounds of rain and the feeling of warm wind on our cheeks help us understand and enjoy the weather even as we engage in other activities. Similarly, we are aware of the activity of neighbors through passing sounds and shadows at the periphery of our attention. Cues like an open door or lights in an office help us subconsciously understand the activities of other people and communicate our own activity and availability.

Current personal computing interfaces, however, largely ignore these rich ambient spaces, and squeeze vast amounts of digital information into small rectangular screens. Information is presented as "painted bits" on flat screens that must be in the center (foreground) of a user's focus to be processed. The interactions between people and digital information are currently almost entirely confined to a conventional GUI (Graphical User Interface) comprised of a keyboard, screen, and mouse.

We are trying to broaden the concept of "display" to make use of the entire physical environment as an interface. We are moving information off the screen into the physical environment, where it is manifested as subtle changes in form, movement, sound, color, smell, temperature, or light. We call such displays "ambient displays." We expect ambient displays are well suited as a means to keep users aware of people, weather, or general states of large systems [21].

This paper will introduce two specific ambient display designs: the Water Lamp and the Pinwheels. We call these standalone displays "ambient fixtures" to separate them from their precursor "ambientROOM" [22].

Ambient fixtures are standalone ambient media displays. We have taken concepts developed within the ambientROOM, and have moved them out of a small room into an open space. In the ambientROOM, the user is "inside the computer," while ambient fixtures allow us to externalize the displays and distribute them throughout an open architectural space. Ambient fixtures thus allow ambient displays to be used by several people at once.

We have implemented two ambient fixtures: the Water Lamp and Pinwheels. The Water Lamp is an extension of the ceiling water ripples of the ambientROOM and the Pinwheels explore the ideas of physical movement caused by invisible information flow. Both are designed based on the metaphor of natural physical phenomena.

13.6.2   The Water Lamp

The first ambient fixture we developed is the Water Lamp. Water ripples created by raindrops on the surface of still water were the starting point of this Water Lamp design. Instead of physical raindrops, we envisioned that "bits" (digital information) falling from cyberspace could create physical water ripples. The raindrops of "bits" have been realized with computer-controlled solenoids tapping the water. Figure 13.13 illustrates the prototype of the Water Lamp.

The Water Lamp is composed of a wooden base, 3 aluminum support tubes and an acrylic water pan. There are 3 small solenoids mounted above the water tray. These solenoids are controlled through a single circuit board. When actuated, the solenoids tap on the surface of the water in the tray, causing ripples in the water surface. Various digital information sources can drive this circuit board to actuate these three solenoids. A light shines upward through a pan of water, and produces changing patterns of light and shadow projected onto a ceiling.

The Ambient fixtures, both Water Lamp and Pinwheels are based upon a common control platform- the iRX 2.0 PIC Micro-controller Board designed at the MIT Media Lab by Robert Poor. The iRX board accepts commands over a serial line from a computer to control each fixture. This allows us to distribute fixtures throughout our research space. TCL-based software sends commands to the fixtures. Information can be relayed from the internet or other networked information source and be routed to the appropriate fixture.

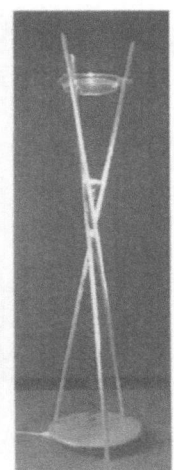

(a)Reflection of water ripple on a ceiling. (b)Water pan with computer-controlled solenoids.

Figure 13.13   Water Lamp.

13.6.3   Pinwheels

The Pinwheels evolved from the idea of using airflow in the ambientROOM. We found that the flow of air itself was difficult to control and to convey information. As an alternative, we envisioned that a visual/physical representation of airflow based on the "spinning pinwheels" could be legible and poetic. The Pinwheels spin in the "bit wind" at different speeds based upon their input information source.

The Pinwheels are made from folded fiberglass mounted on the shaft of a small DC motor. Four Pinwheels are connected to each iRX control board. Pulse width modulation controls the speed at which the motors spin. Figure 13.14 shows the Pinwheels spinning in the bit wind based on the Ethernet traffic of the Media Lab building.

13.6.4   Applications of Ambient Media

Ambient fixtures are envisioned as being all around and suited to the display of a) people's presence (awareness of remote people's status / activities), b) atmospheric and astronomical phenomena, or c) general states of large and complex systems (e.g. atomic power plant, computer networks).

For instance, an atmospheric scientist might map patterns of solar wind into patterns of Pinwheel spins in a room. Other users might want to be aware of tension in the Earth's fault-lines, giving an indication of earthquake activity through an array of Water Lamps.

People have a strong desire to feel connected to others, especially loved ones. The Water Lamp may aid in this sense of connection by displaying the heartbeat of a significant other picked up by a special wristwatch.

Figure 13.14   Array of Pinwheels spinning in the bit wind.

We have been using the Water Lamp and Pinwheels mainly as displays of activities in cyberspace. We have mapped the wireless LAN traffic onto the Pinwheels and the web hits to the Water Lamp.

13.6.5   Design Issues of Ambient Media

Our explorations have given us insight into many research issues that arise in designing and testing ambient display media.

1) Mapping of information to the physical motion
   A designer of ambient displays must transform the digital data into a meaningful pattern of physical motion that successfully communicates the information.

2) Persistence in the space
   Due to the physical persistence of the fixtures in a space, we learn where to "look" for the information within the environment. Selection of location for those fixtures in an architectural space becomes an important design decision.

3) Thresholds between background to foreground
   Ambient displays are expected to go largely unnoticed until some change in the display or user's state of attention makes it come into the foreground of attention. How to keep the level of display at the threshold of a user's attention is an open design issue that we are now working on from a cognitive science point of view.

13.7   Conclusions

This paper has presented our vision of Tangible Bits, which seeks to bridge the gap between the worlds of bits and atoms through graspable objects and ambient media. To illustrate this vision, we have introduced three Tangible Bits designs: Illuminating Light, inTouch, and Water Lamp / Pinwheels. We believe these design instances begin to illustrate the breadth and depth of this new Tangible User Interface research domain.

Current GUI-based HCI displays all information as "painted bits" on rectangular screens in user's foreground, thus restricting itself to very limited communication channels. GUIs fall short of embracing the richness of human senses and skills people have developed through a lifetime of interaction with the physical world.

Although the new streams of AR/MR research suggest important direction for combining real and virtual spaces, current approaches are display-centric, with their interfaces missing *tangibility*.

Our aim is to transform "painted bits" into "tangible bits" by taking advantage of multiple senses and the multi-modality of human interactions with the real world, without requiring encumbering display devices such as HMD. We believe the use of graspable objects and ambient media as core HCI primitives will lead us to a much richer multi-sensory experience of digital information.

Ishii met a highly successful PDA (Personal Digital Assistant) called the "abacus" when he was 2 years old. This simple abacus-PDA was not merely a computational device, but also a musical instrument, imaginary toy train, and a back scratcher. He was captivated by the sound and tactile interaction with this simple artifact. When his mother kept household accounts, he was aware of her activities by the sound of her abacus, knowing he could not ask for her to play with him while her abacus made its music. We strongly believe this abacus is suggesting to us a direction for the next generation of HCI.

Acknowledgments

We thank TTT (Things That Think) consortium at the MIT Media Lab, for its ongoing support of the Tangible Bits project. We also would like to acknowledge the contribution of many hardworking graduate and undergraduate students at MIT for work on the implementation of variety of Tangible User Interface prototypes. In particular, we thank current and former graduate students Brygg Ullmer, John Underkoffler, Scott Brave, Andrew Dahley, Matt Gorbet, Paul Yarin, Craig Wisneski, Phil Frei, Victor Su, Jay Lee, Seungho Choo.

References

[1] W. Buxton: "Living in augmented reality: Ubiquitous Media and Reactive Environments," in (S. Finn and Wilber, eds.) *Video Mediated Communication*, Hillsdale, N.J., Erlbaum, 1997.

[2] D. Smith: "Designing the star user interface," *Byte*, pp.242–282, April 1982.

[3] Apple: *Human Interface Guidelines: The Apple Desktop Interface*, Addison-Wesley, 1987.

[4] P. Wellner, W. Mackay, and R. Gold: "Computer augmented environments: Back to the real world," *Comm. ACM*, vol.36, no.7, July 1993.

[5] R. Azuma: "A survey of augmented reality," *Presence: Teleoperators and Virtual Environments*, vol.6, no.4, pp.355–385, August 1997.

[6] S. Feiner, B. MacIntyre, and D. Seligmann: "Knowledge-based augmented reality," *Comm. ACM*, vol.36, no.7, pp.52–62, July 1993.

[7] G. Fitzmaurice: "Situated information spaces and spatially aware palmtop computers," *Comm. ACM*, vol.36, no.7, pp.38–49, July 1993.

[8] J. Rekimoto and K. Nagao: "The world through the computer: Computer augmented interaction with real world environments," *Proc. User Interface Software and Technology (UIST '95)*, 1995.

[9] P. Milgram and F. Kishino: "A taxonomy of mixed reality visual displays," *IEICE Trans. on Information Systems*, vol.E77-D, no.12, pp.1321–1329, December 1994.

[10] M. Weiser: "The computer for the 21st century," *Scientific American*, vol.265, no.3, pp.94–104, 1991.

[11] H. Ishii and B. Ullmer: "Tangible Bits: Towards seamless interfaces between people, Bits and Atoms," *Proc. CHI'97*, ACM Press, pp.234–241, March 1997.

[12] G. W. Fitzmaurice, H. Ishii and W. Buxton: "Bricks: Laying the foundations for graspable user interfaces," *Proc. CHI'95*, pp.442–449, 1995.

[13] S. Mann: " 'Smart clothing': Wearable multimedia computing and 'Personal imaging' to restore the technological balance between people and their environments," *Proc. ACM MULTIMEDIA '96*, pp.163–174, November 1996.

[14] D. A. Norman: *Psychology of Everyday Things*, Basic Books, 1988.

[15] H. Ishii, M. Kobayashi, and K. Arita: "Iterative design of seamless collaboration media," *Comm. ACM*, vol.37, no.8, pp.83–97, August 1994.

[16] A. Dunne and F. Raby: "Fields and thresholds. presentation at the doors of perception 2," http://www.mediamatic.nl/Doors/Doors2/DunRab/DunRab-Doors2-E.html, November 1994.

[17] M. Weiser and J. S. Brown: "Designing calm technology," http://www.ubiq.com/hypertext/weiser/calmtech/calmtech.htm, December 1995.

[18] W. Buxton: "Integrating the periphery and context: A new model of telematics," *Proc. Graphics Interface '95*, pp.239–246, 1995.

[19] J. Underkoffler and H. Ishii: "Illuminating light: An optical design tool with a Luminous-Tangible interface," *Proc. CHI '98*, pp.542–549, April 1998.

[20] S. Brave, H. Ishii, and A. Dahley: "Tangible interfaces for remote collaboration and communication," *Proc. CSCW '98*, Seattle, Washington USA, ACM Press, pp.169–178, November 1998.

[21] C. Wisneski, H. Ishii, A. Dahley, M. Gorbet, S. Brave, B. Ullmer, and P. Yarin: "Ambient displays: Turning architectural space into an interface between people and digital information," Lecture Notes in Computer Science, Springer Publishing, 1998.

[22] H. Ishii, C. Wisneski, S. Brave, A. Dahley, M. Gorbet, B. Ullmer, and P. Yarin: "ambientROOM: Integrating ambient media with architectural space (video)," *Conf. Summary of CHI'98*, ACM, 1998.

Part IV

Communication and Collaboration

In 1989, VPL, the first company to use the term *virtual reality* (VR), developed RB2 (Reality Built for Two). This VR system allowed two users to experience the same virtual space at the same time through a computer network. It is well known, however, that the data transmission load is relatively low for VR systems that involve only synthetic data, in comparison with telecommunication systems such as a video conferencing. In that sense, it is thus not so difficult to share a VR cyberspace. However, since a mixed reality (MR) systems relies on raw data from the real world, the data transmission load becomes a crucial problem if one wants to operate it over a network. Attaining the goal of networked MR clearly requires technological and/or conceptual breakthroughs to address this problem.

For Susumu Tachi, an expert in telerobotics at the University of Tokyo, the participant at the other end of the communication link is usually a robot or cyborg. In Chapter 14, Tachi extends his earlier work on "tele-existence", or "telexistence", to address "augmented telexistence", a promising new application area for augmented reality. Tachi is currently trying to integrate VR and robotics through some of the national projects in Japan.

Mark Billinghurst and Hirokazu Kato are presently working on incorporating MR into an environment for multi-user collaboration. In Chapter 15, they describe their wearable system for viewing and interacting with shared virtual information while concurrently viewing the real world. This work is quite different from earlier video-conferencing systems based on shared whiteboards, because the wearable MR interface allows participants to point at 3D objects while walking around the same real space. Billinghurst and Kato also discuss the calibration methods required to realize their system.

Throughout this decade, ATR in Japan has been developing a "virtual space teleconferencing system," that supports remote human-to-human communication through virtual space. Jun Ohya and others at ATR have inherited this legacy, and are currently developing on methods to make the virtual scene more realistic and flexible. They are also working on an interface to transmit human actions through the virtual space. Chapter 16 introduces a variety of projects of Ohya's group. Among those, the "virtual metamorphosis system" is noteworthy in that participants can perform as virtual actors with 3D models by real-time projection of their body motion onto the virtual world . It should be noted that this projection is achieved by computer vision technology, not by commercially available motion capture systems.

Chapter 14

Augmented Telexistence

Susumu Tachi
The University of Tokyo, Japan

14.1 Telexistence

It has long been a desire of human beings to project themselves in the remote environment, i.e., to have a sensation of being present or existing in a different place other than the place they really exist at the same time. Another dream has been to amplify human muscle power and sensing capability by using machines while reserving human dexterity with a sensation of direct operation.

In the late 1960s research and development program was planned on a powered exoskeleton that a man would wear like a garment. A concept of Hardiman was proposed by General Electric Co., for example, that a man wearing the Hardiman exoskeleton would be able to command a set of mechanical muscles that multiply his strength by a factor of 25, yet in this union of man and machine he would feel object and forces almost as if he were in direct contact.

However, the project was unsuccessful because of the following reasons: (1) it is potentially quite dangerous to wear a powered exoskeleton when we consider potential malfunction of the machine. (2) Space inside the machine is quite valuable to store computers, controllers, actuators and energy source of the machine, which eliminated the space for a human operator. Thus, the design proved impractical in its original form.

With the advent of science and technology, however, it has become possible to challenge for the realization of the dreams again with a different concept. The concept of projecting ourselves by using robots, computers and cybernetic human interface is called telexistence (tel-existence). This concept expands to include projection in a computer-generated virtual environment. Figure 14.1 illustrates how telexistence has evolved and emerged.

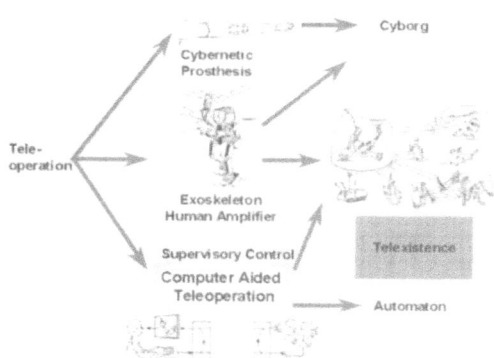

Figure 14.1   Historical diagram on the evolution of telexistence.

The concept of the telexistence was proposed by the author in 1980 and it played the role of the fundamental principle of the eight year Japanese National Large Scale Project of "Advanced Robot Technology in Hazardous Environment," which started in 1983 together with the concept of the Third Generation Robotics. Through this project theoretical consideration has been done and systematic design procedure has been established. Experimental hardware telexistence system haven been made and the feasibility of the concept has been demonstrated [1].

The first prototype telexistence master slave system for remote manipulation experiments was designed and developed, and a preliminary evaluation experiment of telexistence was conducted [2] (see Figure 14.2).

Figure 14.2   Telesar (Telexistence Surrogate Anthropomorphic Robot) developed.

14.2   Augmented Reality in Telexistence

Telexistence can be divided into two categories: telexistence in the real world that actually exists at a distance, and is connected via a robot to the place where the user is located; and telexistence in the virtual world that does not actually exist but is crated by a computer. The former can be called "Transmitted Reality," while the latter is "Synthesized Reality." The synthesized reality can be classified into two, i.e., a virtual environment as a model of the real world and a virtual environment of an imaginary world. Combination of transmitted reality and synthesized reality, which is called mixed reality, is also possible and has a great importance in real applications. This we call augmented telexistence to clarify the importance of harmonic combination of real and virtual worlds in this paper.

Augmented telexistence can be used in several situations. Take for instance, of controlling a slave robot in a poor visibility environment. An experimental augmented telexistence system using mixed reality is constructed as in Figure 14.3. The environment model is also constructed from the design data of the real environment. When augmented reality is used for controlling a slave robot, the modeling errors of the environment model must be calibrated. A model-based calibration system using image measurements is proposed for matching the real environment and a virtual environment. The slave robot has impedance control mechanism for contact tasks and for compensating for the errors that remain even after the calibration. An experimental operation in a poor visibility environment was successfully conducted by using Telesar (Figure 14.2) and the virtual Telesar (Figure 14.4). Figure 14.3 shows the schematic diagram of the augmented telexistence system and Figure 14.4 shows the virtual telexistence anthropomorphic robot used in the experiment [3].

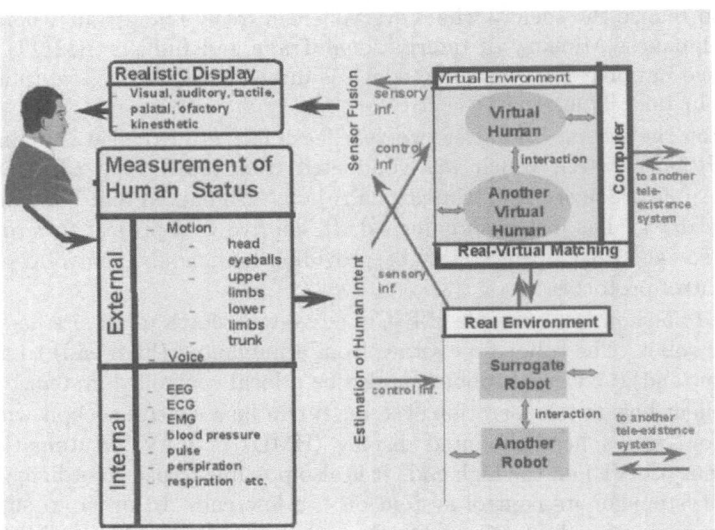

Figure 14.3   Diagram of an augmented telexistence system.

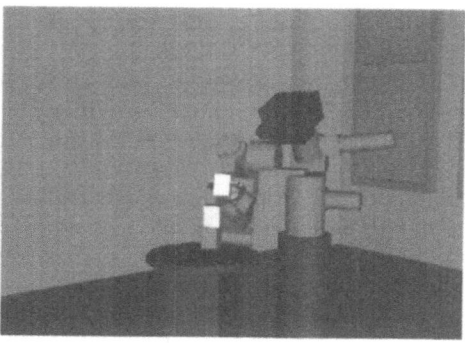

Figure 14.4   Virtual Telesar at work.

Quantitative evaluation of the telexistence manipulation system was conducted through tracking tasks by using a telexistence master slave system designed and developed. Through these experimental studies, it has been demonstrated that a human being can telexist in a remote environment and/or a computer-generated environment by using the dedicated telexistence system [4].

However, it is difficult for everyone to telexist freely through commercial networks like the Internet or the next generation worldwide networks.

14.3   R-Cubed & HRP

In order to realize the society where everyone can freely telexist anywhere through network, Japanese Ministry of International Trade and Industry (MITI) proposed a long-range national R&D scheme, which is dubbed R-Cubed (Real-time Remote Robotics) in 1995 [5] [6].

Based on the scheme and after two-year feasibility study called Human Friendly Network Robot (FNR), which was conducted from April 1996 till March 1998, National Applied Science & Technology Project, "Humanoid and Human Friendly Robotics (HRP)," has just been launched. It is a five-year project toward the realization of so-called R-Cubed Society by providing humanoids, control cockpits and remote control protocols.

Figure 14.5 shows an example of R-Cubed system. Each robot site has its server of its local robot. The robot type varies from a humanoid (high end) to a movable camera (low end). A virtual robot can also be a local controlled system.

Each client has its teleoperation system. It can be a control cockpit with master manipulators and a head mounted display (HMD) or CAVE Automatic Virtual Environment (CAVE) on the high end. It is also possible to use an ordinary personal computer system for its control system on the low end. In order to support the low end users to control remote robots through networks, RCML/RCTP (R-Cubed Manipulation Language / R-Cubed Transfer Protocol) is now under development [7].

To standardize the following control scheme, a language called RCML, which

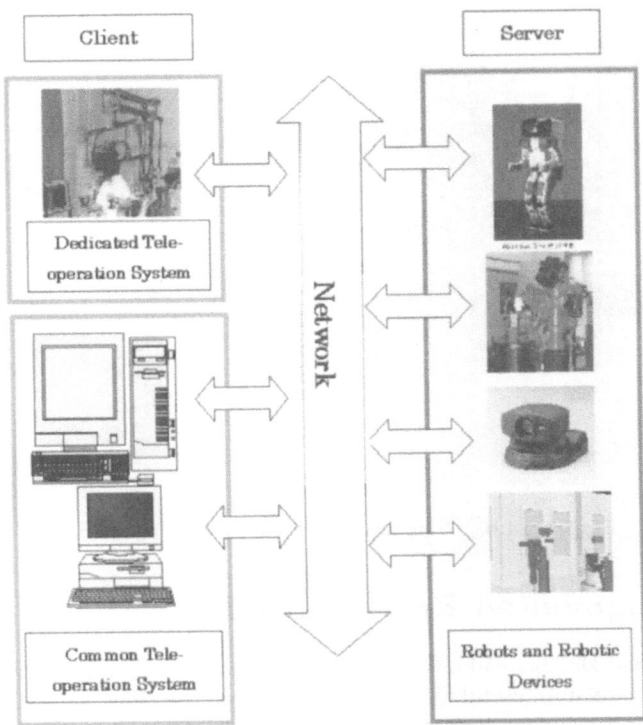

Figure 14.5   R-Cubed concept.

describes a remote robot's features and its working environment, has been proposed. A communication protocol RCTP, which is designed to exchange control commands, status data, and sensory information between the robot and the user, has also been developed.

With a Web browsers a user accesses a Web site describing information of a robot in the form of hypertext and icon graphics using WWW browser. Clicking on an icon downloads the description file, which is written in RCML format, to the user's computer and launches the RCML browser. The RCML browser parses the downloaded file to process the geometry information, including the arrangement of the degrees of freedom of the robot, controllable parameters, available motion ranges, sensor information, and other pertinent information. The browser decides what kind and how many devices are required to control the remote robot. It then generates a graphical user interface (GUI) panel to control the robot, plus a video window that displays the images "seen" by the robot and a monitor window that lets users observe the robot's status from outside the robot. If the user has a device such as 6 degrees-of-freedom (DOF) position/orientation sensor to indicate the robot-manipulator's endpoint, the user can employ that instead of the conventional GUI panel (See Figure 14.6).

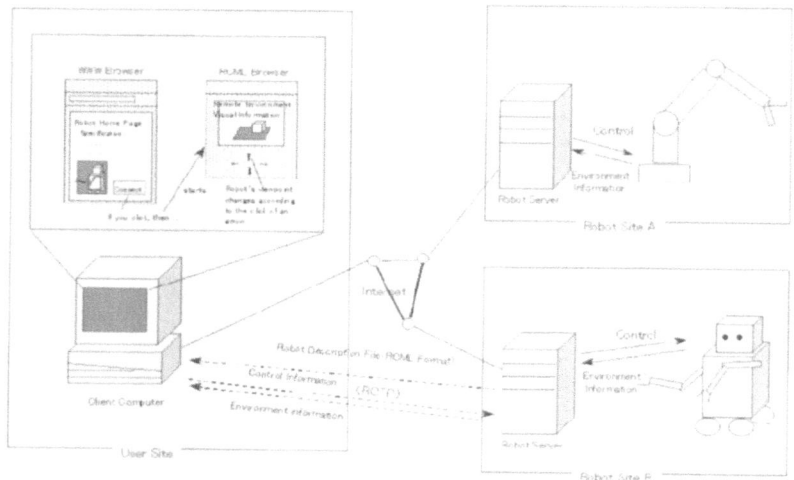

Figure 14.6   Diagram for RCML and RCTP process.

14.4   Augmented Reality in HRP

A Head Mounted Display and a CAVE are two typical virtual reality visual displays. Although they are quite useful displays, it is also true that they have some demerits. The former has a problem of tradeoff of high resolution and wide field of view, and the latter has a problem of a shadow of user's body on a virtual environment and interaction of user's virtual body with their real body.

In our laboratory at the University of Tokyo, a new type of visual display is being developed [7]. It is called X'tal vision, and it uses retro-reflective material as its screen. Figure 14.7 illustrates the principle of the display.

A projector is arranged at the conjugate position of a user's eye, and an image is projected on a screen made of, painted with, or covered with retro-reflective material.

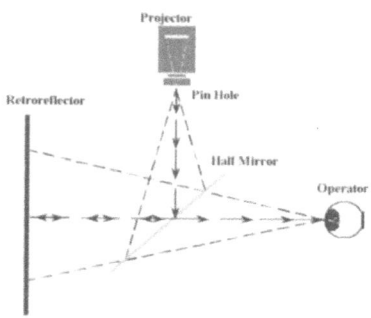

Figure 14.7   Principle of X'tal vision system.

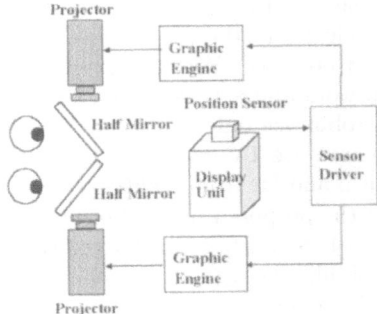

Figure 14.8   Principle of Head Mounted Projector (HMP).

A pinhole is placed in front of the projector to secure adequate depth of focus.

The retro-reflector screen together with the pinhole assures that the user always sees images with accurate occlusion relations. It means that if the user's body has retro-reflector on it, their body becomes a part of the virtual environment and it disappears, and their virtual body replaces it, while if it does not have retro-reflector on it, it will occlude the virtual environment without any troublesome hindrance shadow on the virtual environment.

In the construction of X'tal vision, screen shapes are arbitrary, i.e., any shape is possible. It is due to the characteristics of the retro-reflector and the pinhole in the conjugate optical system. By using the same characteristics of X'tal vision, binocular stereo vision becomes possible using only one screen with an arbitrary shape as in Figure 14.8. This can be mounted on a head of a user, which we call HMP (Head Mounted Projector) System.

Figure 14.9 shows an example of projecting a virtual cylinder on a shape approx-imation device (SAD), which enables the user to touch as if it were a cylinder. The use of SAD as the retro-reflective screen enables us to feel as we see by using a HMP.

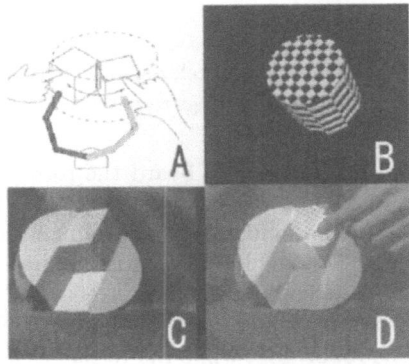

Figure 14.9   Projected image on a spherical retro-reflective screen.

Almost eighteen years have past since our first idea and concept of telexistence, and it is now possible to telexist in the remote environment and/or virtual environment with a sensation of presence. We can have feelings that we are present in several real places and can work and act. However, those people in the place where someone telexists using a robot see only the robot but they can not feel that the person presents. It is useless to use TV display on board the robot to show the face of the user. It is just comical and far from reality.

Figure 14.10 illustrates the proposed method of mutual telexistence using X'tal vision HMP (Head Mounted Projector) in order to solve the above problem, i.e., to make a telexisted robot look like the user of the robot. A human user "A" uses his telexistence robot "A" at the remote site where another human user "B" is present. The user "B" also uses another telexistence robot "B", which exists in the site where the user "A" works. Both robots are painted with retro-reflective material and can act as screens, and they are controlled by their users as conventional telexistence robots.

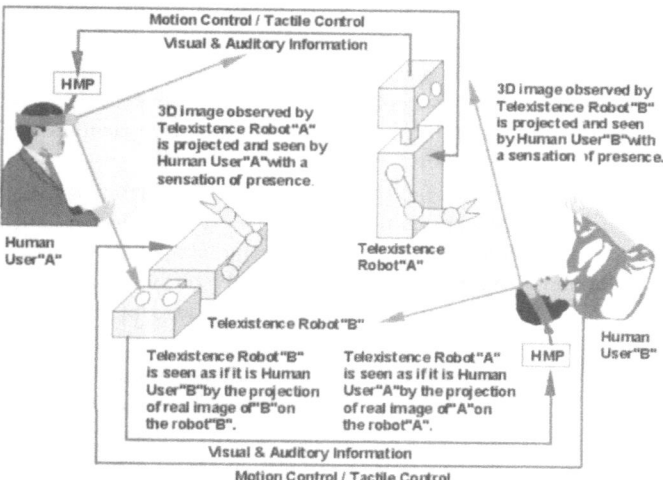

Figure 14.10 Proposed method of mutual telexistence using X'tal vision in HMP (Head Mounted Projector). This enables human beings to telexist with real figures of themselves.

Remote scenery sensed by cameras on board the robots "A" and "B" are sent to HMDs of human users "A" and "B", respectively. 3-D image observed by the telexistence robot "A" is projected and seen by the human user "A" with a sensation of presence, while 3-D image observed by the telexistence robot "B" is projected and seen by the human user "B" with a sensation of presence.

The telexistence robot "B" is seen as if it is the human user "B" by the projection of real image of "B" on the robot "B", while the telexistence robot "A" is seen as if it is the human user "A" by the projection of real image of "A" on the robot "A".

Thus mutual telexistence becomes possible by using X'tal vision method, i.e., not only the user sees other people naturally but also the user of the robot can be observed naturally by other people. We are now in the process of feasibility study of the proposed method using Telesar.

Figure 14.11 shows an example of how a robot can be seen by a human being who wears a HMP. It can be seen as if the robot is a human being telexisting in the robot. Figure 14.11(A) shows an miniature of the HONDA Humanoid Robot, while Figure 14.11(B) illustrates the robot painted with retro-reflective material. Figures 14.11(C) and (D) show how they are seen by a human being wearing a HMP. The telexisted robot just looks like the human operator of the robot, and telexistence can be naturally done.

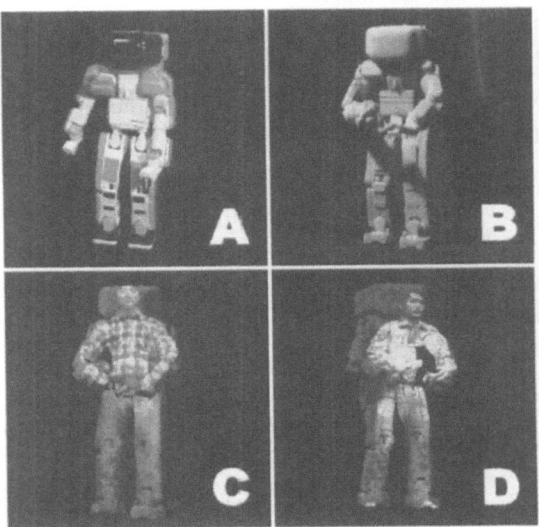

Figure 14.11   (A) Miniature of the HONDA humanoid robot, (B) Painted with retro-reflective material, (C) An example of projecting a human image on it, (D) Another example.

14.5   Conclusion

Virtual reality must have the essence of the reality in its computer-generated environment or a transmitted remote environment so that it is effectively the reality itself. One of the most promising technologies today is the integration of virtual realty and robotics on the network. It is called networked robotics in general and R-Cubed (Real-time Remote Robotics) in particular. R-Cubed is a Japanese national R&D scheme toward the realization of the augmented telexistence through various kinds of networks including the Internet. Mixed reality is expected to play an important role in the realization of augmented telexistence through networks. Japanese Ministry of International Trade and Industry (MITI) launched the "Humanoid and

Human Friendly Robotics (HRP)" 5-year Project in April 1998. This is the first step toward the realization of R-Cube and the results are quite much expected.

References

[1] S. Tachi, K. Tanie, K. Komoriya, and M. Kaneko: "Tele-existence (I): Design and evaluation of a visual display with sensation of presence," *Proc. 5th Symp. on Theory and Practice of Robots and Manipulators (RoManSy '84)*, pp.245–254, Udine, Italy, Kogan Page London, June 1984.

[2] S. Tachi, H. Arai, and T. Maeda: "Tele-existence master slave system for remote manipulation," *Proc. Int'l Workshop on Intelligent Robots and Systems (IROS'90)*, pp.343–348, 1990.

[3] K. Oyama, N. Tsunemoto, S. Tachi, and T. Inoue: "Experimental study on remote manipulation using virtual reality," *Presence*, vol.2, no.2, pp.112–124, 1993.

[4] S. Tachi and Y. Yasuda: "Evaluation experiments of a teleexistence Manipulation system," *Presence*, vol.3, no.1, pp.35–44, 1994.

[5] S. Tachi: "Real-time remote robotics- Toward networked telexistence," *Computer Graphics and Applications*, vol.18, no.6, pp.6–9, 1998.

[6] Y. Yanagida, N. Kawakami, and S. Tachi: "Development of R-Cubed Manipulation Language - Accessing real worlds over the network," *Proc. 7th Int'l Conf. on Artificial Reality and Tele-Existence (ICAT'97)*, pp.159–164, Tokyo, December 1997.

[7] N. Kawakami, M. Inami, T. Maeda, and S. Tachi: "Media X'tal -Projecting virtual environments on ubiquitous object-oriented retro-reflective screens in the real environment-," *SIGGRAPH'98*, Orlando, FL, July 1998.

Chapter 15

Collaborative Mixed Reality

Mark Billinghurst
Hirokazu Kato

University of Washington, U.S.A.

15.1   Introduction

Computers are increasingly used to enhance collaboration between people. As collaborative tools become more common the Human-Computer Interface is giving way to a Human-Human Interface mediated by computers. This emphasis on collaboration adds new technical challenges to the design of Human Computer Interfaces. There are also many social factors that must be addressed before collaborative tools will become common in the workplace.

These problems are compounded for attempts to support three-dimensional Computer Supported Collaborative Work (CSCW). Although the use of spatial cues and three-dimensional object manipulation are common in face to face communication, tools for three-dimensional CSCW are still rare. However new interface metaphors may overcome this limitation. In this paper we describe how Mixed Reality techniques can be used to enhance remote and face to face collaboration, particularly 3D CSCW.

Mixed Reality (MR) environments are defined by Milgram as those in which real world and virtual world objects are presented together on a single display [1]. Single user Mixed Reality interfaces have been developed for computer aided instruction [2], manufacturing [3] and medical visualization [4]. These applications have shown that Mixed Reality interfaces can enable a person to interact with the real world in ways never before possible. For example, the work of Bajura et al. overlays virtual ultrasound images onto a patients body, allowing doctors to have "X-Ray" vision while performing a needle biopsy task [4].

Although Mixed Reality techniques have proven valuable in single user applications, there has been less research on collaborative applications. We believe that Mixed Reality is ideal for collaborative interfaces because it addresses two major issues in CSCW: *seamlessness* and *enhancing reality*. In the next section we describe these issues in depth. We then review approaches for 3D CSCW and describe examples from our work and others of how Mixed Reality can be used to support local and remote collaboration. Finally we conclude with a description of new computer vision techniques for collaborative Mixed Reality interfaces.

15.2   Motivation: Why Collaborative Mixed Reality

15.2.1   Seamless Computer Supported Collaborative Work

When people talk to one another in a face-to-face conversation while collaborating on a real world task there is a dynamic and easy interchange of focus between the shared workspace and the speakers' interpersonal space. The shared workspace is the common task area between collaborators, while the interpersonal space is the common communications space. In face-to-face conversation the shared workspace is often a subset of the interpersonal space, so there is a dynamic and easy change of focus between spaces using a variety of non-verbal cues. For example, if architects are seated around a table with house plans on it, it is easy for them to look at the plans while simultaneously be aware of the conversational cues of the other people.

In most existing CSCW tools this is not the case. Current CSCW interfaces often introduce seams and discontinuities into the collaborative workspace. Ishii defines a seam as a spatial, temporal or functional constraint that forces the user to shift among a variety of spaces or modes of operation [5]. For example, the seam between computer word processing and traditional pen and paper makes it difficult to produce digital copies of handwritten documents without a translation step. Seams can be of two types:

- *Functional Seams:* Discontinuities between different functional workspaces, forcing the user to change modes of operation.

- *Cognitive Seams:* Discontinuities between existing and new work practices, forcing the user to learn new ways of working.

One of the most important functional seams is that between shared and interpersonal workspaces. However, most CSCW systems have an arbitrary seam between the shared workspace and interpersonal space; for example, that between a shared white board and a video window showing a collaborator (Figure 15.1). This prevents users who are looking at the shared white board from maintaining eye contact with their collaborators, an important non-verbal cue for conversation flow [6].

A common cognitive seam is that between computer-based and traditional desktop tools. This seam causes the learning curve experienced by users who move from

Figure 15.1   The functional seam between a shared whiteboard and video window.

physical tools to their digital equivalents, such as the painter moving from oils to digital tools. Grudin [7] points out that CSCW tools are generally rejected when they force users to change the way they work; yet this is exactly what happens when collaborative interfaces make it difficult to use traditional tools in conjunction with the computer-based tools.

Functional and cognitive seams in collaborative interfaces changes the nature of collaboration and produces communication behaviors that are different from face-to-face conversation. For example, even with no video delay, video-mediated conversation doesn't produce the same conversation style as face-to-face interaction [8]. This occurs because video cannot adequately convey the non-verbal signals so vital in face-to-face communication introducing a functional seam between the participants [5]. Thus, Sharing the same physical space positively affects conversation in ways that is difficult to duplicate by remote means.

Ishii et al. showed that it is possible to build seamless collaborative interfaces through their work on the *TeamWorkStation* and *ClearBoard* projects [9] [10]. *Team-WorkStation* reduces the seam between the real world and collaborative workspace by combining video and computer based tools. Video overlay on the collaborative workspace allows collaborators to use pencil and paper together with their computer-based tools in remote collaboration. However there is still a seam between the collaborative workspace and video images of the collaborators. The *ClearBoard* series of interfaces address the seams between the individual and shared workspace. The interfaces are based on the metaphor of users drawing on a glass between them. By using surfaces made of large mirrors, and video projection techniques, users can look directly at their workspace and see their collaborator directly behind it (Figure 15.2). This removes the seam between the interpersonal and shared workspace. Users can effectively and easily change focus, maintain eye contact, and use gaze awareness in collaboration; the result is an increased feeling of intimacy and copresence.

These projects show that in order for an interface to minimize functional and cognitive seams it should have the following characteristics:

- It must support existing tools and work techniques;

- Users must be able to bring real-world objects into the interface;

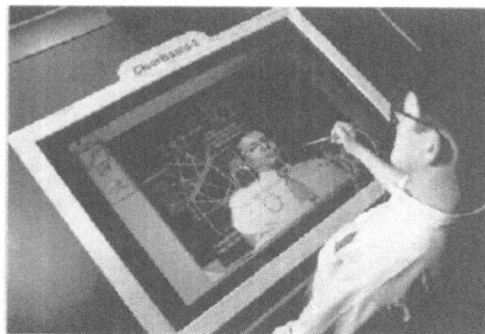

Figure 15.2   The ClearBoard seamless interface (Image courtesy of H. Ishii, MIT media laboratory).

- The shared workspace must be a subset of the interpersonal space;

- There must be audio and visual communication between participants;

- Collaborators must be able to maintain eye contact and gaze awareness.

15.2.2   Enhancing Reality

Removing the seams in a collaborative interface is not enough. As Hollan and Stornetta point out [11], CSCW interfaces may not be used if they provide the same experience as face-to-face communication; they must enable users to go "beyond being there" and enhance the collaborative experience. When this is not the case, users will often stop using the interface or use it differently that what it was intended for. For example, studies of use of the Cruiser video conferencing system between users in the same building found that most people used the system for brief conversations and setting up face-to-face collaboration rather than for replacing face-to-face meetings [12].

The motivation for going "beyond being there" can be found by considering past approaches to CSCW. Traditional CSCW research attempts to use computer and audio-visual equipment to provide a sense of remote presence. Measures of social presence [13] and information richness [14] have been developed to characterize how closely CSCW tools capture the essence of face-to-face communication. The hope is that collaborative interfaces will eventually be indistinguishable from actually being there.

Hollan and Stornetta suggest this is the wrong approach. Considering face-to-face interaction as a specific type of communications medium, it becomes apparent that this approach requires one medium to adapt to another, pitting the strengths of face-to-face collaboration against other interfaces. Mechanisms that are effective in face-to-face interactions may be awkward if they are replicated in an electronic medium, often making users reluctant to use the new medium. In fact, because of the nature of the medium, it may be impossible for mediated collaborations to provide the same experience as face-to-face collaboration [15].

Hollan and Stornetta argue that a better way to develop interfaces for telecommunication is to focus on the *communication* aspect, not the *tele-* part. Rather than using new media to imitate face-to-face collaboration, researchers should be considering what new attributes the media can offer that satisfy the needs of communication so well that people will use it regardless of physical proximity. So one way to develop effective collaborative interfaces is to identify unmet needs in face-to-face conversation and create interface attributes that address these needs.

In this section we have described the need for collaborative interfaces to support seamless interaction and enable collaborators to go beyond being there. Mixed Reality interfaces are ideal for CSCW because they meet both these needs. We demonstrate this in the next section by reviewing other types of interfaces for three-dimensional CSCW, and comparing them to a Mixed Reality approach.

15.3 Collaborative Interfaces for Three Dimensional CSCW

There are several different approaches for facilitating three-dimensional collaborative work. The most obvious is adding collaborative capability to existing screen-based three-dimensional packages. However a two-dimensional interface for three-dimensional collaboration can have severe limitations. For example, Li-Shu [16] developed a workstation based collaborative CAD package but users found it difficult to visualize the different viewpoints of the collaborators making communication difficult. Communication was also restricted to voice and pointing with a graphical icon, further compounding the problem.

Alternative techniques include using large parabolic stereo projection screens or holographic optical systems to project a three-dimensional virtual image into space. CAVE-like systems [3] and the responsive workbench [17] allow a number of users to view stereoscopic 3D images by wearing LCD-shutter glasses. These images are projected on multiple large screen projection walls in the case of the CAVE, or a large opaque tabletop display for the responsive workbench. Unfortunately in both cases the images can be rendered from only a single user's viewpoint, so only one person will see true stereo. This makes it impossible for users to surround the Responsive Workbench table, or to spread themselves throughout the CAVE and see the correct stereoscopic image. The devices are also need bulky hardware such as a projection screen or large beam splitter, are not portable and require expensive optics.

Mechanical devices can also be used to create volumetric displays. These include scanning lasers onto a rotating helix to created a three-dimensional volumetric display [18] or using a rotating phosphor coated plate activated with electron guns [19]. These devices are also not portable, do not permit remote collaboration, and do not allow direct interaction with the images because of the rotating display surface.

Multi-user immersive virtual environments provide an extremely natural medium for three dimensional CSCW; in this setting computers can provide the same type of collaborative information that people have in face-to-face interactions, such as communication by object manipulation and gesture [20]. Work on the DIVE project [21], GreenSpace [22] and other fully immersive multi-participant virtual environments has shown that collaborative work is indeed intuitive in such surroundings. Gesture, voice and graphical information can all be communicated seamlessly between the participants. However most current multi-user VR systems are fully immersive, separating the user from the real world.

15.3.1 Collaborative Mixed Reality

Unlike the other methods for three-dimensional CSCW, Mixed Reality interfaces can overlay graphics and audio onto the real world. This allows the creation of MR interfaces that combine the advantages of both virtual environments and seamless collaboration. Information overlay may be used by remote collaborators to annotate the user's view, or may enhance face-to-face conversation by producing shared interactive virtual models. In this way Mixed Reality techniques can be used to enhance communication regardless of proximity. Thus the use of Mixed Reality facilitates the development of collaborative interfaces that go "beyond being there". Mixed Reality also supports seamless collaboration with the real world, reducing the functional and cognitive seams between participants. These attributes imply that Mixed Reality approaches would be ideal for many CSCW applications.

Despite this, there are few examples of multi-user Mixed Reality systems. Amselen [23] and Rekimoto [24] have explored the use of tracked hand held LCD displays in a multi-user environment. Amselen uses LCD panels as portable windows into a shared multi-user immersive environment, while Rekimoto attaches small cameras to LCD panels to allow virtual objects to be composited on video images of the real world. These displays have the advantage that they are small, light weight, portable and higher resolution than head mounted displays. Unfortunately they do not support a true stereoscopic view, and are not hands free. Users must also hold the LCD panel in front of their face - obscuring their facial expressions.

Klaus et al. [25] also use video compositing techniques to superimpose virtual image over a real world view. Their system is also multi-user, but is monitor and workstation based so users get the impression that the virtual objects are superimposed on a remote real environment rather than their local environment. Their architecture is designed to support distributed users viewing the same real environment remotely rather than local users interacting in the same real environment.

Unlike these systems, our approach is to use see-through head mounted displays with head and body tracking in a collaborative interface. These types of Mixed Reality interfaces allow multiple users in the same location or remote to work in both the real and virtual world simultaneously, facilitating CSCW in a seamless manner. This approach is most closely related to that of Schmalsteig et al. [26]. They use see-through head mounted displays to allow users to collaboratively view 3D models of scientific data superimposed on the real world. They report users finding the interface very intuitive and conducive to real world collaboration because the groupware support can be kept simple and mostly left to social protocols. The AR^2 Hockey work of Ohshima et al. [27] is also very similar. In this case two users wear see-through head mounted displays to play a Mixed Reality version of the classic game of air hockey. Like Schmalsteig, they report that users are able to naturally interact with each other and collaborate on a real world task.

From their work Schmalsteig et al. [26] identify five key advantages of collaborative MR environments:

- *Virtuality:* Objects that don't exist in the real world can be viewed and examined.

- *Augmentation:* Real objects can be augmented by virtual annotations.

- *Cooperation:* Multiple users can see each other and cooperate in a natural way.

- *Independence:* Each user controls his own independent viewpoint.

- *Individuality:* Displayed data can be different for each viewer.

Compared to immersive virtual environments, MR interfaces allow users to refer to notes, diagrams, books and other real objects while viewing virtual images, and they can use familiar real world tools to interact with the images, increasing the intuitiveness of the interface. More importantly, users can see each other's facial expressions, gestures and body language, increasing the communication bandwidth. Finally the entire environment doesn't need to be modeled, considerably reducing the graphics rendering requirements.

15.4 Our Work

In this section we present two of our prototype MR interfaces:

WearCom: An interface for multi-party conferencing that enables a user to see remote collaborators as virtual avatars surrounding them in real space. Spatial cues help overcome some of the limitations of current multiparty conferencing systems.

Collaborative Web Space: An interface which allows people in the same location to view and interact with virtual world wide web pages floating about them in space. Users can collaboratively browse the web while seeing the real world and use natural communication to talk about the pages they're viewing.

MR interfaces can be distinguished by how they present information in the information space. In a MR interface with a head mounted display, information can be presented in a combination of three ways:

- *Head-stabilized* - information is fixed to the user's viewpoint and doesn't change as the user changes viewpoint orientation or position.

- *Body-stabilized* - information is fixed relative to the user's body position and varies as the user changes viewpoint orientation, but not position.

- *World-stabilized* - information is fixed to real world locations and varies as the user changes viewpoint orientation and position.

Each of these methods require increasingly complex head tracking technologies; no head tracking is required for head-stabilized information, viewpoint orientation tracking is needed for body-stabilized information, while position and orientation tracking is required for world-stabilized. The registration requirements also become more difficult; none are required for head-stabilized images, while complex calibration techniques are required for world stabilization.

Body- and world-stabilized information display is attractive for a number of reasons. A body-stabilized information space can overcome the resolution limitations of head mounted displays. For example, Reichlen [28] tracks only head orientation to give a user a "hundred million pixel" hemispherical information surround. World-stabilized information allows annotating the real world with context dependent data and creating information enriched environments [29]. Spatial information displays enable humans to use their innate spatial abilities to retrieve and localise information and to aid performance.

In a Mixed Reality setting, spatial information display can be used to overcome the resolution and field of view limitations of the HMD and provide information overlay on the surrounding environment. This is important because the information presented in a MR interface is often intimately linked to the user's real world location and task. The prototype *WearCom* and the Collaborative Web Space interfaces both use body-stabilized information, however at the end of this chapter we present some more recent applications which use a world-stabilized information display.

15.4.1 Mixed Reality Interfaces for Remote Collaboration

We can use previous research in teleconferencing and CSCW interfaces to suggests attributes of the ideal collaborative MR interface. Research on the roles of audio and visual cues in teleconferencing has produced mixed results. There have been many experiments conducted comparing face-to-face, audio-and-video, and audio-only communication conditions, as summarized by Sellen [30]. While people generally do not prefer the audio-only condition, they are often able to perform tasks as effectively or almost as effectively as in the face-to-face or video conditions, suggesting that speech is the critical medium [31].

Based on these results, it may be thought that audio alone should be suitable for a creating a shared communication space. However attempts to build audio-only communication spaces, such as Thunderwire [32], have found that while audio can be sufficient for a usable communication space, there are several shortcomings. Users are not able to easily tell who else is present and they can't use visual cues to determine other's willingness to interact and discriminate between speakers. These problems suggest that while audio-only may be useful for small group interactions, it is less usable the more people present.

These shortcomings can be overcome through the use of visual and spatial cues. In face-to-face conversation, speech, gesture, body language and other non-verbal cues combine to show attention and interest. Visual cues are present in videoconferencing applications, however the absence of spatial cues in most video conferencing systems means that users often find it difficult to know when people are paying attention to them, to hold side conversations, and to establish eye contact [33].

Virtual reality can provide an alternative medium that allows groups of people to share the same communications space. In collaborative virtual environments (CVEs) spatial visual and audio cues can combine in natural ways to aid communication. Users can freely move through the space setting their own viewpoints and spatial relationships; enabling crowds of people to inhabit the same virtual environment and interact in a way impossible in traditional video or audio conferencing [34]. The well known "cocktail-party" effect shows that people can easily monitor several spatialized audio streams at once, selectively focusing on those of interest [35]. Even a simple virtual avatar representation and spatial audio model enables users to discriminate between multiple speakers [36].

These results suggest that an ideal MR interface for remote collaboration should have high quality audio communication, visual representations of the collaborators and an underlying spatial metaphor.

WearCom

The WearCom project explores how wearable computers can be used to support remote collaboration. Wearable computers are the most recent generation of portable machines. Worn on the body, they provide constant access to computing and communications resources. In general, a wearable computer may be defined as a computer that is subsumed into the personal space of the user, controlled by the wearer and has both operational and interactional constancy, i.e. is always on and always accessible [37]. Wearables are typically composed of a belt or back pack PC, see-though or see-around head mounted display (HMD), wireless communications hardware and an input device such as touchpad or chording keyboard. The use of a see-through display means that wearable computers are an ideal platform for portable MR interfaces.

Many of the target application areas for wearable computers are those where the user could benefit from expert assistance, such as vehicle maintenance or emergency response. Network enabled wearable computers can be used as a communications device to enable remote experts to collaborate with the wearable user. In such situations the presence of remote experts have been found to significantly improve task performance [38]. The question WearCom addresses is how a portable MR interface can be used to support collaboration between multiple remote people. This is becoming increasingly important as telephones incorporate more computing power and portable computers become more like telephones.

In WearCom we use the simplest form of body-stabilized display; one with one degree of orientation to give the user the impression they are surrounded by a virtual cylinder of visual and auditory information (Figure 15.3). We track the user's head orientation so as they as look around they can see different portions of the information space. The cylindrical display is very natural to use since most head and body motion is about the vertical axis, making it very difficult for the user to become disoriented. With this display configuration a Mixed Reality conferencing space could be created that allows remote collaborators to appear as virtual avatars distributed about the user (Figure 15.4).

A wearable computer can be used to provide spatialized 3D graphics and audio cues to aid communication. The result is an augmented reality communication space with audio enabled avatars of the remote collaborators surrounding the user. The user can use natural head motions to attend to the remote collaborators, can communicate freely while being aware of other side conversations and can move through the communication space. In this way the conferencing space could support many simultaneous users. The user could also see the real world, enabling remote

Figure 15.3   Our body-stabilized display.

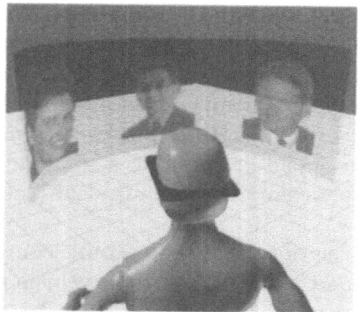

Figure 15.4   A MR spatial conferencing space.

collaborators to help them with real world tasks.

The WearCom prototype implements the wearable communications described above. Our research is initially focused on collaboration between a single wearable computer user and several desktop PC users. The wearable computer we use is a custom built 586 PC 104 based computer with 20mb of RAM running Windows 95 (Figure 15.5). A hand held Logitech wireless radio trackball with three buttons is used as the primary input device. The display is a pair of Virtual i-O iglasses! converted into a monoscopic display by the removal of the left eyepiece. This head-mounted display can either be used in see-through or occluded mode, has a resolution of 262 by 230 pixels and a 26-degree field of view. They also have a sourceless two-axis inclinometer and a magnetometer used as a three degree of freedom orientation tracker. A BreezeCom wireless LAN is used to give 2mb/s Internet within 500 feet of a base station. The wearable also has a soundBlaster compatible sound board with headmounted microphone. The desktop PCs are standard Pentium class machines with Internet connectivity and sound capability.

The wearable computer has no graphics acceleration hardware and limited wireless bandwidth so the interface is deliberately kept simple. The conferencing space runs as a full screen application that is initially blank until remote users connect.

When users join the conferencing space they are represented by blocks with static pictures of themselves on them (Figure 15.6). Although the resolution of the images is crude it is sufficient to identify who the speakers are and their spatial relationships. A radar display shows the location of the other users in the conferencing space, enabling users to find each other easily.

Figure 15.5   Wearable hardware.

Figure 15.6   The user's view of the conferencing space (see color pages).

Each user determines the position and orientation of their own avatar in space, which changes as they move or look about environment. The wearable user has their head tracked so they can simply turn to face the speakers they are interested in. Users can also navigate through the space; by rolling the trackball forwards or backwards their viewpoint is moved forwards or backwards along the direction they are looking. Since the virtual images are superimposed on the real world, when the user rolls the trackball it appears to they are moving the virtual space around them, rather than navigating through the space. Users are constrained to change viewpoint on the horizontal plane, just as in face-to-face conversations. The two different navigation methods (trackball motion, head tracking), match the different types of motion used in face to face communication; walking to join a join a group for conversation, and body orientation changes within a conversational group. The interface was developed using Microsoft's DirectX suite of libraries.

The wearable interface also supports 3D spatialized Internet telephony. When users connect to the conferencing space their audio is broadcast to all the other users in the space. This is spatialized according to the distance and direction between speaker and listener. As users face or move closer to different speakers the speaker volume changes due to the sound spatialisation. Since the speakers are constrained to remain in the same plane as the listener the audio spatialisation is considerably simplified. The conferencing space uses custom developed telephony libraries and the Microsoft DirectSound libraries.

Preliminary informal trials with WearCom have found that users are able to easily discriminate between three simultaneous speakers when their audio streams are spatialized, but not when non-spatialized audio is used. Participants preferred seeing a visual representation of their collaborators over just hearing their speech because it enabled them to see who is connected and the spatial relationship of the speakers. This allowed them to use some of the non-verbal cues commonly used in face-to-face communication such as gaze modulation and body motion. Lastly, users found that they could continue doing real world tasks while talking to collaborators in the conferencing space and it was possible to move the conferencing space with the trackball so that collaborators weren't blocking critical portions of the users field of view.

15.4.2   Co-Located Collaboration

Mixed Reality interfaces can enable co-located users to view and interact with shared virtual information spaces while viewing the real world at the same time. This preserves the rich communications bandwidth that humans enjoy in face-to-face meetings, while adding virtual images normally impossible to see. In this section we present a collaborative web browser that enables users to load and place virtual web pages around themselves in the real world and to jointly discuss and interact with them; users can see both the virtual web pages and each other, so communication is natural and intuitive.

We have developed a three-dimensional Web browser that enables multiple co-located users to collaboratively browse the World Wide Web. Users see each other and virtual web pages floating in space around them (Figure 15.7). The effect is a body-centered information space that the user can easily and intuitively interact with (Figure 15.8). The Shared Space browser supports multiple users who can communicate about the web pages shown, using natural voice and gesture.

Users wear see-through Virtual i-O stereoscopic headmounted displays, and head orientation is tracked using a Polhemus Fastrak electro-magnetic sensor. The interface is designed to be completely hands-free; pages are selected by looking at them and, once selected, can be attached to the user's viewpoint, zoomed in or out, iconified or expanded, or additional links loaded with voice commands. Speaker-independent continuous speech recognition software (Texas Instrument's DAGGER system) allows users to load and interact with web pages using vocal commands. To support this, an HMTL parser parses the web pages to extract their HTML links and assign numbers to them. In this way users can load new links with numerical commands such as "Load link one". Each time a new page is loaded a new browser object is created and a symbolic graphical link to its parent page is displayed to

Figure 15.7   The shared space interface.

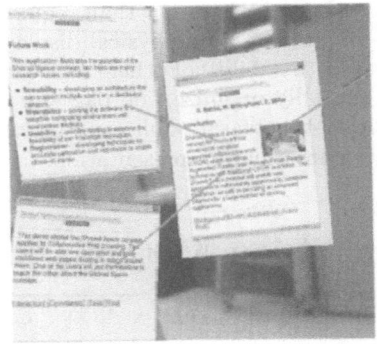

Figure 15.8   The participant's view.

facilitate the visualisation of the web pages. The voice recognition software recognises 46 command phrases with greater than 90% accuracy. A switched microphone is used so participants can carry on normal conversation when not entering voice commands.

Two important aspects of the interface are gaze awareness and information privacy. Users need to know which page they are currently looking at as well as the pages their collaborators are looking at. This is especially difficult when there are multiple web pages close to each other. The Virtual i-O head mounted display has only a 26 degree field of view so it is tempting to overlap pages so that several can be seen at once. To address this problem, each web page highlights when a user looks at it. Each page also has gaze icons attached to it for each user that highlights to show which users are looking at the page. In this way users can tell where their collaborators are looking. When each web page is loaded it is initially visible only to the user that loaded it. Users can change page visibility from private to public with vocal commands; users can only see the public web pages and their own private objects.

The collaborative web interface uses a body-stabilized information space similar to that in the WearCom interface. However, in this case all three degrees of head orientation are used, providing a virtual sphere of information. Even though the head-mounted display has only a limited field of view, the ability to track head orientation and place objects at fixed locations relative to the body effectively creates a 360-degree circumambient display. Since the displays are wearable users can collaborate in any location, and because interface objects are not attached to real world locations the registration requirements are not as stringent.

The Mixed Reality interface facilitates a high bandwidth of communication between users as well as natural 3D manipulation of the virtual images. The key characteristic of this interface is the ability to see the real world and collaborators at the same time as the virtual web pages floating in space. This means that users can use natural speech and gesture to communicate with each other about the virtual information space. In informal trials, users found the interface intuitive and communication with the other participants seamless and natural. Collaboration could be left to normal social protocols rather than requiring mechanisms explicitly encoded in the interface. Unlike sharing a physical display, users with the wearable information space can restrict the ability of others to see information in their space. They were able to easily spatially organize web pages in a manner that facilitated rapid recall, and the distinction between public and private information was found to be useful for collaborative information presentation.

15.5   Computer Vision Methods for Collaborative Mixed Reality

In the previous sections we have described collaborative mixed reality interfaces which use magnetic or inertial trackers to create collaborative spaces. Wired position sensors such as these have been used effectively for immersive virtual reality systems. However, in many MR applications the need for high resolution, large scale position and orientation tracking make the use of these sensors impractical. Computer vision methods do not require any wired sensors and so play an important role in MR applications, particularly interfaces which use world-stabilized information display. In this section we describe our computer vision techniques for accurate world-stabilized image registration, and some collaborative applications that use these technique.

Computer vision techniques have previously been applied with great success in MR interfaces. A common approach is to use physical markers to aid with the registration. For example, Rekimoto proposes a registration method using square markers each with a unique 2D-matrix code [29]. Other methods using fiducial markers include those of State [39] and Kutulakos [40]. All these cases are video based MR systems in which virtual objects are superimposed on video images captured by the camera. This considerably simplifies the camera calibration requirements. In video-based Mixed Reality, only the relationship between camera coordinates and 3D world coordinates is needed. However, in optical see-through MR systems, stereoscopic views of virtual objects appear in the physical space, so the eye and HMD screen coordinate systems must also be known.

In a MR system using optical see-through HMD, stereo images have to be provided to both eyes. In order to do that, accurate measurement of position of the camera with respect to each eye is required. This is because the scene captured by a head-mounted camera is different from the view of each eye. This type of camera calibration is difficult and some systems such as AR^2Hockey [27] ignore the parallax between the camera and eyes entirely. However, one of the application areas we are interested in is a collaborative MR system for wearable computers. In this case, we can expect that the 3D space in which virtual objects appear will be close to the user. The parallax between the eyes and the camera therefore can not be ignored. In order to display a virtual object in a position close to the user, an accurate calibration is required.

15.5.1 Calibration Method

We calibrate the head-mounted camera using the calibration frame shown in Figure 15.9. This is a simple cardboard frame with a ruled grid of lines on it that is attached to the front of the HMD as shown. By attaching the calibration frame to the head mounted display, the head position doesn't change relative to the grid of lines, so we can find the relationship between the camera coordinate system and the screen coordinate system, by way of the calibration frame coordinate system. There are two critical transformations we need to find; that between the screen coordinate frame and calibration frame coordinate frame, and that between the calibration frame coordinate frame and camera coordinate frame. Figure 15.10 shows these transformations and the various coordinates systems used.

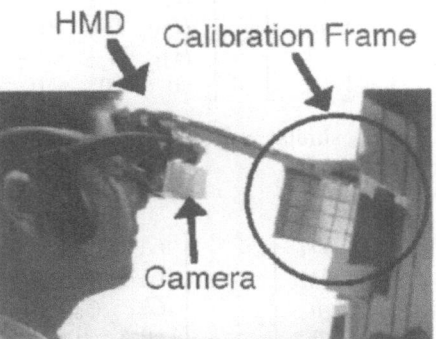

Figure 15.9   The calibration frame used in our calibration method.

We assume all rays from a physical object reach the focal point of the eye through the HMD screen. Then the relationship between the HMD screen and the focal point of the eye makes a perspective camera model shown by the equation below. T_{sh} is the calibration frame to screen coordinate frame transformation matrix, while (x_s, y_s) is the screen coordinates of the point being transformed and (X_h, Y_h, Z_h) the location of the same point expressed in the calibration frame coordinate frame. Using T_{sh} as the transformation between the virtual 3D space and screen

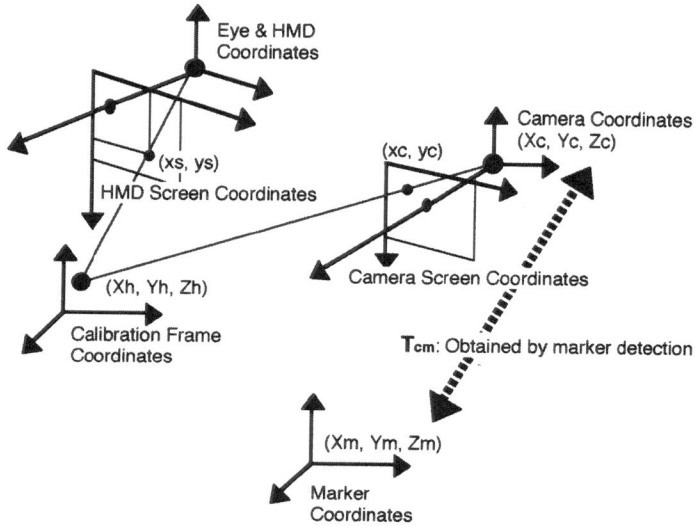

Figure 15.10   The coordinate systems used in our calibration procedure.

coordinates enables the virtual 3D space to coincide with the physical 3D space on the HMD Screen.

$$
\begin{bmatrix} hx_s \\ hy_s \\ h \\ 1 \end{bmatrix} = \begin{bmatrix} H_{11} & H_{12} & H_{13} & H_{14} \\ H_{21} & H_{22} & H_{23} & H_{24} \\ H_{31} & H_{32} & H_{33} & H_{34} \\ 0 & 0 & 0 & 1 \end{bmatrix} \begin{bmatrix} X_h \\ Y_h \\ Z_h \\ 1 \end{bmatrix} = \mathbf{T_{sh}} \begin{bmatrix} X_h \\ Y_h \\ Z_h \\ 1 \end{bmatrix}
$$

In a similar way, the relationship between the camera and calibration frame coordinate is:

$$
\begin{bmatrix} hx_c \\ hy_c \\ h \\ 1 \end{bmatrix} = \begin{bmatrix} C_{11} & C_{12} & C_{13} & 0 \\ C_{21} & C_{22} & C_{23} & 0 \\ C_{31} & C_{32} & C_{33} & 0 \\ 0 & 0 & 0 & 1 \end{bmatrix} \begin{bmatrix} X_c \\ Y_c \\ Z_c \\ 1 \end{bmatrix}
$$

$$
= \mathbf{C} \begin{bmatrix} R_{11} & R_{12} & R_{13} & T_1 \\ R_{21} & R_{22} & R_{23} & T_2 \\ R_{31} & R_{32} & R_{33} & T_3 \\ 0 & 0 & 0 & 1 \end{bmatrix} \begin{bmatrix} X_h \\ Y_h \\ Z_h \\ 1 \end{bmatrix} = \mathbf{CT_{ch}} \begin{bmatrix} X_h \\ Y_h \\ Z_h \\ 1 \end{bmatrix}
$$

where C is a matrix representation of the inner camera parameters, and $\mathbf{T_{ch}}$ is the transformation matrix from calibration frame coordinates ($\mathbf{X_h}$, $\mathbf{Y_h}$, $\mathbf{Z_h}$) to Camera coordinates ($\mathbf{X_c}$, $\mathbf{Y_c}$, $\mathbf{Z_c}$).

The grid of the calibration frame is used to find the exact values of the matrices $\mathbf{T_{sh}}$, C and $\mathbf{T_{ch}}$. When a user wears this frame they see the view shown in Figure

15.11. To calibrate the HMD, while wearing the calibration frame the user fits virtual lines drawn on the HMD screen to the corresponding line segments on the physical calibration tool. The virtual lines can be moved and rotated by keyboard operation. The positions of all of the intersections of the real line segments are known in the calibration frame coordinate system. This user operation finds the corresponding positions in the HMD screen coordinate system. By using this data, the transformation matrix $\mathbf{T_{sh}}$ can be estimated. The user carries out this process for each of eyes, generating two matrices. The resultant $\mathbf{T_{sh}}$ matrices are used as the transformation between the virtual 3D coordinate system and the HMD screen coordinate system. To find the camera calibration matrices, $\mathbf{C}$ and $\mathbf{T_{ch}}$, the same process is used, however in this case video from the camera is displayed on a computer monitor and the user aligns the virtual lines on-screen.

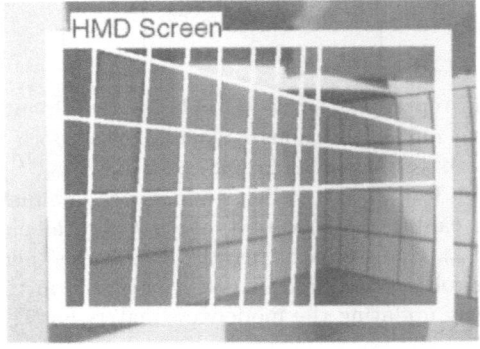

Figure 15.11   Simulated view of the image the user sees in the calibration frame.

15.5.2   Registration Method

Once the camera and eye transformation matrices have been found they can be used to register virtual images with the real world. In order to do this we use square fiducial markers attached to real world objects. These markers are detected using basic image processing techniques and virtual objects are drawn relative to the marker coordinates system.

Displaying a virtual object on markers in the physical 3D space requires that $\mathbf{T_{hm}}$, the relationship between the marker coordinate system and calibration frame coordinate system is known. If this is known then transformation between the marker coordinate system and screen coordinates can be found from $\mathbf{T_{sm}} = \mathbf{T_{sh}} \cdot \mathbf{T_{hm}}$. In Figure 15.10 the matrix $\mathbf{T_{cm}}$ represents the transformation between the marker coordinate system and the camera coordinate system. This transformation can be estimated by image analysis, using the same computer vision methods that Rekimoto [29] and others use to identify fiducial location in MR systems. The transformation matrix $\mathbf{T_{ch}}$, representing the relationship between the camera and the calibration frame coordinate systems, is found from the camera calibration, so the transformation matrix from marker coordinate system to the calibration frame coordinate system can be calculated easily from the following:

$$\mathbf{T_{hm}} = \mathbf{T_{ch}}^{-1}.\mathbf{T_{cm}}$$

The perspective transformation matrix from calibration frame coordinate system to HMD screen coordinates system, $\mathbf{T_{sh}}$, is known from the system calibration. So the screen coordinates $(\mathbf{x_s}, \mathbf{y_s})$ of the virtual image that is to appear at the physical coordinates of $(\mathbf{X_m}, \mathbf{Y_m}, \mathbf{Z_m})$ in the marker coordinate frame can be found from the following equation:

$$\begin{bmatrix} hx_s \\ hy_s \\ h \\ 1 \end{bmatrix} = \mathbf{T_{sh}}.\mathbf{T_{ch}}^{-1}.\mathbf{T_{cm}} \begin{bmatrix} X_m \\ Y_m \\ Z_m \\ 1 \end{bmatrix}$$

The above matrix representation is suitable for the OpenGL graphics libraries we use for developing our applications. OpenGL uses both a modelview matrix and a projection matrix to render 3D graphics, so $\mathbf{T_{sh}}.\mathbf{T_{ch}}^{-1}$ can be regarded as a projection matrix and $\mathbf{T_{cm}}$ as the modelview matrix. The matrix $\mathbf{T_{sh}}.\mathbf{T_{ch}}^{-1}$ does not change after the calibration so it only needs to be calculated once. However, the transformation $\mathbf{T_{cm}}$ changes each time the user moves their head (and camera) position. So a virtual object can be accurately registered in the marker coordinate system by continuously updating the modelview matrix for $\mathbf{T_{cm}}$.

15.5.3   Examples of Vision Based Collaborative Interfaces

The calibration and registration methods we describe above allow us to developed collaborative Mixed Reality interfaces that use world-stabilized images. In this section we briefly describe two applications that are currently under development. Many other world stabilized MR collaborative applications are possible.

Our first application is a prototype of a video conferencing system for wearable computers. This extends the WearCom work by allowing the virtual images of remote participants to be attached to real world locations. In this case unique fiducial markers are used to represent each of the people a user may wish to call. When the user places a marker in view of the head mounted camera the system initiates an audio and visual connection to the remote user and attaches a virtual representation of them to the marker. Figure 15.12 shows the view through the user's head mounted display when they are conferencing with two other people. The user can arrange the layout of participants as they want and can continue to do a real world task without the interference of video display.

The interesting aspect of this approach is that it is the opposite of traditional video conferencing. Our goal is to put a virtual representation of the remote user into the local user's real world location, enabling them to have a videoconference regardless of where they are. In contrast, current video conferencing requires the user to move to a desktop computer or videoconferencing suite, often removing them from their workplace. As with WearCom, the interface also restores the spatial cues lost in traditional videoconferencing. However, having remote users represented as objects in the physical environment means that their virtual avatars can also gesture and interact visually with other objects in the user's space. This is potentially a powerful new interaction technique for collaborative MR interfaces.

A second area we are looking at is how our vision technique can be used to aid co-located MR collaboration on a wearable computer platform. For wearable computers, there are many problems remaining with the input interface. We have made a prototype of input interface using a paper tablet and a pen that enables a user to draw virtual lines on the tablet surface. Figure 15.13 shows a tablet with 6 fiducial markers and virtual annotations aligned with the markers. The user can take this interface anywhere, and more importantly other people with them that have wearable computers can also see the virtual annotations, so this can be used as a powerful tool for co-located collaboration. Since some markers may be occluded by a hand or a pen and missed by the camera, the 3D position of the tablet is estimated from whatever markers are visible. This ensures robust tracking of the tablet interface.

Figure 15.12   MR world stabilized video conferencing.

Figure 15.13   MR tablet writing interface.

15.6   Conclusions

Mixed Reality interfaces enable the development of innovative CSCW applications that are seamless and enhance face-to-face and remote collaboration. They are seamless because they allow the users to use traditional tools and workplace practices while overlaying virtual images onto the real world. Thus MR interfaces enhance the real world rather than replacing it entirely as do immersive VR environments. These MR enhancements can be used to support face-to-face and remote collaboration in ways otherwise impossible, enabling users to go "beyond being there".

In this chapter we have shown several examples of collaborative body and world-stabilized information spaces. User experiences with these interfaces have shown that they facilitate collaboration in a natural manner, enabling people to use normal gestures and non-verbal behavior in face-to-face collaboration, and to have access to their traditional tools and workplace in both face-to-face and remote collaboration.

Although these results are promising, they just scratch the surface of possible applications and there are several important research directions that need to be addressed in the future. First, rigorous user studies must be conducted to identify the unique characteristics of collaborative MR interfaces. The few papers that have been published in the field have shown some of the possible benefits of using these types of collaborative tools, but these benefits will only be realized when applications are developed based on solid interface design principles and user studies. These studies should compare user performance in MR interfaces with the equivalent immersive VR interfaces to examine how the seamlessness between the real and virtual world affects performance. They should also measure the effects of registration errors and system latency on collaborative performance, and provide guidelines for the acceptable latency for a range of tasks. Many of these types of studies have been conducted for single user MR interfaces and need to be replicated for collaborative applications.

A second area for future work is exploration into the types of unique interface metaphors that collaborative MR interfaces make possible. MR interfaces allow users to use real world objects to interact with virtual images, and enhance existing real world objects. However it remains an unanswered question as how to best use these capabilities. In order to explore this there needs to be better supporting technologies developed such as improved techniques for image registration and object and body tracking.

Finally, promising application areas need to be identified. One possibility is applications in the field of wearable computing. This is an area where MR head-mounted displays are commonly used and current user interface needs are not being well met. The traditional WIMP interface is not appropriate for the wearable computing platform, both because of the inherent assumptions that it makes about the input and output devices and the nature of the tasks that wearable computers are being used for. However the ability of MR interfaces to enhance rather than replace the users real world task suggest that MR techniques could be used to develop a more appropriate interface metaphor. Wearable MR interfaces for collaboration is a largely unexplored field that promises to change the way people collaborate with wearable computers.

References

[1] P. Milgram and F. Kishino: "A taxonomy of mixed reality visual displays," *IEICE Trans. on Information and Systems, Special issue on Networked Reality*, vol.E77-D, no.12, pp.1321–1329, Dec. 1994.

[2] S. Feiner, B. MacIntyre, and D. Seligmann: "Knowledge-based augmented reality," *Comm. ACM*, vol.36, no.7, pp.53–62, 1993.

[3] C. Cruz-Neira, D. J. Sandin, T. A. Defanti, R. V. Kentyon, and J. C. Hart: "The CAVE: Audio visual experience automatic virtual environment," *Comm. ACM*, vol.35, no.6, pp.65, 1992.

[4] M. Bajura, H. Fuchs, and R. Ohbuchi: "Merging virtual objects with the real world: Seeing ultrasound imagery within the patient," *Proc. SIGGRAPH 92*, New York: ACM Press, pp.203–210, 1992.

[5] H. Ishii, M. Kobayashi, and K. Arita: "Iterative design of seamless collaboration media," *Comm. ACM*, vol.37, no.8, pp.83–97, Aug. 1994.

[6] C. L. Kleinke: "Gaze and eye contact: A research review," *Psychological Bulletin*, vol.100, pp.78–100, 1986.

[7] J. Grudin: "Why CSCW applications fail: Problems in the design and evaluation of organizational interfaces," *Proc. CSCW'88*, Portland, Oregon, New York: ACM Press, pp.85–93, 1988.

[8] C. Heath and P. Luff: "Disembodied conduct: Communication through video in a multimedia environment," *Proc. CHI'91*, New York, NY: ACM Press, pp.99–103, 1991.

[9] H. Ishii and N. Miyake: "Toward an open workspace: Computer and video fusion approach of teamworksation," *Comm. ACM*, vol.34, no.12, pp.37–50, Dec. 1991.

[10] H. Ishii, M. Kobayashi, and J. Grudin: "Integration of inter-personal space and shared workspace: Clearboard design and experiments," *Proc. CSCW'92*, pp.33–42, 1992.

[11] J. Hollan and S. Stornetta: "Beyond being there," *Proc. CHI'92*, New York: ACM Press, pp.119–125, 1992.

[12] R. S. Fish, R. E. Kraut, R. W. Root, and R. Rice: "Evaluating video as a technology for informal communication," Bellcore Technical Memorandum, TM-ARH017505, 1991.

[13] J. Short, E. Williams, and B. Christie: *The social psychology of telecommunications*, London, Wiley, 1976.

[14] R. L. Draft and R. H. Lengel: "Organizational information requirements, media richness, and structural design," *Management Science*, vol.32, pp.554–571, 1991.

[15] W. Gaver: "The affordances of media spaces for collaboration," *Proc. CSCW'92*, Toronto, Canada, New York: ACM Press, pp.17–24, 1992.

[16] Li-Shu and W. Flowers: "Teledesign: Groupware user experiments in three-dimensional computer aided design," *Collaborative Computing*, vol.1, no.1, pp.1–14, 1994.

[17] W. Kruger, C. Bohn, B. Frohlich, H. Schuth, W. Strauss, and G. Wesche: "The responsive workbench: A virtual work environment," *Computer*, vol.28, no.7, pp.42–48, 1995.

[18] P. Soltan, J. Trias, W. Dahlke, M. Lasher, and M. McDonald: "Laser-based 3D volumetric display system: Second generation," in *Interactive Technology and the New Paradigm for Technology*, IOP Press, pp.349–358, 1995.

[19] B. G. Blundell and A. J. Schwarz: "A graphics hierarchy for the visualization of 3D images by means of a volumetric display system," *Proc. the IEEE Region 10's Ninth Annual Int'l Conf.*, Singapore, IEEE, New York, NY, vol.1, pp.1–5, 22-26 Aug. 1994.

[20] A. Wexelblat: "The reality of cooperation: Virtual reality and CSCW," in (A. Wexelblat, ed.) *Virtual Reality: Applications and Explorations*, Boston, Academic Publishers, 1993.

[21] C. Carlson and O. Hagsand: "DIVE - A platform for multi-user virtual environments," *Computers and Graphics*, vol.17, no.6, pp.663–669, Nov./Dec. 1993.

[22] J. Mandeville, J. Davidson, D. Campbell, A. Dahl, P. Schwartz, and T. Furness: "A shared virtual environment for architectural design review," *Proc. CVE'96 Workshop*, Nottingham, Great Britain, 19-20 Sep. 1996.

[23] D. Amselen: "A window on shared virtual environments," *Presence*, vol.4, no.2, pp.130–145, 1995.

[24] J. Rekimoto: "Transvision: A hand-held augmented reality system for collaborative design," *Proc. Virtual Systems and Multimedia (VSMM'96)* , Gifu, Japan, 18-20 Sep. 1996.

[25] A. Klaus, A. Kramer, D. Breen, P. Chevalier, C. Crampton, E. Rose, M. Tuceryan, R. Whitaker, and D. Greer: "Distributed augmented reality for collaborative design applications," *Proc. Eurographics'95*, pp.C-03–C-14, Sep. 1995.

[26] D. Schmalsteig, A. Fuhrmann, Z. Szalavari, and M. Gervautz: "Studierstube - An environment for collaboration in augmented reality," *Proc. CVE'96 Workshop*, Nottingham, Great Britain, 19-20 Sep. 1996.

[27] T. Ohshima, K. Sato, H. Yamamoto, and H. Tamura: "AR² Hockey: A case study of collaborative augmented reality," *Proc. VRAIS'98*, IEEE Press: Los Alamitos, pp.268–295, 1998.

[28] B. Reichlen: "SparcChair: One hundred million pixel display," *Proc. IEEE VRAIS'93*, Seattle WA, IEEE Press: Los Alamitos, pp.300–307, 18-22 Sep. 1993.

[29] J. Rekimoto: "Matrix: A realtime object identification and registration method for augmented reality," *Proc. Asia Pacific Computer Human Interaction 1998 (APCHI'98)*, Japan, 1998.

[30] A. Sellen: "Remote conversations: The effects of mediating talk with technology," *Human Computer Interaction*, vol.10, no.4, pp.401–444, 1995.

[31] S. Whittaker and B. O'Connaill: "The role of vision in face-to-face and mediated communication," in (K. Finn, A. Sellen, and S. Wilbur, eds.) *Video-Mediated Communication*, Lawerance Erlbaum, New Jersey, pp.23–49, 1997.

[32] D. Hindus, M. Ackerman, S. Mainwaring, and B. Starr: "Thunderwire: A field study of an audio-only media space," *Proc. CSCW'96*, New York, NY: ACM Press, 16-20 Nov. 1996.

[33] A. Sellen: "Speech patterns in video-mediated conversations," *Proc. CHI'92*, New York, ACM Press, pp.49–59, 3-7 May 1992.

[34] S. Benford, C. Greenhalgh, and D. Lloyd: "Crowded collaborative virtual environments," *Proc. CHI'97*, Atlanta, Georgia, New York: ACM Press, pp.59–66, Mar. 1997.

[35] C. Schmandt and A. Mullins: "AudioStreamer: Exploiting simultaneity for listening," *Proc. CHI 95 Conf. Companion*, Denver Colorado, ACM: New York, pp.218–219, 7-11 May 1995.

[36] H. Nakanishi, C. Yoshida, T. Nishimura, and T. Ishida: "FreeWalk: Supporting casual meetings in a network," *Proc. CSCW'96*, New York, NY: ACM Press, pp.308–314, 16-20 Nov. 1996.

[37] S. Mann: "Smart clothing: The wearable computer and WearCam," *Personal Technologies*, vol.1, no.1, Springer-Verlag, Mar. 1997.

[38] R. Kraut, M. Miller, and J. Siegel: "Collaboration in performance of physical tasks: Effects on outcomes and communication," *Proc. CSCW'96*, New York, NY: ACM Press, 16-20 Nov. 1996.

[39] A. State, G. Hirota, D. Chen, W. Garrett, and M. Livingston: "Superior augmented-reality registration by integrating landmark tracking and magnetic tracking," *Proc. SIGGRAPH 96*, pp.429–438.

[40] K. Kutulakos and J. Vallino: "Calibration-free augmented reality," *IEEE Trans. on Visualization and Computer Graphics*, vol.4, no.1, pp.1–20, 1998.

[41] T. P. Caudell and D. W. Mizell: "Augmented reality: An application of heads-up display technology to manual manufacturing processes," *Proc. the Twenty-Fifth Hawaii Int'l Conf. on Systems Science*, Kauai, Hawaii, vol.2, pp.659–669, 7-10 Jan. 1992.

[42] C. O'Malley, S. Langton, A. Anderson, G. Doherty-Sneddon, and V. Bruce: "Comparison of face-to-face and video-mediated interaction," *Interacting with Computers*, vol.8 no.2, pp.177–192, 1996.

Chapter 16

Virtual Reality Technologies for Multimedia Communications

Jun Ohya
Tsutomu Miyasato
Ryohei Nakatsu

ATR Media Integration & Communications
Research Laboratories, Japan

16.1 Introduction

ATR Media Integration & Communications Research Laboratories are aiming at realizing virtual communication environments in which remotely located people can communicate with each other using multiple virtual reality technologies so that the people can feel as if they are physically co-located. To achieve such a virtual communication environment, the following component technologies are very important.

(1) Generating human images in virtual scenes
(2) Generating virtual (background) scenes
(3) Stereo scopic displays
(4) User interface techniques to the virtual scenes

For human to human communications via a virtual scene, human images need to be synthesized in the virtual scene. Here, the virtual scenes should be able to allow people to appear in the scenes in arbitrary forms, which could be different from the people's original forms. The authors have proposed a virtual metamorphosis system in which anyone can change his/her form into any other form [1]. In Section 16.2, the virtual metamorphosis system is described.

Background image generation from arbitrary viewpoints is as important as human image generation for the creation of the virtual communication environment. The background image can be created manually or semi-automatically by computer graphics utility softwares, but usually this is a time-consuming and painful job for designers. Our view generation technique is image based, which is presently a topic of active research. Section 16.3 describes our affine coordinate based approach. Merging computer graphics images and real scene based images is also an important problem, which we address using the same framework [2].

The virtual scenes generated by the methods described above need to be presented to viewers as 3D images. 3D displays have been studied by many people, but one of the most serious problems of conventional 3D displays is that the displays cause fatigues of viewers. Section 16.4 describes the authors' head mounted display (HMD) that does not cause any fatigue even if the viewer uses it for a long time.

In addition, techniques for user interfaces between a human and virtual environments are very important. Section 16.5 describes the authors' user interface method that utilizes hand gesture recognition technologies. Moreover, our haptic interface such as a palm top display having force feedback mechanisms is explained.

16.2   Virtual Metamorphosis System

16.2.1   Concept and Outline

To realize a virtual metamorphosis system, the facial expressions and body postures of a person who wants to metamorphose need to be reproduced in the different form in real-time. Contact type methods, in which tape marks and sensors are attached to human faces and bodies [3], can achieve quite accurate measurements, but they are cumbersome and have limited applications. Therefore, the authors' have developed non-contact, real-time, computer vision based methods for estimating facial expressions and body postures. The virtual metamorphosis system consists of three main modules: (1) 3D modeling of characters (prior to metamorphosis), (2) Real-time estimation of facial expressions and body postures by a non-contact method, and (3) reproduction of estimated expressions and postures in the 3D model. As the first version of the virtual metamorphosis system, the authors have developed the Virtual Kabuki system, in which anyone can metamorphose into a Kabuki actor [1]. Figure 16.1 shows a block diagram of the system. Note that by using a model for a different character, anyone can metamorphose into any other form. In the following, the three main modules are detailed.

16.2.2   3D Modeling of Kabuki Actors

3D models of characters (such as Kabuki actors) to which a person changes his/her form need to be created prior to metamorphosis. Note that in Figure 16.1, characters' models are stored in an Onyx RE2 workstation. A character's model is created by a wire frame model, which is a set of small triangular patches. Color texture is mapped to the wire frame model. According to facial expressions and human motions

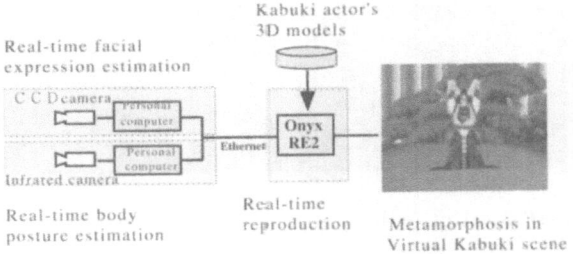

Figure 16.1   Principle of virtual metamorphosis system.

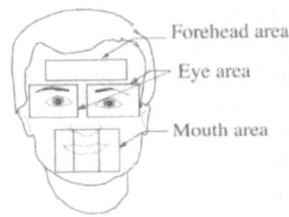

Figure 16.2   Windows for estimating facial expressions.

estimated by the methods described in Section 16.2.3, the wire frame model can be deformed so that the estimated expressions and postures are reproduced in the 3D model (Section 16.2.4).

16.2.3   Estimating Facial Expressions and Body Postures

The authors developed a computer vision based method for estimating facial expressions, more specifically deformations of facial components such as the eyes and mouth, using a frequency domain transform [4]. The person, who metamorphoses into a Kabuki actor, wears a helmet to which a small CCD camera is fixed so that the camera stays at the same position relative to the face regardless of head movements. As shown in Figure 16.2, in the face image acquired by the camera, a window is applied to each facial component, and each window is converted to frequency domain data by DCT (Discrete Cosine Transform), which is featured by its fast processing speed. More specifically, each window is divided into sub-blocks, where a sub-block consists of 8×8 pixels. Then, DCT is applied to each sub-block, and the summations of DCT energies in the horizontal, vertical and diagonal directions in each sub-block are calculated. DCT feature for a direction among the three directions is obtained by adding the summations in that direction. The authors have confirmed that the DCT features represent changes in shapes of facial components [4]. The DCT features are used to reproduce the facial deformations in the face model as described in Section 16.2.4.

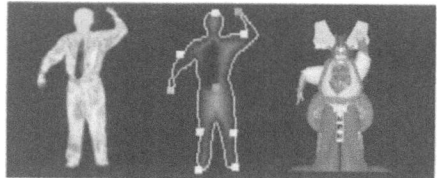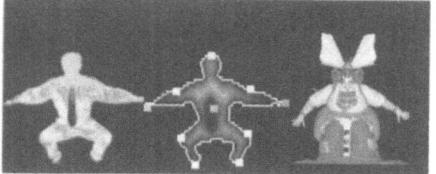

Figure 16.3   Human body posture estimation and reproduction.

The authors have developed a method for estimating the posture of a person at each time instant from the thermal image acquired by an infrared camera that observes the person [5], where thermal images are robust against changes in lighting conditions and colors of clothes and backgrounds. In a thermal image (Figure 16.3, Left row), the silhouette of the human body is extracted from the background by a simple thresholding for temperature values. First, some significant points such as the top of the head and finger tips are located by analyzing the contour of the extracted silhouette. To reproduce the entire posture of a human body, in addition to those significant points, we need to estimate the locations of the main joints such as elbows and knees. These are difficult to estimate by a simple contour analysis, because the main joints do not produce salient features on the contour. Therefore, we developed a learning based method that estimates the positions of the joints using the located positions of the significant points. That is, polynomials that calculate the coordinates of the main joints from the coordinates of the located significant points are constructed in advance by a learning procedure and are used for the estimation. Since estimating the values of the coefficients of the polynomials is a combinatorial optimization problem, a GA (genetic algorithm) [7] is utilized to determine the coefficients values from a sample data set. In the middle row of Figure 16.3, the located positions of the significant points and joints are indicated by the small squares.

16.2.4   Reproducing Estimated Expressions and Postures

To reproduce facial expressions in a 3D face model such as Kabuki actor's face model, the DCT features estimated in Section 16.2.3 are utilized to deform the face model. The authors developed a reproduction method based on anatomy for artists or plastic anatomy [8]. Anatomy for artists, which was proposed in Italy in the 15th century, is an artistic principle for deformation or emphasis that makes viewers feel that works (such as paintings and sculptures) are realistic.

The authors' method consists of a modeling process and reproduction process. In the modeling process, the 10 reference facial expressions chosen by an artist are generated in a 3D face model, where the 3D displacement vector from a neutral facial expression is recorded at each vertex of the face model for each reference expression. The reference expressions are chosen according to the principle of anatomy for artists; therefore, some expressions include deformations that humans cannot actually display. To generate intermediate facial expressions, the 10 reference expressions need to be mixed. Therefore, in the modeling process, at each vertex of the 3D face model, the artist decides the mixing rate of the 3D displacement vectors of each reference facial expression for each image of a sample image set that contains many different facial expressions by comparing each real expression with the generated expression. Then, at each vertex, a linear combination of the 10 reference displacement vectors is constructed, where each coefficient (mixing rate) of the linear combination is represented by a linear combination of the DCT features. A GA based learning procedure determines the coefficients of the linear combination of the DCT features. In the reproduction process, the DCT features obtained from the estimation process (Section 16.2.3) are input into the linear combination for the mixing rates so that the displacement vector of each vertex is calculated.

To reproduce the estimated 2D body postures in a 3D Kabuki actor's model, polynomials that convert the 2D coordinates of the significant points and joints to 3D coordinates are constructed in advance by a learning procedure based on GA. When body postures are reproduced, the estimated 2D coordinates are inserted into the polynomials so that 3D postures are reproduced in the Kabuki actor's model in a reasonable manner. The right row of Figure 16.3 shows the results of the reproductions.

16.2.5   Implementation

The algorithms described in this paper are implemented as outlined in Figure 16.1. For estimating facial expressions, a Gateway 2000 personal computer was used. Thermal images are acquired by an infrared camera (Nikon Thermal Vision LAIRD3), and human body postures are estimated from the thermal images by an SGI Indy workstation. A Kabuki actor's 3D model is rendered together with a Kabuki scene by an SGI Onyx Reality Engine 2 and displayed using a large projector screen (Figure 16.1). With this implementation, the process speed is approximately 20 frames/second for facial expression estimation and body posture estimation, and 12 frames/second for the rendering. The speed for the estimation is good, but the rendering speed needs improvements. The Virtual Kabuki system was demonstrated at Digital Bayou in SIGGRAPH96 held in New Orleans, USA in 1996. The demo system was quite popular; the system attracted the attention of mass media because of its extremely stable performance.

16.3   Novel View Generation

16.3.1   Affine Coordinate Based Theory

To generate views from arbitrary points in the real space, the authors developed a direct approach to scene reprojection, contrary to the traditional 3D reconstruction and rendering strategy. It is based on Jacobs' work [9], where Jacobs shows that the set of 2D images produced by a group of 3D point features can be optimally represented by two lines in the high dimensional affine space.

The authors assume a regular pin hole camera geometry, instead of the para perspective method described in [9], to project the points in 3D into a plane. This is followed by an affine transform of these projected points. Let $(P_1, P_2, P_3, \ldots, P_n)$

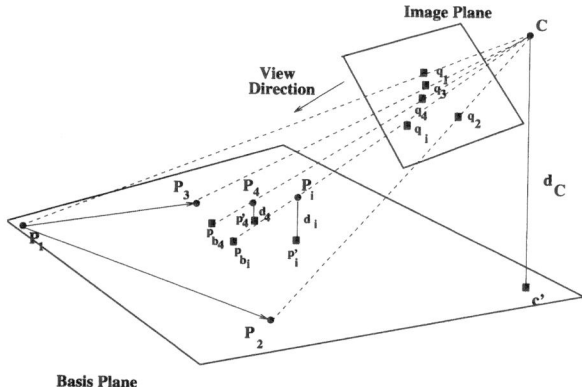

Figure 16.4   Affine transform based projection model.

be the set of 3D points not necessarily lying on a plane. We construct a hypothetical plane passing through points P_1, P_2 and P_3 as shown in Figure 16.4. We call it the basis plane. The point P_4 is projected perpendicularly into the basis plane, and we call this projected point as p_4'. The affine coordinates of p_4' with respect to the basis (P_1, P_2, P_3) are (a_4, b_4). Similarly, for the ith point P_i, its projection on the basis plane is p_i' (with affine coordinates (a_i, b_i)), and for C (view point), c' with (a_c, b_c). Let d_4, d_i and d_c be the distances of point P_4, P_i and C from the basis plane, respectively.

As in [9], for affine coordinates (α_4, β_4), it can be shown that there is a viewpoint C in which the projection of the point P_4 has those affine coordinates. Let p_{b_4} be a point on the basis plane with affine coordinates (α_4, β_4), for the basis (P_1, P_2, P_3). The line passing through p_{b_4} and P_4 sets this viewing direction. This line meets the image plane (whose normal is parallel to the line) at a point q_4. That is, q_4 is the image of P_4. In a similar manner, we project P_1, P_2, P_3 into q_1, q_2 and q_3, respectively on this image plane. With (q_1, q_2, q_3) as the basis, one can easily observe that q_4 has the affine coordinates (α_4, β_4), even when we subject the points on the image plane to an affine transformation (which includes translation, rotation, and scaling, to name a few).

We assume a pin-hole projection to a plane (containing P_1, P_2 and P_3, say), followed by an affine transform. The geometry is illustrated in Figure 16.4, and is intermediate between para perspective and perspective projections. Using the geometry of similar triangles, we can show that:

$$(\alpha_i - a_i) = \frac{d_i}{d_4}\alpha_4', \tag{16.1}$$

where $\alpha_4' = \left(\dfrac{\alpha_i - a_c}{\alpha_4 - a_c}\right)(\alpha_4 - a_4)$.

Thus the plot of (α_4', α_i) over all possible images fall on one straight line. The expression for the β coordinate can be written similarly. We use the straight line property along with the dense point match information between stereo pairs to generate novel views.

16.3.2   Reprojection, Novel View Generation and Merging

Let the two input images be I_1 and I_2. For novel view generation, we assume the knowledge of dense point correspondence between these two images. For a point p_i^1 in image I_1, let the corresponding point in I_2 be p_i^2. We need four reference points (three points to create the basis and a fourth point) to generate the lines in α and β space. Let the reference points be $p_1^j, p_2^j, p_3^j, p_4^j$ in the image I_j ($j = 1, 2$.). To make things simpler, we choose these points as the images of the points P_1, P_2, P_3 and P_4, where $P_1 P_2$ is perpendicular to $P_1 P_3$, and $P_4 P_1$ is perpendicular to the plane containing (P_1, P_2, P_3). Also, let $|P_1 P_4| = |P_1 P_2| = |P_1 P_3| = |P_2 P_3| = 1$. We show this structure (simultaneously) to the two cameras before the experiment, and record the coordinate values of their projections $p_1^j, p_2^j, p_3^j, p_4^j$ ($j = 1, 2$). Now, for points p_4^j and p_i^j in image I_j, let its affine coordinates be (α_4^j, β_4^j) and (α_i^j, β_i^j), respectively. The line in the (2 dimensional) α space in Eq.(16.1) passes through the points $(\alpha_4^{1'}, \alpha_i^1)$ and $(\alpha_4^{2'}, \alpha_i^2)$. To obtain $\alpha_j^{1'}$ and $\beta_j^{1'}$, we need to compute a_c and b_c. For this, we use a fifth control point P_5 collinear with P_1 and P_4, and $|P_5 P_1| = k|P_4 P_1|$. We record its position in I_1 and I_2, respectively.

For merging real and synthetic objects, the polygonal (wavefront) representation is first defined with respect to the axis system of the reference points, followed by the simultaneous rendering of the reprojected points and the synthetic object. The views are stitched by reprojecting one of the images into the other.

16.3.3   Experimental Results

The images shown in Figure 16.5(a) are two images from a multiple baseline stereo configuration. We use a correlation based stereo matching algorithm to obtain the dense point match information. Example of novel views generated are shown in Figure 16.5(b). An example of merging real and virtual objects is shown in Figure 16.6, in which a CAD model for a face is merged in the real scene. Similarly, stitching results are good.

16.4   Fatigueless Head Mount 3D Display

In this section we describe the 3DDAC (3-D Display with Accommodative Compensation) [10]. The principle feature of 3DDAC is compensating accommodation with binocular disparity 3-D image representation by using real-time display screen movement.

16.4.1   Visual Fatigue and 3-D Display

The binocular disparity (stereoscopic) type of 3-D display system is simple and practical in actual applications. However, the binocular disparity type has the problem that it makes uncomfortable feeling and a visual fatigue for users. This is mainly

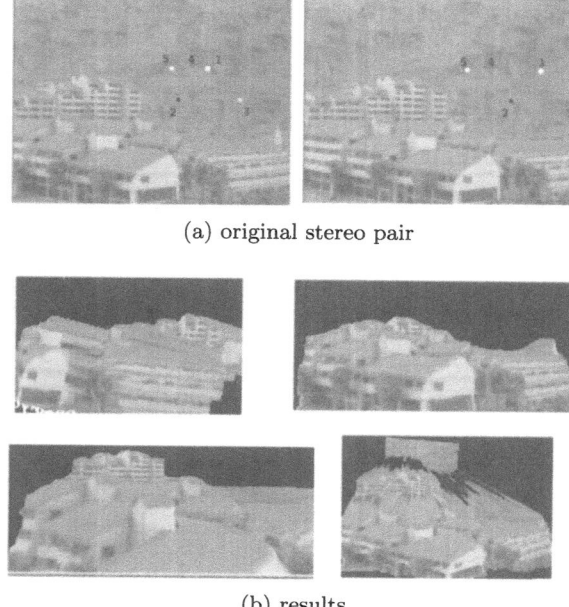

(a) original stereo pair

(b) results

Figure 16.5   Novel view generation.

Figure 16.6   Merging CG and real images.

caused by the mismatching of convergence and accommodation. Known as the convergence accommodation and accommodative convergence [11], both convergence and accommodation cooperate in eye-ball motion, it is natural behavior of a sense organ. There is a difference between the vergence distance and the accommodation distance with a simple binocular disparity type. Thus this mismatching interrupts a natural physiological mechanism, and invokes a physical and a mental work load which causes serious fatigue. Another problem is the inaccuracy of depth perception.

16.4.2   Principle Idea of Accommodative Compensation

We intended to improve the previously mentioned inherent weak point of the binocular disparity methodology. We made a compensation mechanism for that mismatching, called the accommodative compensation. The principle idea of compensation is very simple. There is a gaze detection mechanism and a movable display screen within the same system. The gaze detector detects the viewpoint of the image, and the system calculates the distance of that viewpoint and controls an actuator that drives a movable screen to provide correspondence between the vergence distance and the accommodation distance.

The basic mechanism of screen movement is to move a virtual image of the displayed image on the screen using optical components, which is equivalent to moving the screen unit itself [10].

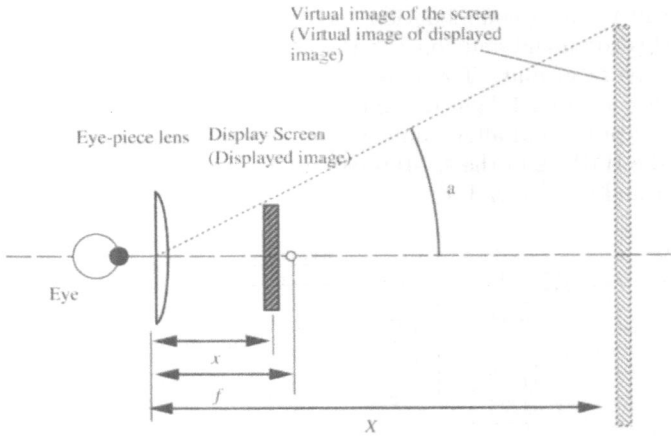

f: Focal distance of a field lens
x: Distance between eye-piece lens and screen
X: Distance between eye-piece lens and virtual image of the screen
a: Field of view

Figure 16.7   Relationship between the displayed image and its virtual image.

In the 3DDAC system, there is a movable screen unit and an eye-piece lens unit, designed to reduce the size and weight and to simplify the optical structure for practical use. Figure 16.7 shows the relationship between the image displayed on the screen and its virtual image. In this optical configuration, the eye-piece lens is a magnifier of the image displayed on the screen and makes its virtual image at a far point. Here, the focal distance of the eye-piece lens is f, and the distance between the eye-piece lens and the screen is x. The distance between the eye-piece lens and the virtual image of the displayed image is X, with following equation:

$$X = \frac{fx}{f - x}.$$

16.4.3 System Implementation of 3DDAC System

Figure 16.8 is a block diagram of the 3DDAC system, consisting of the head unit, image control unit, accommodation control unit, and host CPU interface. Here, the image control unit receives each left-eye image and right-eye image (3-D mode) or monocular image (2-D mode) as an NTSC video signal (Y/C or component RGB) from the host CPU and transfers it to the head unit. The head unit is a head mounted binocular viewer that gives a disparity image to each eye. The accommodation control unit is an actuator controller with a host CPU interface. The host CPU calculates the distance of the displayed image and issues a control command for screen movement to the accommodation control unit. The accommodation control unit receives this command and controls the actuator unit that drives the movable screen unit in the head unit. The movable screen unit consists of two wide-angle TFT LCD panels with back-lighting units. The optical component is a monolithic block of lens and prism and allows outside real images to pass through via an LCD shutter when the HMD is in the see-through mode. The actuator unit consists of a stepping motor and gear assembly.

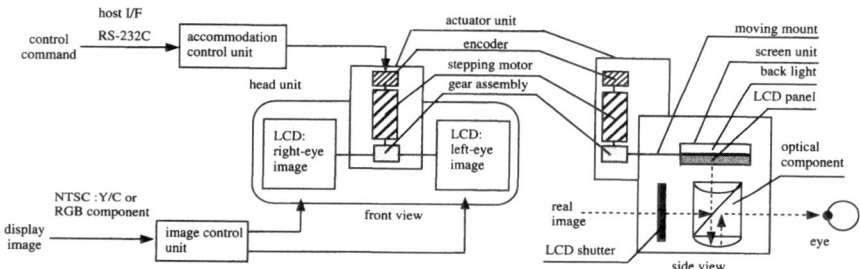

Figure 16.8　Block diagram of 3DDAC.

This system can drive a movable screen at less than 0.3 s to move a displayed image from a distance of 0.25m to infinity, which is a sufficient response for human perception. We have also conducted a simple subjective assessment for evaluation. We made a stimulus CG image from 3D objects and placed them at various depths in a 3D CG world. We then showed those stimulus images to subjects under different conditions of accommodation compensation control, and interviewed them about their impressions of the relative position of each stimulus object. Every subject reported that it was easy to distinguish the relative position when accommodation was correctly compensated and the result shows the efficiency of the accommodative compensation [12].

We are also thinking of using the 3DDAC as an Eye-through HMD [13] to enhance face-to-face conversation through the use of augmented reality and surreal techniques. The Eye-through HMD enables the wearer's eyes to be visible to other people despite the HMD, and as such, prevents the HMD from hindering interaction. It also features the ability to control eye movement according to the content of a conversation, emotions, etc.

16.5   User Interface in Virtual Environments

Applications of virtual worlds generated by VR techniques are expected in many areas, e.g., teleconferencing, cooperative work spaces, remote control, and CAD. Generally, VR systems provide their participants with an interface for direct manipulation, and they are able to handle virtual objects generated by computer graphics. In this section we introduce an effective user interface. In this paper, we focus on user interface technologies related to manipulation by hand.

16.5.1   Hand Gestures Estimation Using Multiple-camera [14]

Hand gestures are a useful interface for humans to interact with not only other humans but also machines. Major applications of hand gesture recognition include 2D and 3D pointing, object grasping, and sign language understanding. Such high-potential applications led to many gesture-based systems using glove-type sensory devices in the early days of virtual reality research. Contact-type devices, however, are troublesome to put on and take off and wearing such devices continuously for a long time can tire their users.

To overcome these disadvantages, vision researchers have been trying to develop non-contact type systems for detecting human hand motion [15]–[19]. However, some instabilities peculiar to vision-based systems have been found. The most significant problem is occlusion. As for the moving of nonrigid objects like a human hand, the detection of feature points and the matching of these points between images are difficult to do correctly.

To avoid such instabilities, our system utilizes only one feature point (COG, center of gravity) from each of the cameras employed to reconstruct 3D pose information, and detects the point using a distance transformation based method that is not affected by occlusion. Detailed information, like hand gestures, can be detected in a camera image selected based on the 3D pose information. By using this camera selection, we can always get a hand image from a particular view. This simplifies the description and recognition process of hand shapes.

Distance transformation is applied on each hand silhouette image and the COG point is extracted as the maximum point in the distance transformed image (Each pixel has a distance value from the closest boundary pixels; Figure 16.9). The hand orientation is also detected as the primary axis of edges in each silhouette image simultaneously. The 3-D position and direction are calculated from the above information from two camera images. The rotation angle of the hand plane about the hand orientation is estimated based on distance transformation values for the hand position [20]. To simplify the estimation, we employ an ellipsoidal model and represent a human palm by an ellipsoid. By using this approach, stereo-matching is no longer necessary for estimating the pose. After the hand plane has been determined, the camera that has its axis most perpendicular to the hand plane is selected. The gesture recognition process is performed for the image. Figure 16.10 shows seven hand shapes which can be recognized in the current implementation.

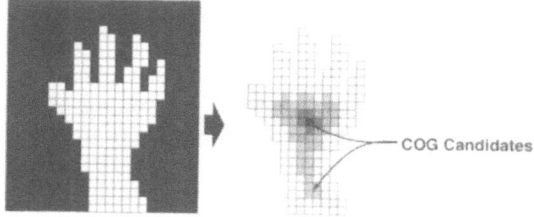

Figure 16.9　COG detection.

Figure 16.10　Seven hand shapes and extracted contours.

16.5.2　Target Prediction in Grasping Hand Motion [21] [22]

While reaching for an object, we usually prepare a grasping pose, which is called "preshaping" and depends on the object's size, shape, and other properties [23]. Analyzing of the preshaping makes it possible to predict a target in advance, and this leads to a quick and subconscious selection of the target without a direct method such as pointing or speech input.

We propose the following hypothesis: When attempting to grasp an object beyond reach, we move a hand toward the object's center of gravity (COG) to overlap the COG of the finger tips (finger's COG) with the object's COG.

Another hypothesis is that we attempt to grasp the long axis from among the principal axis of inertia of an object and adopt our hand orientation to it.

We performed experiments that investigated the motion of grasping an object generated in a virtual space. The results showed that a hand tends to be moved in the direction of an object's COG, independent of the object's shape or size, and that the hand orientation is subconsciously arranged according to the object's shape. According to the experimental results, we considered a simple prediction method based on a stochastic model. The method enables a high target recognition rate and reduces the time required for operation.

16.5.3　See-through Hand [24]

In a typical VR system, using a glove-like device with a magnetic positional sensor, the user grabs and manipulates virtual objects in the VR environment. When reaching for an object, the user's hand is located between the himself/herself and the screen. Therefore, the target object displayed on the screen is often occluded by the user's hand.

The See-through Hand is a palm-coupled display system, and allows a user not only to handle objects directly but also to see through his/her hand and thereby avoid becoming annoyed by the occlusion from his/her own hand while observing the object he/she is grasping.

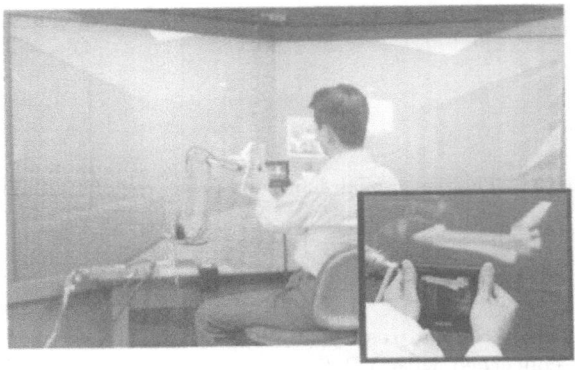

Figure 16.11   Scene using PDDM for manipulating a virtual object.

As our previous work, we integrated force-feedback with our palmtop VR system named PDDM (Palmtop Display for Dextrous Manipulation) [25] shown in Figure 16.11. The PDDM consists of a palmtop display, a position and orientation sensor, and a force display, and can act as a visual display and as a motion input device with haptic sensations in one unit. The PDDM has two phases: an observing phase and a handling phase. The user holds both sides of the display and makes button presses to shift between the two phases. When the user locks the object in the center of the palmtop display, he/she can manipulate the object following the motion of the palmtop display. The motion of the PDDM is directly reflected on the grabbed object as if the user is grasping and moving it. The change in the copy is immediately reflected in the original in the VR space on the large screen. Problems of our PDDM include its lack of intuitiveness. As the display has two buttons for input and phase shift, pushing or holding the buttons is cumbersome. In our daily lives, we use our own hands to grab objects. Therefore, it is difficult to manipulate objects while holding and moving a display with our two hands. The See-through Hand is an attempt to use a palmtop display more intuitively.

Figure 16.12 shows the concept of our See-through Hand. The user has a palmtop size display, and looks at objects in the VR world through the display; the user's hand appears to be see-through. When the user wants to grab an object, he/she reaches for the target object. The object is selected and grabbed, as with most VR systems. The See-through Hand allows the user to ignore all mode shifts, which exist and interrupt the phase flow in manipulations with the former PDDM. Therefore, even a novice user can use it well.

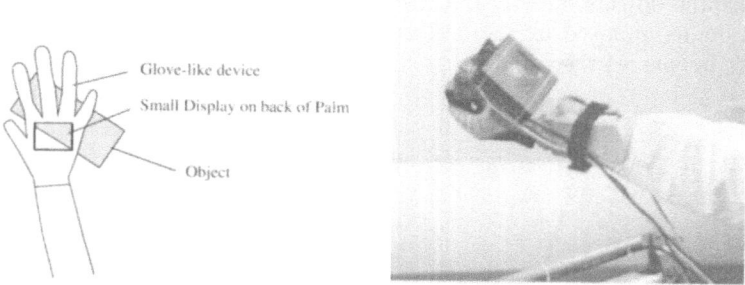

Figure 16.12   The see-through hand and prototype of see-through hand.

16.6   Conclusion

In the applied fields of virtual reality (VR), communications are very important. The authors are concerned with VR technologies needed for realizing multimedia communications. In this paper, the authors briefly introduced research works on VR based multimedia communications that are being conducted at ATR Media Integration & Communications Research Laboratories; that is, virtual metamorphosis system based on real-time human image analysis and synthesis methods, novel view generation from a real image set, a new head mounted 3D display, and user interface techniques based on hand gesture recognition and haptic devices.

In order to construct a virtual space that offers a highly natural sense of immersion, simulation of all the human senses is definitely needed. We will continue our investigation towards creating suitable communication environments.

References

[1] J. Ohya et al.: "Virtual kabuki theater: Towards the realization of human metamorphosis systems," *Proc. 5th IEEE Int'l Workshop on Robot and Human Communication*, pp.416–421, Nov. 1996.

[2] K. Sengupta et al.: "Novel scene generation, merging and stitching views using the 2D affine space," *Proc. IEEE Int'l Conf. on Multimedia Computing and Systems*, pp.602–603, Jun. 1997.

[3] J. Ohya et al.: "Virtual space teleconferencing: Real-time reproduction of 3D human images," *J. Visual Communication and Image Representation*, vol.6, no.1, pp.1–25, Mar. 1995.

[4] K. Ebihara et al.: "Real-time facial expression detection based on frequency domain transform," *Proc. Visual Communications and Image Processing'96*, SPIE vol.2727, pp.916–926, Mar. 1996.

[5] S. Iwasawa et al.: "Real-time estimation of human body posture from monocular thermal images," *Proc. CVPR'97*, pp.15–20, Jun. 1997.

[6] K. Ebihara et al.: "Real-time 3-D facial image reconstruction for virtual space teleconferencing system," *IEICE Trans.*, vol.J79-A, no.2, pp.527–536, Feb. 1996 (in Japanese).

[7] D. E. Goldberg: *Genetic algorithms in search, optimization, and machine learning*, Addison-Wesley Publishing Company Inc., 1989.

[8] K. Ebihara et al.: "Real-time facial expression reproduction based on anatomy for artist for virtual metamorphosis system," *IEICE Trans.*, vol.J81-D-II, no.5, pp.841–849, May 1998 (in Japanese).

[9] D. W. Jacobs: "Space efficient 3D model indexing," *Proc. CVPR*, pp.439–444, Jun. 1992.

[10] S. Shiwa et al.: "Proposal for a 3-D display with accommodative compensation: 3DDAC," *J. SID*, vol.4,no.4, pp.255–261, 1996.

[11] J. Semmlow and D. Heerema: "The synkinetic interaction of convergence accommodation and accommodative convergence," *Vision Research*, vol.16, pp.1237–1242, 1979.

[12] T. Sugihara and T. Miyasato: "A lightweight 3-D HMD with accommodative compensation," *Digest of SID'98*, 32.4, pp.927–930, 1998.

[13] T. Miyasato: "An eye-through HMD for augmented reality in face-to-face communication," *Proc. IEEE RO-MAN'98*, pp.15–20, Sep. 1998.

[14] A. Utsumi and J. Ohya: "Direct manipulation interface to virtual scenes: Estimating Hand Gestures using Multiple-camera Images," *PUI Workshop'97*, pp.22–23, Oct. 1997.

[15] B. Moghaddam and A. Pentland: "Maximum likelihood detection of faces and hands," *Proc. Int'l Workshop on Automatic Face and Gesture Recognition*, pp.122–128, 1995.

[16] P. A. Hadfield et al.: "Uncalibrated stereo vision with pointing for a man-machine interface," *Proc. IAPR Workshop on Machine Vision Applications*, pp.163–166, 1994.

[17] M. J. Rehg and T. Kanade: "Visual tracking of high DOF articulated structures: An application to human hand tracking," *Computer Vision-ECCV'94*, LNCS vol.801, pp.35–46, 1994.

[18] Y. Iwai et al.: "Estimation of hand motion and position from monocular image sequence," *Proc. ACCV'95*, vol.II, pp.230–234, 1995.

[19] J. Davis and M. Shah: "Determining 3-d hand motion," *Proc. Asilomar Conf. in Signals, Systems and Computers*, pp.1262–1266, 1994.

[20] A. Utsumi et al.: "Hand gesture recognition system using multiple cameras," *13th Int'l Conf. on Pattern Recognition*, pp.219–224, 1996.

[21] Y. Nakamura et al.: "Target prediction in grasping hand motion," *Proc. ACCV'95*, pp.II-694–II-698, 1995.

[22] Y. Nakamura et al.: "Target prediction in grasping hand motion," *CHI'96 Interactive posters*, pp.113–114, Apr. 1996.

[23] M. Jeannerod: "The timing of natural prehension movements," *J. Motor Behavior*, vol.16, pp.235–254, 1984.

[24] T. Miyasato, and R. Nakatsu: "A palm-coupled display system for direct manipulation of virtual objects," *Proc. CGIM'98*, pp.242–244, 1998.

[25] H. Noma et al.: "A palmtop display for dextrous manipulation with haptic sensation," *Proc. CHI'96*, pp.126–133, 1996.

Part V

Systems: Design Considerations and Future Trends

The final part of the book comprises a collection of reports about the development of prototype mixed reality (MR) systems. The last three chapters are devoted to outdoor applications, which is a key area for expanding the field of MR beyond the laboratory.

The research of Stephen R. Ellis and Brian M. Menges reported in Chapter 17 is noteworthy as one of the few user-centered empirical studies. For several years, they have been studying the perceptual effects of viewing virtual objects on stereoscopic displays. This chapter presents the results of four experiments about the relative localization accuracy of virtual objects, presented via various viewing conditions using a custom see-through HMD. It also discusses the practical implications of their results for the design of HMDs and virtual objects presentation schemes.

Gudrun Klinker and her colleagues discuss comprehensive design considerations for augmented reality (AR). They have developed a variety of AR systems, and their experience indicates that there are many tradeoffs in AR system design. Chapter 18 explains two opposite approaches: a real-time immersive system that can be implemented using today's technology, and a high quality, future oriented system in which current real-time operation is sacrificed in anticipation of future improvements in processing power and data transmission bandwidth. After presenting the applications and demonstrations they have built in the recent past, as well as the underlying hardware and software architecture, they discuss the trade-offs they made in developing user-tracking and scene-augmentation algorithms.

Kosuke Sato and his group are making efforts to realize practical support systems for inspection of real electronic parts. Chapter 19 describes two types of computer-aided maintenance-support systems based on MR. The first system tracks the positions of parts on a printed circuit board. It prompts the operator about where and how to inspect the board by superimposing graphical instructions. The second system is a wearable backpack computer that uses an HMD and an inertial navigation system to support the inspection of wall-mounted power-system components. Currently, this work is directed toward indoor use such as in factories, but AR/MR systems are also expected to be used outdoors.

Steven Feiner, a pioneer in the field of see-through AR displays with his renowned KARMA system [1], has once again assumed the role of pioneer with his research on the development of a mobile AR system that can be used outdoors. This theme is currently attracting the interest of other researchers as an especially important challenge for wearable user interfaces. In Chapter 20, Feiner and his colleagues at Columbia University describe the architecture and user interface of two experimental systems. The first, which uses commercial indoor tracking technology, is designed to assist a construction worker in assembling a spaceframe structure. It overlays 3D graphics, text, and sound to show where to install the next part. The second system presents information about the Columbia campus through a "hybrid user interface" that combines a head-tracked, see-through HMD, driven by a backpack computer, with an untracked, hand-held, stylus-based computer. This system performs submeter-accurate position tracking using a differential global positioning system receiver.

Ronald T. Azuma is well known as the author of an excellent survey of AR technology [2]. After completing his Ph.D research on AR at the University of North Carolina at Chapel Hill, Azuma moved to HRL Laboratories, where he became a central figure in the development of outdoor AR systems. In Chapter 21, he discusses the motivation and challenges underlying his research in outdoor AR, starting from his background in AR technology.

The final chapter of this book, Chapter 22, describes the conception and technological development of the long-range outdoor AR system being developed by Donghyun Kim and his group in South Korea. They propose a monitor-based AR system to perceive remote geographical characteristics. Their system includes a radio-controlled helicopter with a wireless CCD camera and GPS to determine the camera attitude in real time.

References

[1] S. Feiner, B. MacIntyre, and D. Seligmann: "Knowledge-based augmented reality," *Comm. ACM*, vol.36, no.7, pp.52–62, 1993.

[2] R. Azuma: "A survey of augmented reality," *Presence*, vol.6, no.4, pp.355–385, 1997.

Chapter 17

Operator Localization
of Virtual Objects

Stephen R. Ellis
NASA Ames Research Center, U.S.A.

Brian M. Menges
San Jose State University Foundation, U.S.A

17.1   Introduction

The following four experiments explore such phenomenon by examining subjects'
ability to adjust the distance of a physical pointer to match that of a nearby virtual
object. The object is generated by a high performance, computer graphics system
and presented by a head-mounted, see-through display. This localization task was
selected since it is close to the visual-manual manipulation expected of users of virtual objects in numerous possible applications from surgery to mechanical assembly
on a production line. In these applications virtual objects are proposed to become
3D guides that instruct and aid manual interaction with the world by dynamically
helping the user position his hand. Thus, these virtual objects must be accurately
and reliably localized by the user. For this reason, and because its precision and
accuracy can be easily and reliably measured, the following experiments study the
localization of virtual objects with a real pointer. Preliminary testing showed that
our subjects could set a mechanically displaced, physical pointer to match the distance of physical targets with several millimeter accuracy and that this accuracy

corresponded to their ability to match target distances with their fingers. Most previous work has examined system configuration to optimize display performance (e.g. [1] [2]) but only Rolland's [3] laboratory has systematically investigated user localization performance.

The initial experiment below examines the effects of three different viewing conditions on the subjects' accuracy of placement of the physical pointer under a virtual object. The conditions are: 1) monocular, 2) biocular or 3) stereoscopic viewing. These represent a range of cost and image fidelity, i.e. the completeness with which an objects physical characteristics are presented. The monocular condition presents a virtual image of an object much as monocular helmet mounted sights do and represents a minimal hardware/software rendering cost for such virtual image displays. But the monocular virtual images it presents are subject to visual suppression due to binocular rivalry. The biocular condition, which presents two identical virtual images to the subjects' left and right eyes, can avoid the rivalry problem. The images it presents are projected from a cyclopean position between the viewers eyes but are shifted to allow unstressed fusion at a selected distance (See Figure 17.1). This condition halves the rendering cost with respect to stereo displays, but doubles the head-mounted display hardware requirements with respect to monocular displays. It presents a pattern of disparity approximating that of a transparent flat display surface and represents an intermediate cost system with potentially more stable image brightness. The third condition, conventional stereo display with parallel viewing vectors, presents the highest spatial fidelity but doubles both the rendering and hardware display cost compared to the monocular condition.

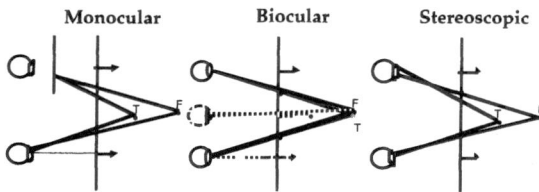

Figure 17.1   Viewing geometry for each viewing condition.

Because the monocular and biocular viewing conditions degrade the fidelity with which distance is presented, we expect that the stereoscopic display should support the most accurate localization. The biocular display may, however, provide a competitive alternative for virtual objects with relatively little internal depth by avoiding the potential problem of binocular rivalry. The first experiment provides a descriptive study of relative localization accuracy of virtual objects presented via the alternative viewing conditions. Since observer age and accommodative demand (required focus) could be expected to interact and influence localization accuracy, these display characteristics are also examined in the first study to develop designer guidelines for the adjustment of focus and selection of personnel to use virtual object displays . Subsequent studies consider the effect of the introduction of a nearby physical surfaces on the localization of the virtual object, identify a phenomenon that introduces errors into such localization, and explore two alternative explanations for the phenomenon.

17.2   Experiment 1

17.2.1   Methods

All experiments reported in this paper used a custom, head-mounted, see-through display, called an electronic haploscope by the authors, capable of presenting a 20-30° diameter circular monocular field to each eye with variable monocular overlap. In the following experiments the system was used at 100% overlap and 20° field of view. The display system used two vertically mounted Citizen 1.5 ' 1000 line miniature Cathode Ray Tube's (CRT) in National Television Standards Committee (NTSC) mode which were driven by a Silicon Graphics (SGI) computer (4D/210GTXB) through custom video conditioning circuits. For the simple 3D imagery used in the following experiments, the computer could maintain a 15 Hz graphics update rate. The CRT images were infinity collimated by standard glass telescope eyepieces (Erfl 32 mm and Ploessl 42 mm) mounted directly under the CRTs. After the signal transformation from the RGB to NTSC, individual pixels which corresponded in the current configuration to 3 arcmin horizontal resolution measured from subjects' eyes were easily discriminated. The collimated light could be modified by lenses and rotating prisms from a standard optometric trial lens set that allowed precise positioning with at least 5 arcmin resolution of the separate left and right images and allowed variation of the accommodative demand for each eye. The images were relayed to the subjects eyes by custom, partially silvered (15%) polycarbonate mirrors mounted at 45° directly in front of each eye. The left and right viewing channels could be mechanically adjusted between 55 mm and 71 mm separations for different subject's interpupilary distances. The video signal conditioning also allowed lateral adjustment of the video frame. Consequently, the display system can precisely position the center of each graphics viewport directly in front of the eyes of all subjects for bore-sight alignment.

The entire display system, built around a snug fitting, rigid headband, is intended to be worn by a freely moving subject and weights between 0.77 and 1.1 kg. depending upon configuration. In the lightest configuration the moments of inertia have been measured when mounted on an erect head to be: 0.0782 kg-m^2 vertical axis, 0.0644 kg-m^2 longitudinal axis, 0.0391 kg-m^2 lateral axis. In all of the experiments described below, the band was fitted to each subject's head and then supported by a special pivoted mount at the end of a 1.8 m table. This mount restricted horizontal movement but allowed some pitch movement. Subjects sat at this end during the course of each experiment. The mount and chair were adjusted so that the virtual objects could be presented at eye level. Lateral head movement was restricted during all the following experiments but a residual pitch of $\pm 10°$ was allowed for subject comfort. In practice the subjects were monitored by the experimenters to keep their heads approximately level at an individually selected orientation during the course of the experiments.

A monocular, biocular or stereoscopic virtual image of an upside down, axially rotating (~ 3 rpm) pyramid was presented at a distance of 58 cm away from the subjects' eyes by a head-mounted see-through display. It was seen against a grey, cloth covered wall 2.2 m from the subjects. Preliminary experiments examining varying the rotation rate of the pyramid for each trial showed that such variation had no effect on the localization of the virtual image. The reference distance of 58 cm was chosen for the experiment because it corresponded to a possible working distance for several industrial applications of interest to the authors. All displays were operated under moderate indoor artificial illumination (approximately 50 lux). The virtual image was presented with either 2 diopters accommodative relief or at optical infinity. The stereo display was, however, calibrated (see below) over a range of 30 to 110 cm. The monocular display was simply the stereo channel that corresponded to the subjects' dominant eye. The biocular display was produced by positioning the graphics eye point midway between the subjects eyes. The left and right images were identical copies of this view but were shifted laterally so that when the subjects eyes converged to the centers of each view-port they would have 0 disparity relative to the reference convergence point 58 cm. The plane of 0 disparity was thus set so that the subjects could easily fuse the images when converged at 58 cm. This technique was used in general for all biocular stimuli at different depths which were experimentally interjected as described below. Though no keystone correction was applied for the distortion caused by the image shift, the disparity pattern produced in this biocular image closely approximates that of a flat image of the target as if it were drawn on a transparent projection surface at the simulated convergence distance.

The wire-frame pyramid had a nominal 10 cm base and 5 cm height. The width of the wire frame lines were about 9 arcmin. The depicted size of the presented virtual object was randomly scaled from 70 to 130% of its nominal size for each trial to interfere with subjects' possible use of angular size as a depth cue. The lines of the wire frame an all other computer generated lines had a luminance of about 65 cd/m^2 and were seen against the gray cloth background of 2.9 cd/m^2 with visible vertical seams. While the presented luminance did not approach that used in aircraft heads up displays which must be visible against a 30,000 cd/m^2 background, it was just adequate for indoor work against most colored surfaces and is at least 3 times brighter than the luminance available in off-the-shelf, see-through head-mounted displays such as the IGlasses[TM] formerly made by VIO.

Four out of every 30 judgments were based on unanalyzed random variations in the depicted depth of the pyramid. This variation was introduced to help insure that the subjects did not notice that the same depicted depth was repeated. However, the major factor masking the repetition of the depicted depth were the perceptual effects causing changes in the apparent depth with different viewing conditions. Since the viewing conditions could be unobtrusively intermixed and no feedback was given to the subjects, there was no way for them to tell that the depicted distance was not in fact changing. In fact, no naive subject in any of the following experiments reported noticing the repetition of the target depth.

The subjects' task was to use a method of adjustment to position the binocularly visible, physical cursor, a yellow-green light emitting diode (LED) (about 20 cd/m^2) pointer, shaped like a pyramid (base 0.5 cm, height 1 cm) into vertical alignment with the apex of the inverted pyramidal virtual object. The physical cursor was moved on a rail by a chain and gear system and positioned about 2 degrees of visual angle below the virtual object. The distance to the pointer was automatically recorded through use of a shaft-encoder interfaced to the display computer. The adjustment was self paced, but subjects were encouraged to take between 15 and 30 seconds for each adjustment and were allowed to take breaks at one half to 1 hours intervals as needed. As part of the standard procedure for use of human subjects, all subjects were informed they could terminate the experiment at anytime if they experienced any undo discomfort and were asked at the end if they experienced any viewing difficulties seeing the virtual objects.

The haploscope display system was adjusted by monocular superimposition of reference virtual images on an 18 cm diameter circumscribed circle presented at a distance of 2.2 m. In addition to position adjustment this allowed adjustment of the field of view angle of the graphics system to match the total magnification of the system. This was done separately for each eye and for each subject before the experiment was started. Thus, we could account for any changes due to variation of accommodative demand and corrective lenses that might be worn by the subjects. The subjects' interpupilary distances (IPD) were measured with a binocular-type viewing device with digital readout (Varilux Model: Digital CRP). All displays and algorithms were adjusted to reflect the measured IPD values.

Preliminary test results for virtual targets placed between 33 and 108 cm showed that within the full range of adjustment used for the experiment subjects using a stereo display could align the cursor within ± 0.3 cm of the depicted virtual object target depth [4]. The distance responses were completely linear, unbiased and unskewed and were conducted in the same full room illumination as the experiment. Similar tests of the biocular viewing condition showed equally linear responses. Tests in the monocular condition showed, expectedly, inconsistent behavior. In further examination of the localization technique pilot subjects were asked to use the pointer to match the depth of physical targets. These tests showed linear, unbiased, unskewed estimates with statistical ranges of ± 0.15 cm about depicted physical distances for targets used in the experiment.

Ten subjects, five young (15 - 29 yrs.) and five older (38 - 47 yrs.) participated in the experiment. Subjects in the older group could be presumed by population data to be at least early presbyopes [5]. All but one young and one older subject (i.e. the authors) were naive with respect to the purpose of the experiment. The others were either paid subjects recruited through the Ames Bionetics contractor or were laboratory personnel. All subjects were screened on the Bauch & Lomb Orthorater stereo tests for stereoacuity better than 1 arcmin. Subjects who normally wore prescription spectacles were allowed to wear them during the screening test and during the experiment. During pilot testing for the experiment, inadvertent errors of 0.1 - 0.2 cm in modeling of subjects' interpupilary distances in the graphics simulation produced easily noticeable artifacts. Precise stereo or biocular presentation of virtual objects evidently requires measurement and modeling of interpupilary distance with an accuracy on the order of ± 0.1 cm.

Viewing Conditions (monocular, biocular, and stereoscopic) were crossed with Accommodation (0 or 2 Diopters) and nested within Age groups. The experiment used a blocked design in which blocks of 5 replications of a given condition were presented for each of the three viewing conditions producing uninterrupted 15 judgment sequences. The sequence of viewing conditions were randomly assigned to each subject and thereafter systematically permuted after each set of 3 viewing conditions were presented. In general, it was possible to switch the viewing conditions solely through software. Thus, the subjects were generally unaware which viewing condition was presented and the perceptual variation in apparent depth caused by variation in view condition was readily interpreted by them as variation in depicted depth. The viewing conditions were blocked for a given accommodative demand which was switched by interrupting every 15 trials to change viewing lenses. The order of presentation of accommodative demand was permuted within subjects and balanced across subjects.

17.2.2   Results

Analysis of variance (ANOVA) showed that the viewing conditions had a major effect on the bias of the subjects distance judgments (F= 15.580, df =2, 16, p < .001). The mean stereoscopic and biocular localizations were almost completely correct, but a judgment bias appeared as an overestimate when the stereo depth cues associated with the virtual object were removed by monocular viewing. This effect interacted with accommodative demand and age, as indicated in Figure 17.2 (F=7.76, df = 2, 16, p < .004). All other effects are related to this three way interaction and will not be discussed individually. No subjects reported any difficulties seeing the virtual objects during the experiment.

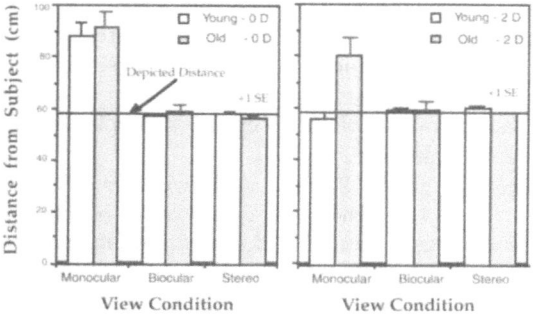

Figure 17.2   Effect of Age/Accommodative on localization.

17.2.3   Discussion

The results of the ANOVA plotted in Figure 17.2 show that when depth cues to the virtual object are degraded to monocular conditions, judgments of its distance drop back towards the distance of the background wall at 2.2 m from the subjects. A phenomenon that could explain the increased judged target distance in the monocular condition is the specific distance effect which causes visual targets of unknown physical size presented with weak or ambiguous depth cues to tend to appear in visually impoverished environments about 2-3 meters away [6]. This distance effect is also associated with tonic accommodation and vergence which relax to approximately 1-2 meters in the absence of distance cues [7]. Changes of convergence to a more distant resting position could cause the localization of the virtual object to recede from the viewer. But since these effects are generally seen when targets appear against featureless backgrounds they are not likely important for the present results.

Another phenomenon probably more relevant is the direct effect of change in ocular convergence on perceived depth [8] [9]. It is the kind of result to be expected if the subjects' actual convergence was driven or attracted by the wall which provided a visually sharp, textured cloth with vertical seams at 2.2 m. In the monocular condition, the only source of information that the virtual object is located any particular distance in front of the background is provided by accommodative demand. It is therefore not surprising that when 2 diopters of accommodative demand was provided, only the subjects young enough to respond to this cue were able to localize the virtual object approximately correctly. In these subjects the accommodative-convergence reflex allowed them to maintain convergence on the virtual object. Older subjects unable to respond to accommodative cues only have the disparity information provided by the background cloth to control their vergence. Thus, they would still diverge to fuse the background and the monocularly presented virtual objects would still appear toward or on the background wall.

That the monocular virtual objects are not judged to be exactly on the wall, reflects the response bias within the experiment originating from constraints on movement of the physical cursor, unavoidable guesses the observers may have made about the approximate size of the object, and the possibility that the observer converged to other closer objects such as the pointer and its supports which, though darkened to be less conspicuous, were still visible against the background. In fact, if the observers' were to hold their eyes completely still, the monocular viewing situation is very similar to that of viewing a monocular afterimage in a demonstration of Emmert's Law, the classic observation that an afterimage often appears at the distance of a physical surface against which it is projected [10]. The readily available correct disparity information in the biocular and stereo conditions, however, provide the missing cue that allowed all observers to correctly judge the distance to the virtual object.

Finally, since no difficulties seeing the virtual objects were reported during the experiment, we find no evidence that binocular rivalry interfered with subjects' ability to see the virtual objects in the monocular condition. The finding is consistent with all observations we have made of the monocular wire-frame virtual objects during preparation for the experiment. Apparently, the high contrast & motion of the objects we have examined easily overcomes any binocular rivalry that might be present.

17.3   Experiment 2

Experiment 1 examined the effect of different viewing conditions on the localization of virtual objects superimposed on a physical surface 2.2 m distant. But since new uses of virtual objects are likely to bring them closer to physical surfaces, Experiment 2 examines the effect of introduction of a much closer physical surface. In view of the interacting roles of accommodative and convergence in the discussion of Experiment 1, one could reasonably expect the introduction of a nearby real surface to cause the observers to localize monocularly viewed virtual objects at the same distance of introduced surface. If the accommodative demand for the virtual object is already matched to its displayed distance, one would expect that this improvement in the accuracy of localization would be larger for older observers than younger ones who already would have accommodative cues to the virtual object distance [11]. Accordingly, subjects of different age groups were used while correct accommodative demand to the virtual object was provided. In order to study the effects of introduction of a physical surface, subjects must first judge the distance of the virtual image by itself. This first judgment is identical to those made with the viewing conditions in Experiment 1 with 2 D accommodative demand and provides a chance to replicate that part of the experiment.

17.3.1   Methods

The virtual image stimuli used in Experiment 2 were identical to those in Experiment 1 for the 2 diopter accommodation condition but a new physical stimulus was introduced. This physical surface was a slowly, irregularly rotating checkerboard ($\sim$ 2 rpm) made of xeroxed paper glued on foam-core and was mechanically introduced along the line of sight to the pyramid as illustrated in Figure 17.3. Motion of the checkerboard was introduced because preliminary testing showed that changes in localization that it produced were enhanced by motion. The checkerboard was a disk 29 cm in diameter with 5 cm black and white checks having either 1.3 cd/m^2 or 17.8 cd/m^2 luminance. It was positioned so that the virtual image of the pyramid could be seen against the lower rim of the disk in order to allow the subjects to adjust the physical cursor to the apparent distance of the virtual image in the presence of the disk. Care was taken to be sure the physical cursor was below the bottom of the virtual object and the edge of the disk as in Experiment 1. As before, subjects viewed the virtual objects with either monocular, biocular or stereoscopic view conditions.

Thirteen subjects, seven young (15 -29 yrs.) and six older (38 - 47 yrs.) participated in the experiment. All but one young and one older subject (the authors) were naive with respect to the purpose of the experiment. The others were either recruited through the Ames Bionetics contractor or were laboratory personnel. All subjects were screened for stereo vision as in Experiment 1.

The first part of the subjects' task was to mechanically place the yellow-green LED pointer under the nadir of the slowly rotating, wire-frame virtual pyramid, which varied randomly in size for each trial as in Experiment 1. The second part of the task involved an adjustment of the pointer to match the pyramid's distance after the slowly, irregularly rotating, opaque checkerboard was introduced along the line of sight to the pyramid. The checkerboard was introduced at the previously judged distance of the apex of the virtual pyramid. This fact was unknown to all the naive subjects and remained unnoticed throughout the experiment. Though the virtual pyramid was also presented a second time at the same distance as the first localization, the experimental variations generally concealed this fact from the naive subjects who were led to believe each trial, with or without the checkerboard, involved a potentially different depicted depth. As in Experiment 1, the occasional introduction of unanalyzed sham targets at different depths enforced the naive sub-

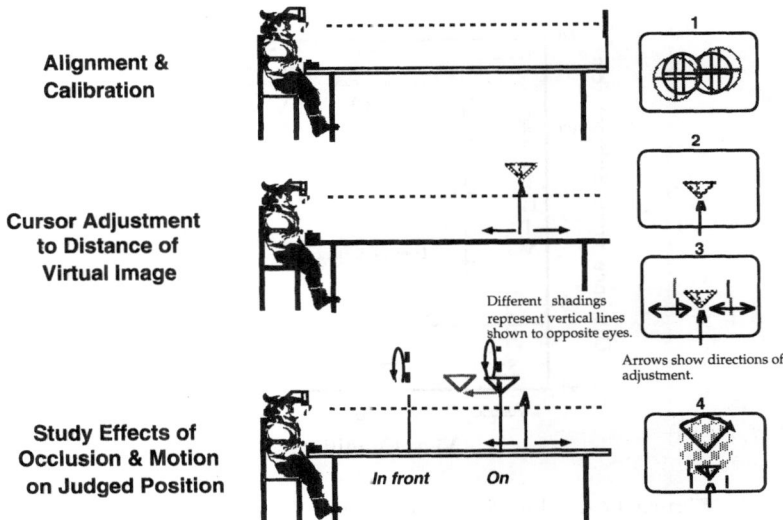

Figure 17.3 Experimental procedure illustration for Experiments 2, 3, and 4. Top: alignment, magnification, and interpupilary adjustment, Middle: Initial localization of virtual object depth, Bottom: Testing conditions representing, the "on" or "in front" placement of the rotating checkerboard and the second localization of the virtual object depth. The rightmost panels 3 and 4 represent use of nonius lines to detect relative convergence in Experiment 3.

jects' belief that the virtual image possibly could be displaced variously in depth for every localization.

17.3.2   Results

Analysis of the subjects' first localization the virtual object under the three viewing conditions with 2 diopters of accommodative demand closely replicate the findings of Experiment 1. The basic result is a significant 2-way interaction of view condition and age (F(2,42)=19.160 p < 0.0009) in which age variation effects the judged distance only for the monocular viewing condition, with the younger subjects (Figure 17.4).

Analysis of the offset of the mean judged distance to the virtual object associated with the introduction of the physical surface also showed a main effect of viewing condition (F(2,26) = 91.340, p <.0001) and a significant interaction between viewing condition and age (F(2,26) = 21.921, p <.0001) (Figure 17.4). These effects modulated the overall significant offset (F(1,13) = 90.623 p <.0001) of the judged distance to the virtual object towards the viewer which was caused by introduction of the physical surface. This effect appears for all viewing conditions as a closer localization of the target after interposition of the physical surface.

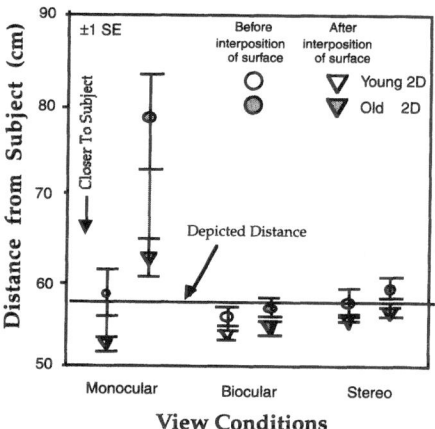

Figure 17.4   Interaction of Age and Viewing Condition.

17.3.3   Discussion

The first virtual object localization shown in Figure 17.4 does replicate the 2 D viewing conditions in Experiment 1 showing that the older subjects are unable to use the accommodative information to estimate the virtual object distance and consequently localize the virtual object erroneously towards the background wall. Interestingly, interposition of the checkerboard at the judged distance of the monocularly viewed virtual object causes a substantial forward movement of the judged virtual object position (Figure 17.4). This change is what would be expected if insertion of the disparity and other cues to nearness of the checkerboard were to cause relative convergence with respect to the position of eyes before its introduction.

It is important to note that the forward movement of the localization is from the initial judged position for each appearance of the isolated virtual object. Since the older subjects tended to initially judge the virtual object to be too far away, the tendency for the checkerboard to bring the judge distance closer was corrective(See Figure 17.4). On the other hand, younger subjects viewing monocularly did not significantly misjudge the virtual object distance, as evident from the error bars. For them introduction of the checkerboard was detrimental, causing the virtual object to be judged too near.

In fact, the checkerboard insertion caused the virtual object to generally appear too near for the other two viewing conditions which otherwise supported correct localization of the isolated virtual object. This was generally true for both age groups and suggests that introduction of the physical surface, the checkerboard, could be causing a small relative convergence under these conditions as well. Though disparity information for correct convergence to the virtual object is available under these two conditions, the virtual object providing this information is not of high visual quality. Its presentation corresponds roughly to 20/100 visual acuity. Under such conditions, convergence based on stereoscopic cues might not be precise and could exhibit a fixation disparity. If this error were an exofixation disparity, i.e. it would tend to the distant side, introduction of the high visual fidelity checkerboard could correct it, causing a relative convergence and associated decrease of the judged distance to the virtual object. If this small corrective convergence were incompletely compensated due to the breakdown of distance constancy, errors of localization could be expected. Thus, the change in localization after introduction of the checkerboard could be due to a change in static convergence.

17.4   Experiment 3

Experiment 3 explicitly tests for such a change associated with the change in judged distance. Attention is focused on the stereoscopic viewing condition to see if the closer judged distance of the virtual object associated with introduction of the physical surface could be associated with an increase in static convergence. Since the amount of expected change is small, for example a change of 3 cm from 58 cm to 55 cm, a sensitive measure of convergence is needed. Angular changes of monocular position of only about 3 arcmin would be expected if the change in localization were explainable by vergence change alone. Such a measurement is difficult to make without encumbering eye tracking technology that preserves a clear visual field. But it is conveniently just at the display resolution of the display configuration used. Therefore, a technique using nonius lines on the display itself was adopted [12]. A nonius line is a line which is broken into two line segments each of which is visible by only one eye. Such lines have typically been used to measure equivalent oculocentric directions to determine the position of the stereoscopic horopter. When the two segments are moved laterally during a period of constant convergence so that they appear to be collinear, their positions may be used to record a specific convergence position.

17.4.1   Methods

Since pilot studies had suggested that longer distances enhance the effect of introduction of the checkerboard, the virtual pyramid was presented at 108 cm rather than 58 cm away from the subjects' eyes. Such an enhancement was deemed helpful to increase the delectability of any convergence. This display, otherwise similar to that used in Exp. 2, was operated under normal room illumination with one diopter accommodative relief for the virtual image. Flanking nonius lines (lower right panels of Figure 17.3) of the same luminance and line width as the pyramid described in the Task section below were also occasionally presented to detect changes in static convergence.

Five men and one woman with measured stereo resolution of better than 1 arcmin participated in the experiment. Some subjects had vision corrected by contact lenses or glasses and were able to wear their corrections during the experiment. Subjects' ages ranged from 17 - 47 and included laboratory personnel as well as paid subjects recruited by a contractor at Ames. Because of the computer control of the experiment it was possible to conduct this experiment double blind.

The subjects task had three basic parts: 1) Localization of an isolated virtual pyramid and measurement of associated static ocular convergence 2) Relocalization of the virtual pyramid in the presence of a real surface either at or in front of the pyramid's apparent distance. 3) Measurement of changes in static convergence associated with the relocalization. The first part of the subject's task was to mechanically place the LED pointer under the nadir of the slowly rotating, wire-frame pyramid. After aligning the pointer, the subjects were presented with two sets of vertical nonius lines just flanking the pyramid (Figure 17.3, right panel #3). These lines were then adjusted to appear vertically collinear, i.e. to have equal visual directions on each side of the pyramid. This adjustment was made by moving the lower left and right segments with a joystick control (See Figure 17.3) and effectively recorded the subjects' static convergence during this part of the experiment. Subsequent brief presentations of the nonius line will accordingly show how static convergence may have changed by revealing vertical misalignments. The second part of the task involved another adjustment of the pointer to the pyramid's depth after the slowly, irregularly rotating checkerboard was introduced along the line of sight to the pyramid. The pyramid was then presented a second time at the same depicted depth in this new configuration. As before, the experimental variations generally concealed this fact from the subjects so that they believed each trial, with or without the checkerboard, involved a potentially different depicted depth. And as before, unanalyzed trials with random variation in distance were introduced to maintain the subjects' uncertainty.

After the second judgment of the pyramid's depth, the nonius lines were flashed briefly (ca 250 msec) next to the pyramid while the subjects fixated it (Figure 17.3, right panel #4). Then the subjects made a forced choice indicating whether the upper or lower pair of the flashed nonius lines appeared closer laterally. The eye assignments of each segment of the nonius lines were randomly selected so that the meaning of the alternative possibilities in terms of convergence or divergence varied randomly across the trials. The assignment of the lower part of the left nonius line and the upper part of the right line to one eye and the other upper-lower pair to the other eye, produced a differential effect doubling the relative misalignment for any given vergence change and increasing the sensitivity of the technique for detecting changes in convergence. The subject reported by a button press which of the paired nonius lines were closer.

In fact, three different experimental conditions were used in the second part of the experiment because of the need for a control case. In the "on" condition the checkerboard was mechanically introduced at the judged depth of the virtual pyramid object so that the pyramid appeared "on" the checkerboard. For the "in front" condition the checkerboard was introduced 30 cm in front of the judged depth. In the control condition the second judgment was a replication of the first judgment in that the subject made a second judgment of the depth of the virtual object. But this time the subject made the forced-choice judgment of the nonius lines alignment without the addition of the checkerboard. Thus, the control was identical to the experimental conditions except the checkerboard was not introduced into the line of sight. Therefore, this control provides an individual baseline for subjects' judgment biases and changes of their convergence during the course of a trial. Each condition was repeated 15 times for each subject in a randomized block design in which blocks of 5 replications of each condition were repeated. The 6 possible orders of the 3 conditions were distributed randomly across the 6 subjects in the experiment.

The change in judged distance of the virtual object was analyzed in a single factor repeated measures ANOVA. Chi-square analyzes were conducted on each individual subject's distribution of judgments of convergence/divergence for each of the 3 experimental conditions. Taking the control condition as a baseline, the relative strength of convergence could be measured by a ratio of the probability of convergence in each experimental condition to the probability of convergence in each subject's individual control. This ratio allows control for the possibility that subjects might have an individual bias to converge or diverge simply because of a repeated presentation of the virtual object.

17.4.2 Results

Single factor repeated measures analysis of the effect of superposition of the checkerboard and virtual images replicated the previous observations that the virtual object was moved closer to the viewer $(F(2,10) = 7.549 \ p < 0.01)$. Individual data are shown in Figure 17.5. This effect was somewhat stronger for the "on" condition than for the "in front" case and varied in strength across the 6 subjects. One subject interestingly showed no major effect.

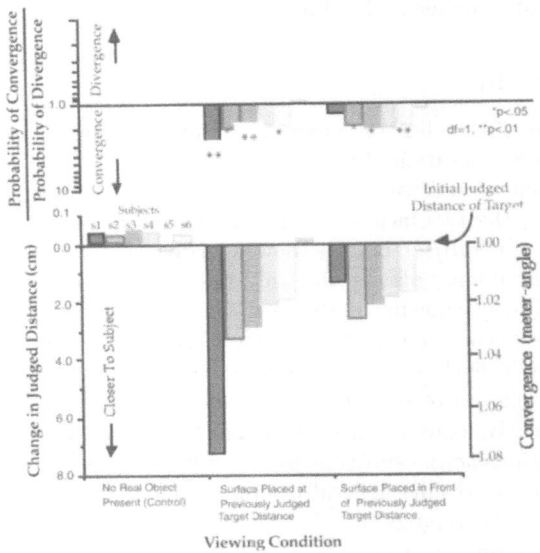

Figure 17.5   Relative convergence vs. judged distance.

The cause of this individual subject's result is illuminated by considering all subjects' tendency to relatively converge during judgment of the depth of the virtual object in the presence of the checkerboard. This tendency is summarized for the experiment in Table 17.1 which displays the frequency of convergence or divergence indicated by the nonius judgments for all subjects in the three experimental conditions (Chi-square = 20.37, df=2, p< .001). The control case shows the expected 50:50 break, collapsing across all subjects, while the other two conditions show clear convergence, the "on" condition being somewhat stronger.

Table 17.1   Frequency of convergence to divergence.

	Convergence	Divergence	Total Judgments	Ratio of Convergence to Divergence
On	84	21	105	1.58
In Front	70	35	105	1.32
Control	53	52	105	1.00

For further analysis each subject's individual tendency to converge was computed separately as the ratio of their probability of convergence in an exp. condition to their probability of convergence in the control. These ratios are plotted in Figure 17.6 for each subject. Since the control was used as reference, all ratios for the control condition are 1. A 2x2 Chi square contingency was also computed to compare the distribution of convergence and divergence for each exp. condition to that of the control. This was done separately for each subject for whom statistically significant differences are indicated by asterisks (Figure 17.5).

17.4.3   Discussion

The individual subject's localization errors in Figure 17.5 are sorted by the size of the change in the judged position of the virtual object for the "on" condition. These results can then be compared with the ratio of the convergence probabilities. As is clear from the figure, the two measurements are almost perfectly correlated across the subjects. The only subject not to show a displacement of the virtual object caused by the checkerboard, also is the only one to show essentially no relative convergence. The subject showing the largest displacement due to introduction of the checkerboard is also the one with the strongest tendency to converge. The results for the "in front" condition show a weaker apparent displacement of the virtual image but also show a correlation of convergence tendencies and changes in localization. The correlation of relative convergence with magnitude of displacement for the "on" and "in front" conditions across subjects and conditions is; in fact, r= 0.894 (df=10; t=6.31, p < .002). These results generally support the supposition that the change in judged depth could be related to a change in convergence, but the mechanism underlying this change remains to be clarified. Correction of a fixation disparity, for example, by introduction of a high resolution physical stimulus, for example, is not the only possible mechanism.

One other possibility is that change in convergence is due to so-called perspective [13] or proximal vergence [14]. These phenomena are changes in convergence due to changes in the apparent nearness of objects. They provide evidence that spatial interpretations of the distance of a visual image themselves can simulate the vergence system. Accordingly, the results from the present experiment, while showing that there is a clear oculomotor response associated with the error in judged depth, does not resolve its cause. A change in the apparent nearness of the virtual object due to its appearing to occlude the nearby checkerboard could be the cause of the measured convergence rather than its consequence. This question can only be resolved experimentally.

17.5 Experiment 4

One approach to analyzing whether the oculomotor effect, i.e. the convergence, observed in Experiment 3 is caused by the superposition of the virtual object on the background is to devise a stimulus condition which on one hand strengthens the oculomotor cues to convergence but on the other hand weakens the visual evidence for occlusion, thus reducing the likelihood of convergence caused by proximal vergence.

We have attempted to create such a stimulus by cutting an annular slot 8 cm wide out of our rotating checkerboard so that the virtual pyramid would be just able to "fall through" the resulting hole (see Figure 17.6) . The outer rim of the checkerboard was supported by thin radial wire matching the color of the background wall and therefore being invisible to the observers. This stimulus triples the number of moving edges that provide the strong disparity discontinuity which could be the stimulus to convergence that could be the cause of the change in static convergence observed in Experiment 3. If the better stimulus to convergence provided by the checkerboard were the cause of the change in static convergence, this stimulus should strengthen the effect. On the other hand, the slotted hole virtually eliminates the visual evidence for occlusion. If proximal vergence triggered by occlusion were the cause of the change in judged distance, one would expect not to find a change in the judged distance of the virtual object when it is presented in the slot.

Figure 17.6   Slotted physical surface.

17.5.1 Methods

Nine subjects aged 23 - 47 who were either laboratory personnel or paid subjects provided by Bionetics were used in Experiment 4.

This experiment was conducted using a methodology equivalent to that of Experiment 2 for stereoscopic virtual objects. Two different depicted distances of the virtual objects, 83 cm and 108 cm, were randomly ordered into blocks of 20 runs. In each block either a solid checkerboard or a slotted checkerboard was introduced along the line of sight of the virtual object. After introduction of the checkerboard, the change in the judged distance to the virtual object was measured. Block types were alternated for all subjects. All but one subject, from whom half of his data was lost, experienced 4 blocks, making a total of 80 judgments per subject. The order of presentation of the two checkerboard types was counterbalanced across subjects.

17.5.2 Results

Analysis of variance showed that while the solid disk caused a previously observed offset of the judged virtual object distance towards the observer (Mean: 2.80 cm; SE: ±0.75 cm), the slotted disk caused only a mean 0.72 cm (SE ±0.49 cm) change. This difference was statistically significant. $F(1,8) = 19.605$ p < 0.002. The offset for the slotted disk condition, though quite close to 0, is statistically significantly less than 0. (t=-2.59, df=8, p< 04)

The size of the offset was larger for the greater depicted distance. The offset for the 83 cm virtual object was 1.49 cm (SE ±0.43) and 2.03 cm (SE ±0.57) for the 108 cm object. This difference was just significant ($F(1,8) = 5.349$ p < 0.05). There was no statistical interaction ($F(1,8) = 3.068$ p > 0.05) so the effects of placing the virtual object in the slot was statistically indistinguishable for the two presentation distances. Accordingly, the data from the two presentation distances may be collapsed as in Figure 17.7. This figure shows the full distribution of all the subjects responses illustrating the effect of the slot on the judged distance to the virtual object.

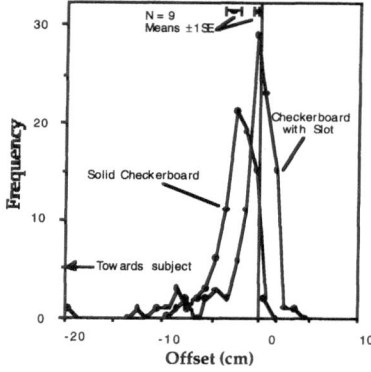

Figure 17.7   Hole effect on judged distance to virtual objects.

17.5.3 Discussion

As is clear from Figure 17.7, introduction of the slot in the checkerboard that removed the occlusion between the virtual object and the checkerboard greatly reduced the offset in judged virtual object position produced by the checkerboard introduction. This reduction makes the oculomotor explanation of the change less likely since the binocular depth cues of the checkerboard would be expected, if anything, to strengthen any vergence response. Since the overlap of the virtual object contours and those of the checkerboard seems to be the key feature causing the shift in its judged position, the occlusion cue placing the virtual object in front of the checkerboard seems to be the best explanation for the nearer localization of the virtual objects. The change in static convergence, thus, appears to be a consequence of proximal vergence and resembles effects in recent reports of convergence being driven by the kinetic depth effects which produce a perception of nearness [15].

There is, however, another possible interpretation of the results which could be based on the proposal that the alternative explanations of the offset of the judged distance of the virtual object are not mutually exclusive [16]. In this view the introduction of the slotted checkerboard stimulus could introduce a proximal vergence tendency to fixate farther away since any objects seen through the slot would necessarily be more distant than the slot. Were this tendency for a more distant fixation to occur, it would oppose any tendency of the visual information to reduce oculomotor bias. The net effect of the opposition could account for the almost negligible forward displacement of the virtual object observed while subjects were exposed to the slotted checkerboard.

One way to dissociate the two influences could be to study the individual oculomotor biases of each subject, measuring for example their phoria with a Maddox Rod test or equivalent. To the extent subjects exhibit exophoria, one could expect a exofixation disparity while viewing the virtual object. Its removal by presentation of the checkerboards could be associated with the change in judged distance. An explanation of the observed effects of the checkerboards solely in terms of proximal vergence would not predict a correlation between the phoria tests and the subjects' individual phoria variations. Accordingly, future investigations could examine individual subject's oculomotor biases as a technique to more precisely determine the cause of the offset of the change in judged distance associated with superimposition of virtual objects against physical surfaces.

17.6   Design Considerations

1) The present results were observed with a static eye point and can be expected to change when significant lateral head-movement producing motion parallax is introduced. Nevertheless, it is important to realize that since many of the new applications of head-mounted see-through displays, in fact, will involve relatively static viewing, the conditions used remain practically relevant.

2) Since weight and cost considerations may argue for the use of monocular displays, they are likely to be initial candidates for many applications. Accordingly, such displays should have a variable focus control to appropriately direct the convergence of prepresbyopic users. Designers and supervisors should be aware that operators over 40 will generally not benefit from the variable focus adjustment.

3) Biocular and stereo displays should be used with a bore-sighting procedure in which focus is adjusted to a reference target so as to correct for any errors in depth due to inappropriate vergence.

4) Computer generated or other targets presented binocularly should have individually tailored stereo disparity to correct their spatial localization so as to compensate

for the tendency of virtual objects to appear to float in front of the surfaces that they are seen against.

Acknowledgments

Earlier reports of the above: 1995, 1996 Proc. HFES,1995 Psychonomics Soc., the 1995 IFAC Proc. A more detailed report will appear in *Human Factors*.

References

[1] R. Azuma and G. Bishop: "Improving static and dynamic registration in an optical see-through HMD," *Proc. SIGGRAPH '94*, July 24-29, Orlando, Fl., New York: ACM, pp.197–204, 1994.

[2] A. L. Janin, D. W. Mizell, and T. P. Caudell: "Calibration of head-mounted displays for augmented reality applications," *Proc. VRAIS '93*, Seattle New York: IEEE, pp.246–255, 1993.

[3] J. P. Rolland, D. Ariely, and W. Gibbon: "Towards quantifying depth and size perception in 3D virtual environments," *Presence*, vol.4, no.1, pp.24–49, 1995.

[4] S. R. Ellis and B. M. Menges: "Judged distance to virtual objects in the near visual field," *Presence*, vol.6, no.4, pp.452–460, 1997.

[5] R. A. Moses: "Accommodation," in (R. A. Moses and W. M. Haret Jr., eds.), *Adler's Physiology of the Eye*, Washington D.C., Mosby, pp.291–310, 1987.

[6] W. C. Gogel and J. D. Tietz: "Absolute motion parallax and the specific distance tendency," *Perception and Psychophysics*, vol.13, pp.284–292, 1973.

[7] D. A. Owens, and H. W. Liebowitz: "The specific distance tendency," *Perception and Psychophysics*, vol.20, pp.2–9, 1976.

[8] D. A. Owens and H. W. Liebowitz: Perceptual and motor consequences of tonic vergence, in (K. J. Ciuffreda and C. M. Shor, eds.), *Vergence Eye Movements: Basic and Clinic Aspects*, p.50, Boston: Butterworths, 1983.

[9] B. Zuber: "Physiological control of eye movements in humans," Ph.D. thesis, p.103, Cambridge, Mass: MIT, 1965.

[10] J. L. Brown: "Afterimages," in (C. H. Graham, ed.), *Vision & Visual Perception*, p.485, New York, Wiley, 1965.

[11] S. R. Ellis and B. M. Menges: "Effects of age on the judged distance to virtual objects in the near visual field," in (W. A. Rogers, ed.), *Designing for an Aging Population*, Santa Monica, CA: Human Factors and Ergonomics Society, pp.15–19, 1997.

[12] S. R. Ellis, U. J. Bucher, and B. M. Menges: "The relationship of binocular convergence to error in the judged distance of virtual objects," *Proc. the Int'l Federation for Automatic Control*, Boston, June 26-27, pp.297–301, 1995.

[13] J. T. Enright: "Paradoxical monocular stereopsis and perspective vergence," in (S. R. Ellis, et al., eds.), *Pictorial Communication in Virtual and Real Environments*, London, Taylor and Francis, pp.567–576, 1991.

[14] K. J. Cuiffreda: "Components of clinical near vergence testing," *J. Behavioral Optometry*, vol.3, no.1, pp.3–13, 1992.

[15] D. L. Ringach, M. J. Hawken, and R. Shapley: "Binocular eye movements caused by the perception of three dimensional structure from motion," *Vision Research*, vol.36, no.10, pp.1479–1492, 1997.

[16] W. Shebilske: personal communication, 1997.

Chapter 18

Augmented Reality: A Balancing Act Between High Quality and Real-Time Constraints

Gudrun Klinker
Technical University of Munich, Germany

Didier Stricker
Dirk Reiners
Fraunhofer Project Group
for Augmented Reality at ZGDV, Germany

18.1 Introduction

Augmented Reality (AR) constitutes a very powerful three-dimensional user interface paradigm for many "hands-on" application scenarios in which users cannot sit at a conventional desktop computer. Users' views of the real world are augmented with synthetic information from a computer. Users can thus continue their daily work involving the manipulation and examination of real objects. At the same time, they receive additional information about those objects and the task at hand, such as up-to-date instructions how to perform the next step of a task. These concepts have been demonstrated for construction and manufacturing scenarios like the computer-guided repair of copier machines [1], the installation of aluminium struts in diamond shaped spaceframes [2], for electric wire bundle assembly before their installation in airplanes [3] [4] and the assembly or repair of machines [5] [6].

Current AR research fans out into several different activities, all of which are essential to generating a system which eventually will be able to sustain a truly immersive AR-experience in extended practical applications rather than short laboratory demonstrations: Virtual objects need to be presented as realistically as possible, integrated physically correctly into the real world. This means that occlusion and light reflection properties between virtual and real objects must be established and maintained, as well as physical laws such as non-penetration, gravity and friction [3] [7]. Furthermore, users must be free to roam an extended area, without being tethered to a stationary system [8]–[10]. Thus, AR-systems must be wearable and mobile, either by carrying all information "on-board" or by being wirelessly connected to distributed sources of information. At the same time, the system should facilitate collaboration with other AR users, allowing them to work together [3] [11]. Finally, in order to make all augmentations worth their while, AR systems must be able to correctly track user motions and even predict future motions ahead of time [12] [13] such that virtual objects are rendered according to the user's changing perspective.

The current state of technology cannot yet provide simultaneous support for an optimal solution to all aspects of AR. Most critical in this respect is the real-time performance of the overall demonstration system. Today's AR systems have to balance a wealth of trade-offs between striving for high quality, physically correct presentations and user modelling on the one hand, and making short cuts and simplifications on the other hand in order to achieve a real-time response. Such trade-offs occasionally involve rather perplexing alternatives: not always is the physically most precise approach the best one since a much coarser approach may be so much easier to compute that it can run in real-time - just fast enough to keep pace with the user's actions whereas the former one never even begins to get a grasp at integrating with the quickly changing real world environment. A beautiful picture is useless, if it's rendered too late. Similarly, a precise model of user motion, involving translational and rotational speeds as well as accelerations is useless, if it cannot adapt in real-time to the erratic head motions of a user assembling a car door.

In our work, we have selected two different positions among many possible trade-offs. In one approach, we emphasize the real-time immersive impression that can be generated with today's technology [14] [15]. Our on-line presentations give the user immediate feedback to his actions and thus generate a very tight, immediate human-computer interaction scheme. Our second approach is intended to present a glimpse of the future, forecasting what quality might be achievable with continuously increasing processing power and data bandwidth. In these demonstrations, we currently use pre-recorded video clips which we analyze and augment semi-automatically off-line [16] [17].

Due to the current need for trade-offs, the optimal configuration of an AR-system depends on the needs and acceptable simplifications of a particular application, as well as on the selected hardware base. Design decisions may change towards including more sophisticated approaches when new applications are chosen or better hardware becomes available. We thus begin by presenting some of the applications and demonstrators we have built in the recent past (Section 18.2), as well as the underlying hardware and software architecture (Section 18.3). Further sections discuss the trade-offs we made towards developing user tracking algorithms (Section 18.4, Section 18.5), and scene augmentations (Section 18.6).

18.2 Applications / Demonstrations

Our two approaches to AR have been the basis of a number of demonstrations. We are using the off-line AR system to shcw how AR can be used to augment outdoor scenes with proposed new buildings, as part of the acquisition and bidding phase of large exterior construction projects (Section 18.2.3). We use our real-time system in all phases of the life cycle of an object, such as object design (Section 18.2.1, Section 18.2.2), object assembly or construction (Section 18.2.4, Section 18.2.5), and object maintainance (Section 18.2.5). Finally, we explore the interactive, world-changing nature of AR applications in a board game scenario, augmented Tic Tac Toe (Section 18.2.6).

18.2.1 Model Presentation and Physical Manipulation

In many industry sectors (e.g., architecture, automotive design, etc.), digital three-dimensional prototypes of a designed object are becoming common place. Viewing such models is a typical application of Virtual Reality, as well as more mundane 3D viewing kits. Augmented Reality provides a new, very intuitive approach towards viewing and manipulating virtual objects [14] [18]. As shown in Figure 18.1, objects, such as a VRML model of St. Paul's Cathedral in London, can be attached to a physical placeholder - a cardboard with a few markers. Users can then manipulate the virtual object simply by moving or rotating the attached physical place holder - without having to deal with complex popup menus full of sliders or dials.

Figure 18.1   St. Paul's cathedral.

18.2.2 Mixed Virtual/Real Mockups

Going one step further, the virtual prototypes can be mixed with partially existing physical prototypes, thus forming mixed virtual/real mockups [17]. Physical prototypes are still essential to evaluating the design of many products, such as cars and buildings. Such physical mockups are time-consuming and expensive to create. They are thus built only at very critical stages of the design process - typically after many of the preliminary decisions have already been made. AR provides the opportunity to build mockups more gradually, using physical prototypes for the already maturing components of the design and inserting virtual models for the currently evolving components. Figure 18.2 shows a real toy house in combination with two virtual buildings. Both the virtual and real objects can be manipulated in the scene by moving associated markers.

Figure 18.2   Manipulation of virtual and real objects (see color pages).

Figure 18.3 shows the relationship between mixed mockups and the already well-known concept of an enhanced desk [19] or Responsive Workbench [20]. In this case, the physical scene is laid out on a planar desktop covered by a (real) map of the city of London. The paper is augmented with a VRML model of St. Paul's cathedral, as well as with a CAD model of a new footbridge across the Thames that is being designed. The new bridge can be moved about interactively until it is placed in the correct spot on the map. The enhanced map is a special case of the more general capabilities of AR. The real world doesn't have to be a flat desk. It can assume arbitrary 3D shapes and become a 3D terrain.

By waving a hand across the virtual camera icon at the river shore, users can request to see a video clip showing this area of London augmented with the proposed new footbridge (see Section 18.2.3). This illustrates that AR provides users access to all kinds of synthetic information, be it three-dimensionally integrated into the scene or presented like a movie or a graph, in a flat, 2D window.

18.2.3   Augmented Landscapes and Cityscapes for Architectural Design

As part of the project design and acquisition process, architects need to convince their prospective customers that the proposed building will fit well into the existing environment. This concern is of particular importance in the context of large objects that will have an impact on the landscape or the city skyline of a town, such as new

Figure 18.3   Augmented map of London (see color pages).

bridges, towers, or major exposition areas.

AR technology can help visualize such new buildings in their eventual environmental setting. In the context of the European CICC project, we have augmented video sequences of several potential construction sites with new buildings to facilitate communications between architects and community leaders [17]. Figure 18.4 (a) shows the shore line of the river Wear in the Sunderland area, Newcastle, UK, augmented with a proposed Millennium bridge. Similarly, Figure 18.4 (b) shows the London Thames area with a millennium footbridge connecting the area near St. Paul's cathedral with the Tate museum. Both objects have been designed by Sir Norman Foster and Ove Arup. In Figure 18.4 (c), the Expo'98 construction site in Lisbon has been augmented with the model of a pavilion being built by Europroject Ingeniera (Spain).

Exterior construction applications impose very demanding challenges on the robustness and usability of evolving AR technologies. Real construction sites are huge. Information has to be integrated into many views, both at close range and from long distances, requiring a significant range of tracking skills. Furthermore, construction environments are not well structured. Information has to be mixed plausibly with existing natural objects such as bushes and trees and heaps of earth, that are likely to change over time.

Since video sequences of such environments are very complex, we currently pre-record the sequences and employ off-line, interactive calibration techniques to determine camera positions for every frame, as well as a scene description. Given all calibrations, the augmentation of the images with virtual objects can be performed live on a high-end graphics computer. These demonstrations thus offer a glimpse of the future, indicating what kind of complexity AR technology needs to be able to deal with automatically in order to become usable in outdoor exterior construction applications.

18.2.4   Augmented Car Door Assembly

The phase following object design involves object construction and assembly. During this phase in the life cycle of an object, AR finds many obvious applications. Hands-on work, such as the assembly of a car door, currently is generally performed without

(a) Sunderland bridge   (b) London bridge

(c) Expo'98 pavilion in Lisbon

Figure 18.4   Augmented landscapes and cityscapes (see color pages).

the benefit of much computer assistance since such work has to take place far away from desktop computers. AR provides the means for bringing a wealth of information into the workplace in the form of up-to-date 3D illustrations of work steps to be performed or objects parts to be manipulated.

We have demonstrated such concepts during the Hanover Industry Fair '98 at the example of an AR-assisted assembly of a door lock for a car [15]. The task of assembling a doorlock is quite challenging, requiring significant planning and dexterity. It is very spatial and three-dimensional in nature. The movement of the hand holding the lock in the small space inside the door requires precise preparation and motion. Since the space inside the door is just big enough for the lock, it has to be held in a very special way for the hand not to get stuck halfway through. Several screws then have to be inserted and fixed in the right order.

In our augmented car door assembly demonstration, the real-time AR-system instructs the user step-by-step how to hold the door lock, where to insert it with what kind of hand motion, what levers to push and what screws to fix. All illustrations are shown as 3D augmentations to the real car door. In his heads-up display, the user sees in stereo how the virtual objects coexist with the real door, being partially hidden inside the door as part of the process. The user controls stepping through the assembly routines via voice input, requesting to proceed to the next part of the illustration whenever he is ready. Figure 18.5 shows a snapshot from the demonstration during the lock insertion stage.

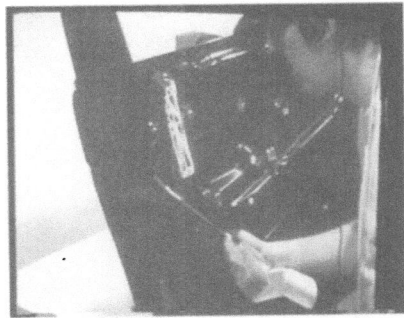

Figure 18.5 Augmented car door assembly: Doorlock insertion (see color pages).

18.2.5 Building Construction and Maintainance

AR is also useful in both the construction and the maintainance phase of buildings [17]. As shown in Figure 18.6, virtual walls can be shown before they are built, thus helping construction workers determine their exact location. Using up-to-date online information, workers are thus guaranteed to work with the latest version of an ever-changing construction plan. Furthermore, the plan can be sequentialized into small construction steps suited to the current schedule of immanent activities. Figure 18.7 shows a set-up for a small bathroom under construction. The water pipes have already been installed, and a dry wall is scheduled to be installed in front of the pipes the next day. Figure 18.7 shows a virtual dry wall in its place.

Figure 18.6 Outdoor construction site, augmented with a virtual wall about to be built (see color pages).

Once the construction has been completed and the building has been taken into commission, the next phase in the life cycle of a building begins: its maintainance, repair and modification. To this end AR can help maintainance crews access all available information about the building in a suitable manner. For example, many walls are photographed before electric and sanitary installations are covered by the final layers of plaster. AR can overlay such visual data on the walls, thus providing people with x-ray vision skill to find (or avoid) electric wires or pipes in the wall (Figure 18.8) when drilling holes.

Figure 18.7   Augmented exterior construction bathroom partially augmented with a virtual dry wall (see color pages).

Figure 18.8   Same room after completion, showing the real dry wall with a semi-transparent augmentation of the piping in the wall.

18.2.6   Augmented Tic Tac Toe

At the example of board games like Tic Tac Toe (Figure 18.9) we are exploring various concepts of mouseless 3D user interaction [14] [17]. To fully exploit the AR paradigm, the computer must not only augment the real world, it also has to accept feedback from it. In truly 3D human-computer interaction, actions or instructions issued by the computer cause the user to perform actions changing the real world - which, in turn, prompt the computer to generate new, different augmentations. Gesture languages, 3D pointers or speech input are all tools with which users can communicate with the computer about their work at an abstract level. If the computer is capable of automatically detecting and correctly interpreting scene changes caused by user actions, much such meta-level communication becomes superfluous.

In the Tic Tac Toe demonstration, the user sits in front of a real game board wearing a head-mounted display with an attached mini-camera. The user and the computer alternate placing real and virtual stones on the board. During the user's turn, he can try out various moves, playing them out on the board. When he

Figure 18.9   Augmented Tic Tac Toe (see color pages).

eventually settles on one, he indicates his choice to the computer by waving his hand across a virtual 3D "GO" button or by speaking a command into an automatic speech recognition system. The computer then scans the image, looking for a new stone. If it finds one, it proceeds by planning its own next move. If it doesn't find a new stone or if it finds more than one, it writes an appropriate comment on the virtual 3D panel placed behind the board game. Note that the user is not requested to indicate by voice, text or other means where he has placed the new stone - the computer interprets and evaluates the user's action automatically.

18.3   System Architecture

18.3.1   Real-Time AR System

Our real-time AR-system (Figure 18.10) uses an SGI O2 with a 180 MHz R5k processor and 128 MB memory. The machine has very good video capabilities and reasonably fast rendering, although not comparable to higher end SGI machines used for virtual reality applications.

The strong spatial nature of AR demands a display that can convey spatial information, such as an HMD. We use Virtual IO! i-glasses with an attached Toshiba IK-M48PK camera using a 7.5 mm lens (Figure 18.1, Figure 18.11 (a)). The headset is a standard affordable piece of equipment allowing see-through and feed-through use. The camera is reasonably small and light enough to be worn on the head without undue strain for the user, while still giving very good quality for a single-CCD camera.

Before settling on the minicamera attached directly to the HMD, we have run our live AR-system in a monitor-based set-up with cameras held in the user's hand or positioned on a tripod (Figure 18.11 (b)). The tracker worked very reliably across a wide range of cameras including high quality Sony 3CCD Color Video Cameras as well as low-end video-conferencing cameras such as an SGI IndyCam.

Two-handed action in many of our demonstrations (e.g., car door assembly) task requires a hands-free interaction technique. In some applications, we thus use a

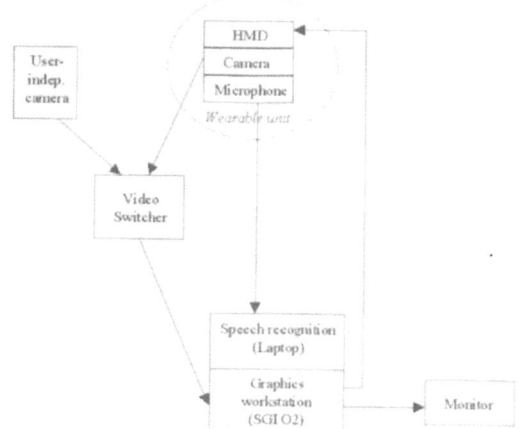

Figure 18.10   Hardware set-up.

(a) HMD with attached mini-camera   (b) Monitor-based AR

Figure 18.11   AR set-ups.

voice-driven interface. It runs on a separate machine, a Laptop running Windows 95 and an IBM Voice-Type based speech recognition software. It is connected to the O2 via RS-232 which is adequate for the transmissions of short, pre-defined commands.

Figure 18.12 shows the software architecture of our live AR system. It revolves around the central tracker loop. The tracker deals with reading and analyzing the video image to calculate the camera calibration parameters. A second, coupled component provides the application dependent augmentations and handles user interaction modalities [15].

Experimentally, the system was distributed between two machines, one for tracking (SGI O2) and one for rendering more complex virtual objects (an SGI Onyx RE2). Only the calibration information was sent via Ethernet between the two machines. But the lag resulting from the system and the network proved to be too

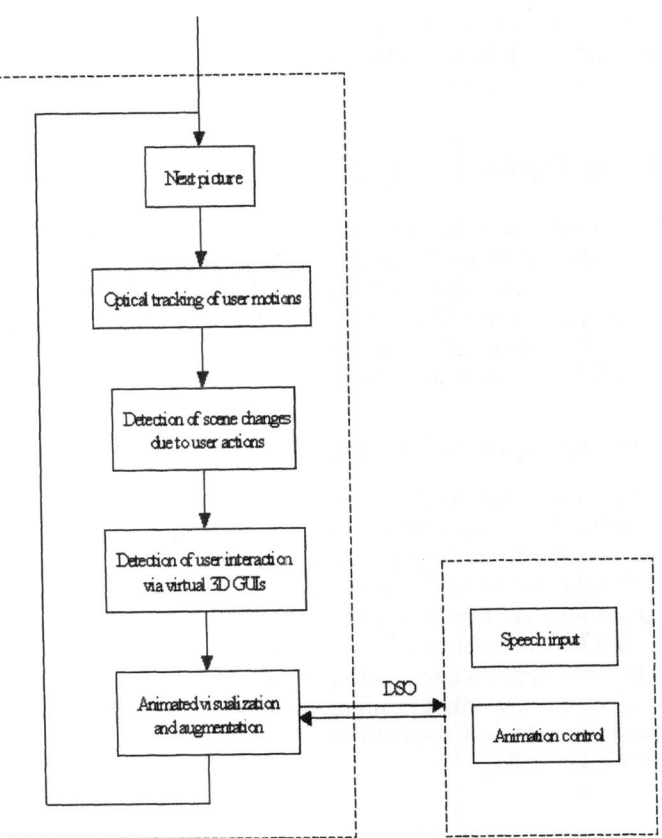

Figure 18.12   Software architecture.

bad to be useful for head-mounted applications. Thus, we now prefer using a single processor system.

The tracking itself runs at 20-25 Hz. Combined with the additional rendering task the speed drops. In most of our demonstrations, we have observed speeds of about 10-15 Hz.

18.3.2   Off-line AR System

For the off-line demonstrations, timing is not very critical. We have developed an interactive calibration system (InCal) which runs on any OpenGL-based machine, such as a low-end SGI workstation. It generates a calibration file for every frame of an image sequence [17].

The subsequent live augmentations of the video sequence require a high-end graphics workstation (e.g., an SGI Oryx RE2) in order to render the very complex virtual buildings at sufficient speed. Our AR-Viewer is a highly optimized SGI-

Performer-based video augmentation system, using InCal's calibration files to insert virtual objects into each frame of the video sequence while simultaneously allowing the user to interact with them.

18.4   Live Optical Tracking of User Motions

Tracking user motions is currently the most intensely investigated aspect of AR - due to its critical importance to the overall performance of an AR system. In the past, quite a few technical approaches have been tried (magnetic, inertial, sound, optical). None works perfectly. Currently, the most successful approaches track optical markers in indoor laboratory demonstrations [9] [13] [21]–[24] sometimes combined with information from other tracking modalities in hybrid approaches.

18.4.1   Use of Optical Markers

In order to achieve real-time optical tracking performance, simplifying assumptions have to be made. It is currently customary for AR labs to place special, easily recognizable markers at carefully measured locations in the scene and make such information available to the tracker for its operation. The placement of such markers certainly is quite a severe restriction to the overall applicability of the system. It is tedious to install a significant number of markers necessary for robust operation of the system. In some applications, such an approach is completely impractical. Yet, more general solutions are only beginning to approach real-time speed. The current demonstrators thus provide a good starting point to begin experimenting with more general concepts [25] [26].

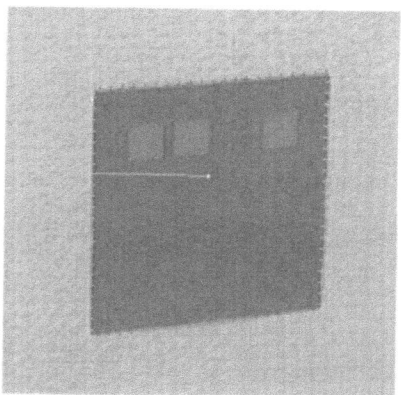

Figure 18.13   Picture of a black rectangular (square) target.

In our approach, we use black rectangles (typically squares) on a white sheet of paper (Figure 18.13) [13]–[15] [17]. The exact size and shape of each rectangle is specified in a scene configuration file. We can place targets of different sizes at more or less confined positions of the scene (e.g., a car door). Large targets can be identified from quite a distance. Small targets in many intricate positions help the tracker when the user comes close. The system doesn't depend much on the exact illumination conditions and the camera sensing parameters since black rectangles on a bright background can be identified very easily from high-contrast edges. We have been able to run demonstrations in many different settings with a variety of cameras (various demonstrations at fairs, conference exhibits, and other people's laboratories or conference rooms).

To uniquely identify each rectangle independently of the current field of view, the rectangles contain a labeling region consisting of 2 rows with 4 positions (bits) each of small red squares. Using a binary encoding scheme, we can define up to $2^8 = 256$ different targets. In any particular image during a demonstration, any subset of two targets suffices for the system to succeed in determining the current user position. Since targets may rotate by more than 45 degrees when they are attached to mobile objects (Figures 18.1, 18.2), we restrict the labelling scheme slightly: label 1 ("0001") can be confused with label 8 ("1000") when rotated right by 90 degrees. We thus discard such potentially misinterpretable labels. In principle, this problem could be alleviated by adding one further unsymmetrically placed symbol to the target. We could also exploit the non-square aspect ratio of rectangles. Yet, any change in the target design and its detection algorithm can have unexpected negative implications on the robustness and speed of the system and thus needs to be thoroughly investigated - which we haven't found time to do yet.

A similar target design based on pentagons rather than rectangles has recently been developed by Mendelsohn, et al. [25]. In their system, the bottom edge is gray rather than black, allowing the system to determine potential target rotations. Furthermore, the identifying labels are placed under the pentagon rather than inside it, playing the trade-off concerning the extent of a target versus the ease of detecting it in a scene differently than our system: the larger the extent of a target, the more likely it is to be partially occluded or out of the field of view. The concept of simplifying the target can be investigated further by making each target even simpler (small dot, or maybe an LED) and determining its identity from its position in unique groupings of several targets. Such approaches are used in commercial tracking systems and also in photogrammetry.

Other optical AR systems currently work with multi-colored, concentric circular targets [22] [23]. When using c different colors in r concentric rings, the multi-coloring scheme can distinguish between up to cr different markers. Typical numbers are $c=4,6$ primary and secondary colors, and $r=2,3,4$ concentric rings, thus allowing between $4^2 = 16$ and $6^4 = 1296$ different targets [22] [23]. There is a trade-off between using a photometric (coloring) scheme versus using a geometric (positional) encoding scheme. Targets distinguished by their colors can be smaller than those using binary codes that have to be discernible from some distance. Colored targets are also less susceptible to partial occlusions. On the other hand, they are less tolerant to changing illumination and sensing conditions than a geometric approach which essentially operates on binary image data. Approaches using multi-colored concentric rings merge the benefits and problems of both approaches, requiring a certain spatial extent for the rings to be distinguishable, as well as being dependent on the illumination.

18.4.2   Initialization of the Optical Tracking System

Our system uses fast image processing techniques to initially find the rectangles in an image [14] [17]. The robustness, simplicity and speed of not only the on-line tracker but also its initialization routine is absolutely essential to the overall acceptability of the AR system. Even the fastest optical tracking systems cannot always keep pace with erratic user head motions. Thus, trackers tend to fail every once a while, requiring a re-initialization. A fast recovery from tracking failure is absolutely essential since users won't stop moving just for the system to re-initialize. Our initialization routine currently runs at about 3 Hz.

To find the black rectangles in an image, we scan every nth line for strong white-to-black image gradients followed by similar black-to-white gradients. A pair of such gradients is hypothesized to indicate the left and right edges of a rectangle. We use the center in-between them as a starting point to vertically seek for a third black-to-white gradient defining one of the horizontal edges. The horizontally and vertically scanned pixels are evaluated with respect to the overall homogeneity and blackness of the potential target. Inhomogeneous or very bright target candidates are discarded.

We next follow the border around each dark blob, starting from one of the three edge points. To fit rectangles to each blob, the edge pixels are classified into four clusters according to their gradient directions, using a standard ISO-data algorithm. After a statistical homogeneity test, we fit straight lines to each of the four edge clusters. The intersection of neighboring lines defines the corner points of the rectangle.

To determine the ID of a target, we scan the image along two lines known to pass through the two rows of red squares. The position of each row is defined by an offset from the top left corner of the rectangle, scaled relative to the length to the top edge. The size of the red squares and their position in each row is also defined relative to the length of the edge. We sample each line and correlate it with templates representing encodings of labels 0000 through 1111, selecting the label yielding the highest score. This method has proven to be very robust, working well even with low quality cameras and under bad illumination conditions.

When a set of markers has been identified in an image, we use one of a suite of calibration routines to determine the current camera position from a mapping between scene co-ordinates and image co-ordinates of those corner points of the targets that have been reliably detected. Using the pin-hole model, we compute the external camera parameters (camera pose), as well as the internal camera parameters (focal length and center, aspect ratio) by the algorithms described in [27] [28]. Whenever possible, we use simplified versions of the algorithms, computing only the external camera parameters, using user-provided data for the internal parameters and thus gaining much more robust calibration estimates.

18.4.3   Camera Tracking

To redetect targets in subsequent images, we predict their position in the next frame. Due to the randomness of user head motion, it is hard to predict future motion from history. Sophisticated motion prediction models that we have experimented with, such as Kalman filters [13], have had serious problems tracking a user-held or user-worn camera. Such prediction of 3D user motion, which includes schemes to model the current translational and rotational speed as well as acceleration of a camera, is too slow to quickly react to users' head rotations (e.g., during a quick glance to one side, or a head shaking motion), generating an effect of "swimming" off track. If the head motion is very abrupt, the system never recovers from its "detour".

Instead, we have now returned to a much simpler, two-dimensional tracking approach. We linearly predict (locally in 2D) the position of a target's corners from their position in the previous two images - ignoring all influence of 3D rotations, acceleration or perspective foreshortening which can result in different 2D motion vectors in different parts of the image. This approach is very fast and gives us a good starting point to search for the precise position of the targets. Due to its speed (close to the real frame rate), we only have to deal with very little head motion between images. We can thus limit the search radius for redetecting targets - thereby further improving tracking performance. Thus, linearizing 2D target motion has had a very positive effect on system performance and robustness - even though it is a very inadequate motion model for describing user head motions from a physical point of view.

When the target locations have been predicted for the next image, we locally search the image for the best match. For every edge of a rectangle, the algorithm scans the new image perpendicularly to the edge along several scan lines for maximal image gradients. The new edge is determined by fitting a line through the set of maximal gradients [14].

The new edge positions are then used to determine the new camera position. We need to account - in principle - for potential changes of all internal and external parameters of the camera. Yet, for the sake of real-time response, we assume that the internal camera parameters are fixed. For the external parameters (camera position and orientation), we use the information from the previous image as a starting point, applying a fast gradient descent technique to settle on a new set of parameters that fit the target locations in the new image [14].

Since the target tracker only requires edge information and not full target descriptions, our tracker can operate in areas of the scene where no rectangular target is visible. To this end, we have extended the scene model to also include descriptions of naturally occurring linear features, such as window sills and edges of walls or tables. The special markers are only required during the initialization phase.

18.4.4 See-through HMD Calibration

For see-through applications, the position of the user's eyes must be established relative to the computed camera (head tracker) position. Unfortunately, no sensor can record exactly what the user sees. We thus cannot determine the eye position without user involvement. To this end, we have developed an interactive augmented computer vision method. The AR-system displays 3D outlines of the rectangular targets on the HMD, using the calibration parameters of the camera that is attached to the HMD. Due to the offset between the camera and each of the user's eyes, the user observes a misregistration between the target outlines and the rectangles on the wall. We then ask the user to indicate the misregistration with the mouse for one eye at a time by drawing lines from the outlined target corners to their true positions, as seen by the user. A specially constructed head rest helps the user keep the head still during this procedure. Using the mappings between projected and actually observed target positions, the offset of each eye can be computed.

Even though this is not a difficult procedure in principle, it depends very much on just how still users can hold their heads during the interactions. During extended mouse motions, users tend to change head position somewhat, resulting in a gradual drift in the mappings of the corners. We are considering using mouseless interfaces, such as cursor control by voice or by pushing the arrow keys on the keyboard. So far, the use of additional constraints regarding the relative position and orientation of two eyes has sufficed to compute camera-to-eye calibrations that were good enough to generate a believable immersive stereo effect for users when wearing the HMD. Yet, the procedure is not sophisticated enough to deal with a level of precision requiring us to recalibrate the system when different people use the HMD. Thus, we use the same calibration parameters for all users - as long as the attachment of the camera to the HMD is not accidentally moved or twisted.

18.4.5   Real-time Lag Reduction via Motion Prediction

When a scene is augmented with virtual objects, lag cannot be avoided entirely since it takes time to draw the virtual objects. This is not a problem for fast monitor-based solutions since the video image used for calibration can be the one also being used for augmentation - with users barely noticing that the video data they see is slightly outdated. The see-through-mode is much more critical since users keep sight of the real world. Time for capturing the video data, processing it, and generating the augmentations all sums up to a misregistration that depends on the speed of the user's head motion. During a continuous head rotation, virtual objects "lag behind" by a constant distance.

This is the classical tracking and motion prediction problem in AR [12]. Kalman filters are one approach for predicting user motion in three dimensions and thus allowing the system to render objects according to expectations where the user's head will be by the time the drawing is done. Yet, discussions in Section 18.4.3 have outlined why Kalman filters currently have not proven suitable in our applications for real-time AR: head motions are just to erratic to be modelled predictably in 3D to date.

In our current system, we use the 2D feature prediction technique of Section 18.4.3 to also predict where the features will be one time step further in the future. A futuristic calibration then provides the rendering parameters to draw virtual objects ahead of time. In order not to loose time, such calibration only computes the rotational components of predicted head motions since those are affecting the visible misregistration most strongly. When the user spontaneously changes the motion direction, only few images are affected since no complex motion model has to learn about the change of course. After two frames only the new direction is relevant to the motion predictor. This approach works well as long as objects can be rendered at frame rate. Predictions into more distant futures will suffer greatly from the linearized 2D feature prediction approach.

18.5   Off-line Calibration of Video Sequences

We are currently not able to augment scenes of landscapes or cityscapes automatically and in real-time with complex virtual objects, as shown in Section 18.2.3. This section describes our semi-automatic, off-line techniques for augmenting pre-recorded video sequences of such scenes. Such approaches need to become more automatic and faster for high-quality AR presentations.

18.5.1   Interactive Calibration

Our interactive calibration system, InCal, provides a user interface to calibrate and track camera positions semi-automatically in image sequences. It superimposes a rough 3D model of the real scene on a start image. The user can interactively manipulate the model to make it fit the scene approximately. Furthermore, the user can interactively indicate correspondences between 3D model features and 2D image features. Such correspondences are then used to automatically compute the current camera position. When the virtual camera is set to the same position, the 3D scene model "snaps" into alignment with the image [17].

When calibrating image sequences InCal exploits inter-frame coherence to automatically propose feature locations in new images from their locations in previous images, using a normalized cross-correlation technique. Users thus barely have to interact with the system once it has been initialized. Most features are nicely tracked across many images. Occasional mismatches can be corrected interactively.

We have been able to successfully calibrate many live and pre-recorded video sequence this way. Even for complex landscapes, we have been able to interactively calibrate sequences of hundreds of images nearly automatically within a few hours. Yet, calibrations are very sensitive to noise and to the shape of the 3D scene model, requiring suitable heuristics in order to achieve good augmentations of images. Tracking stability is another key issue. Virtual objects must be precisely positioned in the picture and keep their position over time despite camera motion and noise. In video sequences, apparent stability within the scene over time is visually more important than the precise calibration of individual images by themselves. Thus, it is important to make stabilizing assumptions. In particular, assuming that the internal camera parameters remain constant throughout the video recording provides significant overall improvements. Using schemes that avoid computing all six external parameters together for every image further stabilizes the system.

18.5.2   Acquisition of 3D Scene Models ("Reality Models")

High quality AR-applications require a very accurate model of the environment (a reality model) to augment the current view seamlessly with synthetic information (the virtual model) such that the virtual objects behave in physically plausible manners: They occlude or are occluded by real objects, they are not able to move through other objects, and they cast shadows or light reflections on other objects. The automatic construction of scene models is a long-standing issue in computer vision research. Recent approaches have suggested using it for the reconstruction of buildings [29] [30]. In our work, we explore a very applied, pragmatic approach which is closely related to the requirements of rather realistic applications in the exterior construction industry [16].

In our interactive system, InCal, we begin with a very sparse model of a scene, constructed from externally provided information such as the known position and height of a few buildings, power poles or bridge pillars. We measure more information, such as the course of rivers and streets, from 2D maps and insert it a zero height into the model. From this model, we generate an initial camera calibration for a few site photos, interactively indicating how features in the image relate to the scene model. Once an image has been successfully calibrated, the model is overlaid on the image, showing good alignment of the image features with the model features. Models of new structures in the landscape are then added to the model, using their two-dimensional position in the city map and estimating their height from their alignment in several images [16] [17].

18.6   Presentation of Virtual Information

Once appropriate scene models and calibrations have been obtained, they form the basis for mixing real and virtual worlds. This section describes the steps necessary to achieve realistic and fast inclusion of virtual information into the scene.

AR thrives on fast, real-time augmentations of the real world. All virtual information thus has to be rendered very quickly. To this end, we carefully tune and prune geometric models to achieve maximal rendering performance while maintaining an acceptable level of realism. For this purpose, our home-built models are very simple, consisting merely of a few polygons and maybe a texture map which can be rendered very quickly even on low-end graphics machines. When collaborating with industry partners, typical geometric models of virtual car prototypes or buildings are much more complex - too large even to be rendered by high-powered graphics supercomputers at an acceptable frame rate. They thus have to be simplified. We employ an interactive in-house tool [31] building on standard algorithms [32] [33]. Except for close-ups, the resulting models are virtually indistinguishable from the originals, albeit at a fraction of the cost.

Mixing virtual objects into a real scene requires that the object obey the basic physical laws - first of all occlusion. Occlusions between real and virtual objects can be computed quite efficiently by the geometric rendering hardware of graphics workstations when provided with a list of the geometric descriptions of all real and virtual models. By first drawing transparently the scene models of the real objects, we initialize the z-buffer to subsequently draw only those virtual objects that are located in front of the real objects. The user thus sees the real scene through the transparent HMD where the real objects are closest to him, and the virtual objects where they are closest.

Real objects cast shadows and reflect light. So should virtual objects. When 3D scene descriptions are available, standard computer graphics algorithms [34] can be used to compute the geometry of shadows cast by virtual objects onto real ones [23]. Given the right hardware, shadows can be blended in real-time with the image of the underlying object, thereby accounting for ambient light. Reflections are more difficult and can only be solved for special cases, such a mirror reflections of virtual objects on planar surfaces of real objects (windows, water). We are using some of the concepts for the off-line augmentations of pre-recorded video sequences of prospective construction sites. We don't handle shadowing and light reflections in the on-line AR-system in order to leave as much time as possible to the optical tracker.

In the final step of actually merging virtual objects into an image we need to consider whether the virtual objects should completely occlude what is behind them or whether a blended presentation of both reality and virtuality is more appropriate. When we want to create the illusion of a physically changed world with as much realism as possible, the opaque inclusion of virtual objects is appropriate. Yet, in many cases, the augmentations are intended to add supporting information to the world rather than change it. For example, users may want to choose a suitable blending ratio to include virtual text panels with data records or instructions (Figure 18.9) or to gain semi-transparent x-ray viewing capabilities to see what's inside a wall (Figure 18.8). Thus, our system provides users with the option to interactively select a percentage level at which virtual objects are blended into the real world.

18.7 Discussion

In this paper, we have emphasized the current need for AR-Systems to trade off perfection for real-time efficiency. Currently, the most critical issue for the system is to be fast enough to keep pace with a user's spontaneous and erratic head motions when being involved in real, hands-on tasks. Our demonstrations show that this is becoming feasible for "reasonable" motions in pre-arranged scenes. Thus, a starting point has been set to begin embarquing on more ambitious goals. Among those, the following seem the most important. First, the tracking approach has to become more general and flexible, getting by with naturally occurring scene features rather than requiring special targets. The system also has to begin learning about the environment while the user works in it, thus incrementally building a knowledge base describing the scene. Second, the system has to become more interactive, responding to real-world changes autonomously without being prompted by the user. We have indicated a first glimpse of such concepts at the example of the Tic Tac Toe game. Yet, much more has to be done to really understand and track the changing real world in which the AR-user is working. Third, the system needs to develop more sophisticated visualization and rendering schemes. Regarding all these issues it is critical to explore from the beginning with real users which approaches are acceptable in real applications and which ones are making the wrong (impractical) simplifying assumptions. Considering the current rate of progress, we expect many of these aspects to be addressed in the foreseeable future. AR has the potential to become a "killer application" for mobile 3D sensing and visualization technology.

Acknowledgements

This work was conducted while the authors formed the Fraunhofer Project Group for AR at ZGDV, in close collaboration with the department for visualization and virtual reality of Fraunhofer-IGD in Darmstadt. Office space was provided by ECRC in Munich. The work was partially funded by the European projects CICC (ACTS) and Cumuli (Esprit).

References

[1] S. Feiner, B. Macintyre, and D. Seligmann: "Knowledge-based augmented reality," *Comm. ACM*, vol.36, no.7, pp.53–61, 1993.

[2] A. Webster, S. Feiner, B. MacIntyre, W. Massie, and T. Krueger: "Augmented reality in architectural construction, inspection, and renovation," *Proc. ASCE Third Congress on Computing in Civil Engineering*, Anaheim, CA, pp.913–919, Jun. 17–19, 1996.

[3] K. H. Ahlers, A. Kramer, D. E. Breen, P. -Y. Chevalier, C. Crampton, E. Rose, M. Tuceryan, R. T. Whitaker, and D. Greer: "Distributed augmented reality for collaborative design applications," *Proc. Eurographics'95*, 1995.

[4] D. Curtis, D. Mizell, P. Gruenbaum, and A. Janin: "Several devils in the details: Making an AR app work in the airplane factory," *Proc. First Int'l Workshop on Augmented Reality*, Nov. 1, 1998.

[5] E. Rose, D. Breen, K. H. Ahlers, C. Crampton, M. Tuceryan, R. Whitaker, and D. Greer: "Annotating real-world objects using augmented reality," *Proc. Computer Graphics: Developments in Virtual Environments*, Academic Press Ltd., 1995.

[6] J. Molineros, V. Raghavan, and R. Sharma: "AREAS: Augmented reality for evaluating assembly sequences," *Proc. First Int'l Workshop on Augmented Reality*, Nov. 1, 1998.

[7] A. Fournier: "Illumination problems in computer augmented reality," *Journee Analyse/Synthese d'Images (JASI)*, pp.1–21, Jan. 1994.

[8] S. Feiner, B. MacIntyre, T. Hoellerer, and A. Webster: "A touring machine: Prototyping 3D mobile augmented reality systems for exploring the urban environment," *Proc. First Int'l Symp. on Wearable Computers*, Cambridge, MA, pp.74–81, Oct. 1997.

[9] T. Starner, S. Mann, B. Rhodes, J. Levine, J. Healey, D. Kirsch, R. W. Picard, and A. Pentland: "Augmented reality through wearable computing," *Presence, Special Issue on Augmented Reality*, vol.6, no.4, pp.386–398, Aug. 1997.

[10] A. Smailagic and R. M. Metronaut: "A wearable computer with sensing and global communications capabilities," *Proc. First Int'l Symp. on Wearable Computers*, Cambridge, MA, pp.116–122, Oct. 1997.

[11] M. Billinghurst, S. Weghosrt, and T. Furness III: "Wearable computers for three dimensional CSCW," *Proc. First Int'l Symp. on Wearable Computers*, Cambridge, MA, pp.108–115, Oct. 1997.

[12] R. Azuma and G. Bishop: "Improving static and dynamic registration in an optical see-through HMD," *Proc. SIGGRAPH'94*, Orlando, pp.197–204, Jul. 1994.

[13] D. Koller, G. Klinker, E. Rose, D. Breen, R. Whitaker, and M. Tuceryan: "Real-time vision-based camera tracking for augmented reality applications," *Proc. ACM Symp. on Virtual Reality Software and Technology (VRST'97)*, Lausanne, Switzerland, Sep. 15–17, ACM Press, 1997

[14] D. Stricker, G. Klinker, and D. Reiners: "A fast and robust line-based optical tracker for augmented reality applications," *Proc. First Int'l Workshop on Augmented Reality*, Nov. 1, 1998.

[15] D. Reiners, D. Stricker, G. Klinker, and S. Muller: "Augmented reality for construction tasks: Doorlock assembly," *Proc. First Int'l Workshop on Augmented Reality*, Nov. 1, 1998.

[16] G. Klinker, D. Stricker, and D. Reiners: "The use of reality models in augmented reality applications," *European Workshop on 3D Structure from Multiple Images of Large-scale Environments (SMILE)*, Freiburg, Germany, Jun. 6–7, 1988.

[17] G. Klinker, D. Stricker, and D. Reiners: "Augmented reality for exterior construction applications," in (W. Barfield and T. Caudell, eds.) *Augmented Reality and Wearable Computing*, Lawrence Erlbaum Press, 1998.

[18] Z. Szalavari and M. Gervautz: "Using the personal interaction panel for 3D interaction," *Proc. Conf. on Latest Results in Information Technology*, pp.3–6, Budapest, Hungry, May 1997.

[19] B. Ullmer and H. Ishii: "The metaDESK: Models and prototypes for tangible user interfaces," *Proc. UIST'97*, Banff, Alberta, Canada, Oct.14–17, pp.223–232, 1997.

[20] W. Krueger, C. -A. Bohn, B. Froehlich, H. Schueth, W. Strauss, and G. Wesche: "The responsive workbench: A virtual work environment," *Computer*, pp.42–48, 1995.

[21] M. Bajura and U. Neumann: "Dynamic registration correction in video-based augmented reality systems," *Computer Graphics and Applications*, vol.15, no.5, pp.52–60, 1995.

[22] J. Park, S. You, and U. Neumann: "Fast color fiducial detection and dynamic workspace extension in video see-through self-tracking augmented reality," *Proc. Fifth Pacific Conf. on Computer Graphics and Applications*, 1997.

[23] A. State, G. Hirota, D. T. Cheng, W. F. Garrett, and M. A. Livingston: "Superior augmented reality registration by integrating landmark tracking and magnetic tracking," *Proc. SIGGRAPH'96*, New Orleans, Aug. 4–9, ACM Press, pp.429–438, 1996.

[24] M. Uenohara and T. Kanade: "Vision-based object registration for real-time image overlay," *Proc. Computer Vision, Virtual Reality and Robotics in Medicine (CVRMed'95)*, Nice, France, pp.13–22, Apr. 1995.

[25] J. Mendelsohn, K. Daniilidis, and R. Bajcsy: "Constrained self-calibration for augmented reality registration," *Proc. First Int'l Workshop on Augmented Reality*, Nov. 1, 1998.

[26] J. Park, S. You, and U. Neumann: "Natural feature tracking for extendible robust augmented realities," *Proc. First Int'l Workshop on Augmented Reality*, Nov. 1, 1998.

[27] J. Weng, P. Cohen, and M. Herniou: "Camera calibration with distortion models and accuracy evaluation," *IEEE Trans. on Pattern Analysis and Machine Intelligence*, vol.14, no.10, pp.965–980, 1992.

[28] R. Y. Tsai: "An efficient and accurate Camera calibration technique for 3D machine vision," *Proc. CVPR*, pp.364–374, 1986. See also http://www.cs.cmu.edu/rgw/TsaiCode.html.

[29] P. E. Debevec, C. J. Taylor, and J. Malik: "Modelling and rendering architecture from photographs: A hybrid geometry- and image-based approach," *Proc. SIGGRAPH'96*, New Orleans, Aug. 4–9, pp.11–20, 1996.

[30] O. Faugeras, S. Laveau, L. Robert, G. Csurka, and C. Zeller: "3D reconstruction of urban scenes from sequences of images," in (A. Gruen, O. Kuebler, and P. Agouris eds.) *Automatic Extraction of Man-Made Objects from Aerial and Space Images*, Birkhauser, 1995.

[31] J. Schiefele: "Methoden der automatischen Komplexitatsreduktion zur effizienten Darstellung von CAD-Modellen," *Diplomarbeit*, TU Darmstadt, 1996.

[32] J. Rossignac and P. Borrel: "Multi-resolution 3D approximation for rendering complex scenes," *Proc. Second Conf. on Geometric Modelling in Computer Graphics*, Genova, Italy, pp.453–465, Jun. 1993.

[33] W.J. Schroeder, J. A. Zarge, and W. E. Lorensen: "Decimation of triangle meshes," vol.26, pp.65–70, Jul. 1992.

[34] J. D. Foley, A. V. Dam, S. K. Feiner, and J. F. Hughes: *Computer Graphics, Principles and Practice, 2nd. Edition,* Addison Wesley, 1989.

[35] D. E. Breen, E. Rose, and R. T. Whitaker: "Interactive occlusion and collision of real and virtual objects in augmented reality," Technical Report ECRC-95-02, ECRC, Arabellastr. 17, D–81925 Munich, 1995.

[36] T. Caudell and D. Mizell: "Augmented reality: An application of heads-up display technology to manual manufacturing processes," *Proc. HICCS'92*.

Chapter 19

MR Aided Engineering: Inspection Support Systems Integrating Virtual Instruments and Process Control

Kosuke Sato
Yoshihiro Ban
Kunihiro Chihara

Nara Institute of Science and Technology, Japan

19.1 Introduction

This chapter discusses how Mixed Reality (MR) will be able to aid engineering in near future. Among several engineering processes, it focuses inspection processes for electronic parts due to its affinity for computer systems. As actual inspections of electronic parts usually have to utilize a set of electric instruments (voltage meter, ohm meter, oscilloscope or logic analyzer etc.) and a set of manuals, an operator has to alternate his/her focus of attention among the parts to be inspected, the instruments and the manuals. In order to alleviate operator's eye fatigue, MR-integrated instruments and manuals are introduced. While the MR-integrated instruments exist virtually and appear near a test-pin to be proved in the MR space by see-through optics, they imitate the original look-and-feel of real electric instruments. Operator's eyes can be fixed on a narrow portion. The MR-integrated manual controls the entire inspection procedures; it switches the appearance and the function of MR-integrated instruments and gathers parameters from on-board real instruments. A combination with MR-integrated instruments and manuals can supervise the completion of each inspection task, so that they prevent operator's human errors. Two prototypes for inspection supporting are implemented, which realize the above scheme. One is a desktop HMD-free system utilizing 45 degree half-mirror, which dedicates the inspection of printed circuit boards (PCBs). The other is a backpacking system utilizing an HMD and an inertia navigation system, which inspects wall-mounted power electric parts. The operator can walk around a wide manufacturing plant with the battery-operated backpacking system. Making MR systems that work outdoors is a natural step in the future development of MR toward the ultimate goal, which is that people can be supported with MR technology anywhere, anytime in any environments.

19.2   MR Technology Applicable for Manufacturing

19.2.1   Virtual Reality Aims Computer-Aided-Design

Over the past few years, a considerable number of studies have been made on Virtual Reality (VR). The frontiers who explore VR technologies creates many new practical fields, such as a medical training system for surgery, aerial training system for jet pilots, molecular modeling for pharmacy, and so on. Unfortunately, few successful industrial solutions using VR have so far been made. Only the field of Computer-Aided-Design (CAD) takes the profit of VR. A pair of an HMD and a glove-like hand sensor instead of a pair of a CRT display and a mouse has changed the way of 3D geometric modeling for industrial design. The reason why VR assists only 3D modeling in industrial processes is that VR manipulates only virtual objects. All of remained industrial processes need to handle real objects; for examples, inspection task in maintenance processes.

19.2.2   Mixed Reality Aims Computer-Aided-Maintenance

One of breakthroughs of VR is to integrate the real scene into the virtual scene. In that case, the real scene could be augmented by computer technologies, such as database, data communication, AI, visualization and so on. Such integration is the true nature of MR. The major different point between VR and MR is that the operator is able to see the real environments. Basically, VR technology aims to show virtual environments to the operator. On the other hand, as for MR technology, the virtual environment coexists in the real environment. Therefore, the real objects to be inspected can be perceived in the real environment. In MR, the role of the virtual environment changes itself. While there is no need to display virtual environment as the entire world, a virtual display is able to concentrate to provide useful and helpful information to the operator directly [1]–[3]. The information includes not only text data but also graphic data such as three-dimensional (3D) geometrical information. For example, when looking for a object to be inspected, a superimposed cursor guides the operators' eyes like a lookup display of a jet-fighter. An innovative point of MR comparing to the conventional lookup display is that the operator is able to perform anything without restrictions; sitting on a fixed seat [4]. However, if the lookup display technology meets Computer Vision (CV) technology which is capable to identify the target objects in realtime, a helpful MR-Based support system would be realized.

In this chapter, two types of new supporting environments for inspection work of electronic parts are proposed; one is a desktop system by combining MR technology and CV technology, the other is a backpacking one by involving MR technology with an inertia navigation. Although most of MR applications aim how to augment the real world, the goal of these proposed systems is to solve problems of real concrete inspection works. For the goal, the following sections discuss how to superimpose visual markers, instruction prompts, and virtual instruments on real electronic/electric parts.

19.3 Paper Manual to MR-Integrated Instruction

19.3.1 Problems of Paper Manual

We know that conventional printed manuals have the following difficulties.

Unavoidable worker's view point frequent change between a manual and a object
Because it is difficult to memorize all the contents of the manuals, an operator has to carry the manuals and refer them by transferring his/her eyes alternately with the manual and the object. However, this lazy referencing may cause the next two problems regarding the progress of the work process.

Duplication/omission of some work process
As the average industrial manuals are simple and it is not fun to understand their contents, the operator tends to stop understanding what process of the manual at present work is going on by the frequent view point change between the manual and object.

Conditional branch of work process depending on inspection results
There are conditional branches depending on measured results in the adjustment or repair of electronic parts; for example, IF (Voltage-P > 1.2V) THEN GOTO Process-145, IF (Voltage-P < 0.3V) THEN GOTO Process-92. In short, they are dynamic procedure to omit/append some part of work processes. Such time, the operator is requested to turn pages of the manual. This kind of dull conditional branches of manual reading may cause human errors.

Difficulty of finding target object
Commonly instructions in printed manuals are written in natural language sentences as illustrated in Figure 19.1. This document instructing DIP switch setting on a PCB is written in Japanese. When the PCB is an international product, nothing can be done about it for operator speaking other language. Even if photographs or illustrations appear on the manuals, the operator needs to find out the real target corresponding to the graphical materials, it is also dull thing for him/her.

例：No. 5 の U 5 と No. 27 の D 12 の抵抗値を計測し、500 KΩ以下なら、No. 18 のスイッチ 1 を On
にしなさい。

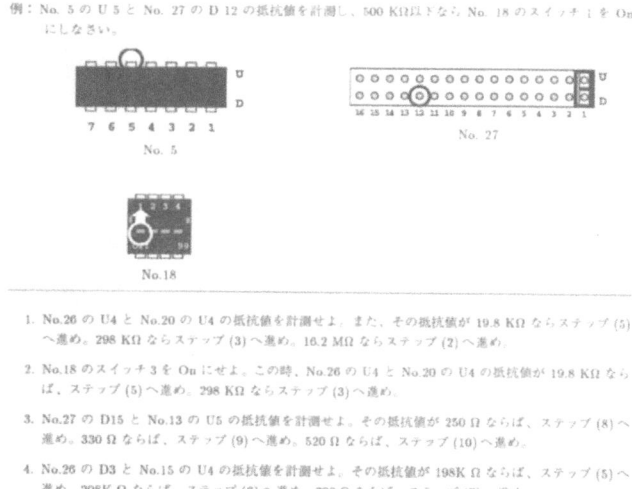

1. No.26 の U4 と No.20 の U4 の抵抗値を計測せよ。また、その抵抗値が 19.8 KΩ ならステップ (5)
   へ進め。298 KΩ ならステップ (3) へ進め。16.2 MΩ ならステップ (2) へ進め。

2. No.18 のスイッチ 3 を On にせよ。この時、No.26 の U4 と No.20 の U4 の抵抗値が 19.8 KΩ なら
   ば、ステップ (5) へ進め。298 KΩ ならステップ (3) へ進め。

3. No.27 の D15 と No.13 の U5 の抵抗値を計測せよ。その抵抗値が 250 Ω ならば、ステップ (8) へ
   進め。330 Ω ならば、ステップ (9) へ進め。520 Ω ならば、ステップ (10) へ進め。

4. No.26 の D3 と No.15 の U4 の抵抗値を計測せよ。その抵抗値が 198K Ω ならば、ステップ (5) へ
   進め。298K Ω ならば、ステップ (6) へ進め。398 Ω ならば、ステップ (7) へ進め。

5. No.26 の D2 と No.8 の D1 の抵抗値を計測せよ。その抵抗値が 480 Ω ～ 520 Ω になるよう No.29
   の可変抵抗を調節し、ステップ (3) へ進め。

Figure 19.1   Sample of paper manual.

19.3.2   Solution in MR Technology

Above problems could be solved by using immersing MR technology. The major
cause of the problems is frequent eye zapping of the operator. The reason why
the frequent eye moving is required is that there is a physical distance between
the work object and the paper manual. Therefore, it is possible to decrease the
eyes moving of the operator by superimposing sufficient information onto the work
object directly, also measured data through the instruments, as shown in Figure
19.3. The graphical superimposing technology is one of key technology of MR,
which enables the fusion of the real world and the virtual world. It gets easier to
find portion to be inspected by superimposing eye-catching markers. For conditional
process branching, an integrated work process management system is able to solve
its problems. As the operator has not to need search and turn the page according
to the work process branching, he/she can concentrate his/her work with graphical
prompts instructing the next task by the process management system.

19.4   MR-Integrated Instruments

Again, this chapter focuses that a target object to be inspected is an electronic
part. Most of electronic inspection needs to use several electronic measurement
instruments. These instruments are having various kinds of user interface designs
because of historic reason. If the operator use them constantly in daily work, it is not
difficult thing to remember how to use them. However, there are lots of beginner

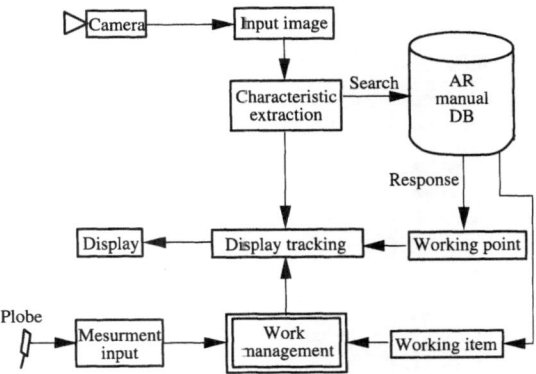

Figure 19.2   Work process management system with AR and online measurement.

or non-expert engineers who are not familiar to how to use them, for example, how to set a multimeter to DC-mode. MR technology also may solve the problem; that is MR-integrated electronic measuring instrument. It is a virtual measuring instruments which is just displayed on the operator's view. Its outlook and online data through an probing interface of an MR system are graphically generated in realtime CG. The MR-integrated instruments are virtually existing in only the MR world with their graphical metaphors. There are several advantages in comparison with usual electronic measuring instruments.

- Free graphical design for displaying measured values: It is possible to set up the display of measured values freely by the computer graphics unlike a usual electronic measuring instrument; analog indicator, bar meter, seven-segment number display and so on.

- Prompt of measuring points visually: For example, as IC parts have many pin leads, it is so severe task to identify a certain pin lead. Instead of written document like 'PIN-23 of IC-8', the MR-integrated electronic measuring instruments indicates the target portion directly by superimposing markers.

- Concealment of work branching depending on measured result: In the electronic inspections, there are lots of tuning tasks; range the voltage of a certain test pin within fixed margin by rotating a variable resister. In this tasks, if the voltage is in the margin, there is no need to tune the resister. At that case, the operator has to skip some portion of the instructions in the procedure. It is a conditional process branching of the instruction. Thereupon, it is possible to search and show the correct work process to be proceeded next by the process management system with an instruction database. While the management system analyzes the output from the measuring instruments and maintains the process control, the worker gets to concentrate on only the work process which is conducted by the management system, without concerning conditional branches.

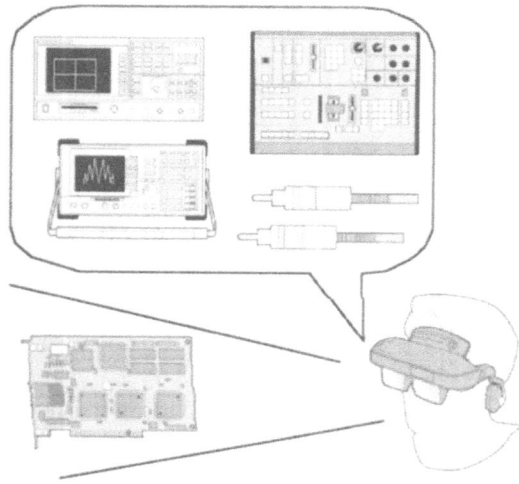

Figure 19.3   MR-integrated instruments.

- Inspection plural targets simultaneously: As the process status is controlled
  by the management system, the status can be temporally stored in anytime.
  It makes a good advantage that the system can manipulate plural number
  of targets simultaneously. The operator can pause and change the inspection
  anytime.

It is possible to overcome the various problems that are conceivable with usual
inspection work by using the MR-integrated electronic measuring instruments.

Figure 19.3 illustrates an example of superimposed figure of an electronic measuring instrument. The operator can know the measured result without turning his/her
eyes away largely from the work object. The measuring instrument (an ohm meter)
is displayed in the upper part. The measured data are displayed not in number, but
in good looking graphics. Sometimes, graphical metaphor of a real instrument is
one of excellent display for expert engineers.

19.5   A Desktop Environment for PCB by MR

In this section, a desktop environment for electronic inspection system by MR is
described. Many electronic products are carried on a conveyor and its inspection is
done on a desktop.

19.5.1   Goal

It is sufficient with the repetition of the simple work by the memory, if it is a simple
routine work that is uniformly arranged like assembly lines in a mass-production
process. However, many works on the real world require specialized knowledge and

force an operator refer a manual even electronic engineers. A work ranges desktop inspections for electronic printed circuit boards. Inspection work for complicated PCBs requires the deep knowledge and observation regarding electronics. The goal of the inspection support system aims the following points.

- Realization of a work environment which requires no special knowledge,

- Prevention of human errors.

As all manual information is stored in a computer, an operator need neither to memorize the contents of the manual nor to refer that takes the correspondence between a figure and a target. The supporting environment prevents workers' failures which come from his/her misunderstanding and mistakes by using visual presentation of instructions and automatic reconfirmation.

Figure 19.4 shows its outlook. This system consists of the following items.

- Cameras: Two cameras are located both side of working area to avoid occluding due to worker's arm,

- PC: It maintains the MR-based manual, manages the current inspection process, gathers current instrumental data, tracking camera images in realtime, and generates superimposed information for the virtual world,

- Display: Graphical instructions are displayed,

- Half mirror: The graphical instructions and the real are merged optically by this half mirror.

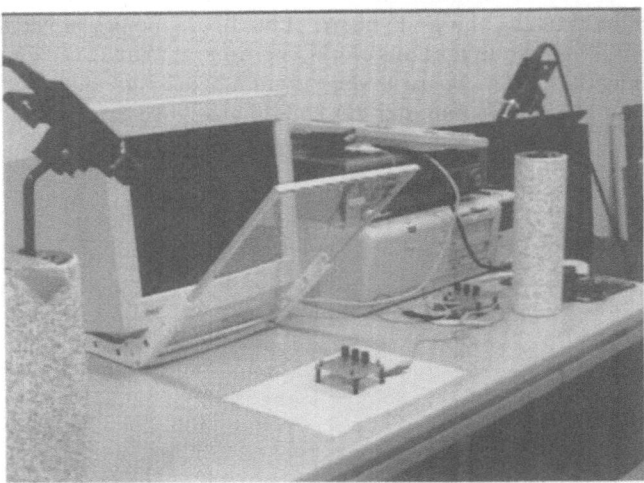

Figure 19.4   MR-based desktop PCB inspection support system with video tracking cameras.

The PCBs to be inspected are planar print circuit boards. Therefore, if it corresponds the optical path length to the display with the optical path length to the real boards, the operator can watch the completely merged MR world from any viewing position without special glasses. All work procedures are stored in the database inside the PC. The type or ID of the object estimated on the characteristic spatial arrangement of LED markers attached on the target PCB by analyzing spot images acquired from two cameras. Beside the prompting markers are superimposed on the target board in accordance with the video tracking information of the characteristic markers, the graphical instructions are given by the operator. The electronic data gathered from measuring devices is inspected automatically and simultaneously.

19.5.2   Calibration

Because the cameras are overlooking it from the right and left upper part like Figure 19.4, a taken image of the real world's coordinate are distorted. Thereupon, a camera calibration between the real world coordinates and camera image coordinates are necessary. We use a lookup table (LUT) method in consideration of the easiness and avoid lens distortion. To do calibration it using distributed flashing infrared LED array that is arranged to the known coordinates. An image sequence while moving this tool in an interval of the grid points of all workspace area is captured into the system.

19.5.3   Object Tracking

The differentiation and the position and also attitude of the work object are detected by performing image. Thereupon, the characteristic makers by infrared LEDs are located on the object, to facilitate the recognition of the position and its ID. Figure 19.5 shows video tracking the PCB board. Five bright points in the figure are the prompting marker of the hypothesis world coordinates that caused to the infrared LEDs. Identification of the board is done from a spatial arrangement information of the LEDs on the object-centered 2D coordinates. An interesting points to be adjusted on the board is described on the object coordinate system.

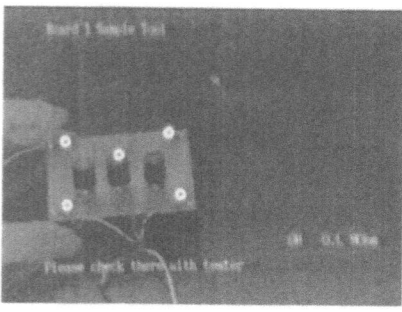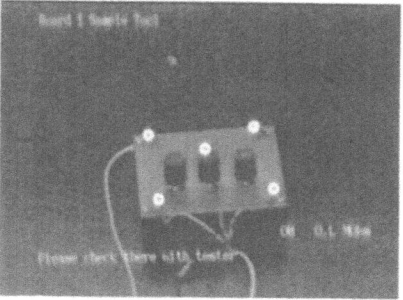

Figure 19.5   Video tracking with infrared LED markers.

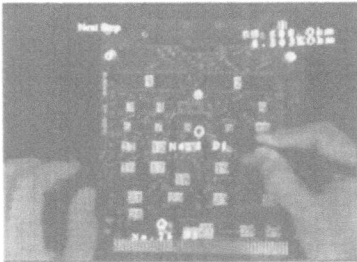
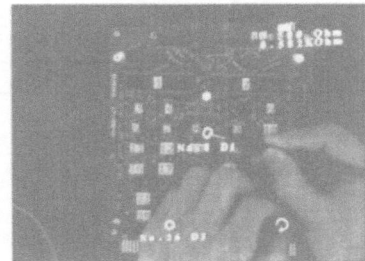

(a) prompting test-pin   (b) prompting variable resistance

Figure 19.6   Operator's views.

- Image acquisition: images are acquired from two cameras that were arranged in the both ends of the workspace.

- Center-of-gravity calculation: Spot image areas due to the infrared LEDs are binarized. Next, for each bright point its accurate center-of-gravity is calculated on the camera coordinates.

- Coordinates conversion: The world coordinates of each bright point are obtained by looking the LUT.

- Estimation of the types: The object type is selected by matching an arrangement relation of the bright points on the world coordinates.

19.5.4   Operator's View

It is possible to make satisfactory the virtual world on the half mirror by using MR technology. The situation that is causing to flatter the work point to the object and tracking of an in Figure 19.6 is being shown. Bright four points of surrounding and middle point of print circuit board are markers of traced the infrared LEDs. Also, it can instruct to the operator that is the measuring point and work place etc. This figure shows the instructions that turn the knob of variable resistance clockwise. And, it show the measuring point that should touch about two points of resistance.

19.5.5   Experiments

Here, we experiment about the work time and errors carried out the work with this system and one with printed manual, to confirm the feasibility of the proposed inspection supporting environment applying MR technology. Test subject people were twenty graduate students who were not familiar to electronic circuits. The same task was carried out by two groups; one using a printed manual and one using the proposed MR system. A PCB shown in Figure 19.6 is a target work object. Lead pins of some ICs are chosen as probe points of an experimental inspection. The circuit board had two pieces of variable resistances and four DIP switches as target parts to be inspected. The experimental inspection procedure included twelve

Table 19.1   Task performance.

Worker	Work time	Error count	Worker	Work time	Error count
A	206	0	K	640	2
B	321	0	L	744	2
C	244	0	M	885	0
D	168	0	N	618	1
E	162	0	O	1001	0
F	298	0	P	527	0
G	243	0	Q	596	0
H	174	0	R	582	1
I	178	0	S	728	0
J	229	0	T	604	0
Average	222.3	0	Average	692.5	0.6

processes; adjusting the resistance which branches according to a measured value of the resistance and turn on/off the DIP switches.

Table 19.1 indicate the summary about the average work time and average number of errors of the two groups of the people. The average work time of the people who used this system was about one third shorter that one of the people who used the printed manual. Also, the average number of total error of the people who used this system was exactly zero. There was 0.6 times error per person who used the printed manual. Major reason was that the process control PC in the proposed system did not proceed the inspection procedure unless the measured value by the MR-Integrated instruments became a designated value.

19.6   A Backpacking Environment for Power Parts by MR with HMD

In this section, we describe to introduce an idea of an MR-based inspection system which augments usual inspection processes [5]. Conventional VR systems have fatal problems when applying industrial inspection processes both indoor and outdoor. The major problems should be concerned as follows. When a human operator wears an HMD which is opaque one and displays only virtual scenes,

- The operator can not avoid dangerous situations, i.e. closing real mobiles, powerful machine tools or real steps of road, because the HMD is not see-through,

- Usual VR systems can not be carried out of laboratory, because the measurable area of popular position/attitude sensors to be attached to the operator's head is so narrow and limited, such as mechanical arms, ultrasonic sensors, magnetic sensors which need transmitter coil modules,

- Conventional VR systems are too heavy and large to be carried, because they are usually implemented on graphic workstations.

In order to solve the above three problems, the following necessary requirements arise.

- In order to walk anywhere and to avoid unexpected dangerous happenings, the operator wearing the HMD needs to see a real environment all the time,

- The position/attitude sensors to be attached to the operator's head should be independent and stand-alone one,

- All equipment to be carried by the operator needs to be implemented in light weight and compact size.

In order to satisfy above conditions, ideal MR-based inspection system should use a) see-through HMD, b) inertia navigation and c) PC-based hardware. Because the inspection data is provided from a host CAD/CAM station with a large database, the system have to be split into a carrying system and a host system with a wireless data-communication channel.

19.6.1   Carrying System

A carrying system includes a complete stand-alone HMD-based MR hardware. In the system, computer-generated display changes itself depending on the operator's head motion. In other words, the carrying system needs to be able to manage the operator's position and orientation and to generate virtual environments according to them. At present, a tradeoff situation exists between weight and size of graphics hardware which can be loaded on a carrying system, and the quality of generated virtual environments. Fortunately, the system using a see-through HMD needs to generate not the entire world but objects to be interested. The see-through HMD we adapt is Shimazu STV-01 which is specially modified to DC operation as shown in Figure 19.7. When opening a visor, the operator can see superimposed stereo color dynamic images onto half-dark transparent glasses.

19.6.2   Host System

A role of a host system is to support the carrying system. It provides and transmits via a wireless ethernet (1.2GHz SS radio, 100Kbps) necessary information for inspection. And also it accepts inspected data entered by the field operator, and then verifies to the CAD/CAM database.

19.6.3   Inertia Sensing Subsystem

Unlike a tele-existence system, this kind of walkable VR system moves itself to any types of cites; in indoor building, open-air outdoor, underground, underwater, on bridge made of iron, etc. In order to achieve independent position/attitude measurement of circumstance, inertia sensing methodology is its unique solution. Two types of inertia sensors are involved; one is a gyro-sensor suitable for measurement of head attitude and the other is an acceleration sensor suitable for one of head position. To reduce the weight of sensors to be attached to the HMD, those two

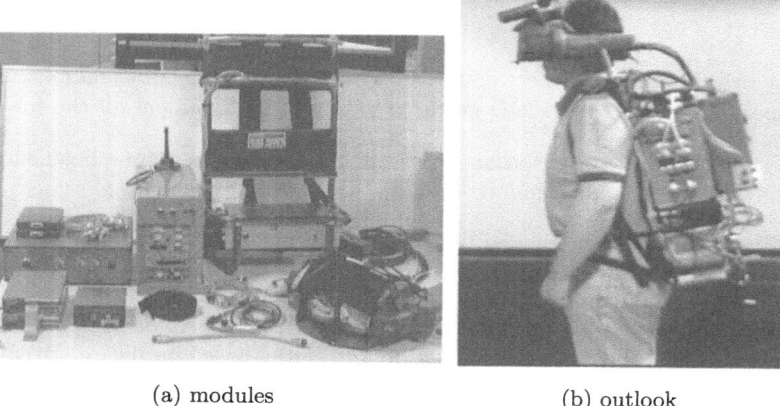

(a) modules   (b) outlook

Figure 19.7   MR-based backpacking electric inspection system.

sensors are split and only the gyro-sensor is mounted to the HMD. The rest acceleration sensor is mounted at the operator's waist position. The position of operator's head is estimated from the position of waist, and is used to generate graphics. As for gyro-sensor, semiconductor type has advantages that it is usually small-sized, light-weight, low consumption of electricity.

19.6.4   Calibration

The initial calibration between the real world and the virtual world is indispensable for inertia sensing based of integral calculus. Unlike PCBs and configurable desktop support environments, it is difficult problem to calibrate coordinates of the worlds. As power electric modules to be inspected are located spread over an industrial plant, special calibration mechanism requiring power supply is not seriously practical. No one can provide the same number of calibration system as target modules. For the backpacking support environment, we design a simple calibration attachment without power supply, which is a couple of bent hinges shown in Figure 19.8(a). Before inspecting the target module, an operator stands at the intersection point of the two hinges and looks forward as illustrated in Figure 19.8(b). As the height of the operator is registered previously, the calibration of head position and attitude is done. This kinds of simple calibration mechanism is practical rather than high-tech gear.

19.6.5   Operator's View

As the desktop system described in 19.5 exists in 2D Mixed Reality world and treats only planar electronic parts, the back-packing system handling complex shaped electric parts has to support fully 3D Mixed Reality world. The stereoscopic HMD connected to a 3D-CG rendering engine displays graphical prompts and virtual instruments in the virtual world three dimensionally. The operator may percept the

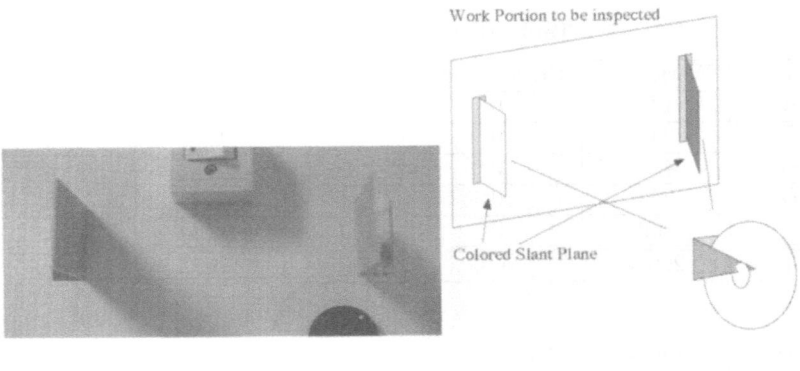

(a) angled hinges   (b) stand positioning

Figure 19.8   Calibration.

(a) prompting terminal   (b) prompting power switch

Figure 19.9   Operator's views.

virtual objects at the fixed point in the 3D world by canceling his/her head motion based of inertia sensing. Unlike Polhemus magnetic sensor, an inertia sensing produces jitter-free comfortable virtual images while a certain period until when the accumulated error gets over a threshold. Figure 19.9 shows frame shots of the operator's view. In (a), the system is prompting turn switch S111 off in natural words while indexing switch S111 where a ring shaped pointer indicates. In (b), it is prompting adjust the voltage to 2.5V by tuning a knob while displaying the measured voltage DC 1.841V in a lower part highlight box.

19.6.6   Experiments

In order to show the feasibility of the system, we experimented about the work time and error counts carried out the work with this system and one with printed manual. An experimental target to be inspected is a wall with four switches and sixteen terminals with nameplate like S101, T14. Test subject people were ten graduate

Table 19.2   Task performance.

Worker	Work time	Error counts	Worker	Work time	Error counts
A	146	0	F	150	1
B	349	0	G	167	0
C	422	0	H	332	3
D	253	0	I	544	5
E	282	0	J	148	1
Average	290.4	0.0	Average	268.2	2.0

students. The same task was carried out by two groups; one using a printed manual and one using the back-packing system. An experimental task consists of nine steps; change a digital multi-meter mode to DC, connect a probe to specified terminal, read the voltage, refer a manual, read a table of repair method, then service separately, etc. Table 19.2 indicate the summary about the average work time and average number of errors of the two subject groups. Unlike the desktop system, the average work time of the subjects who used this system was longer that one of the people who used the printed manual. The major reason is restraint upon the freedom of working due to the heavy HMD and narrow viewing area. On the other hand, as for the total error of the subject who used this system, another error-free operation was accomplished.

19.7   Conclusions

This chapter describes a conceptual idea of MR-based inspection support system and two prototypes; desktop one and backpacking one. The experimental results support that a supporting environment for electronic inspection of PCBs or wall junction boxes applying immerse AR technology is fairly successful. As the MR-Integrated process control maintains inspection process totally, it prevented worker's human error by prompting the certain instructions and the certain points on the target. We gained wonderful finding that AR technology plays an important role in quality control of manufacturing, and in quality improvement of operator's working conditions. We believe that the MR-based inspection support system evolves into AR Aided Engineering in near future. In the desktop case, as the targets are planar print circuit boards, the virtual graphical display can be superimposed without depth discontinuity by using a desktop display and a half mirror. Unlike backpacking one, as a worker does not need to wear a heavy and uncomfortable HMD and PC gear, the desktop MR-based inspection support system is close to practical use. However, even the backpacking MR-based inspection support system is too heavy to carry at present day, its future is so promising. Another kind of immersing technology, Wearable Computer, is drastically growing [6]–[8]. The fusion between Wearable Computer, Mixed Reality and Computer Vision opens a new vista of wearable MR-Based support systems with intelligent computer eyes.

References

[1] S. Feiner, B. MacIntyre, and D. Seligmann: "Knowledge-based augmented reality," *Comm. ACM*, vol.36, no.7, pp.52–62, 1993.

[2] S. Feiner, B. MacIntyre, and T. Höller: "A touring machine: Prototyping 3D mobile augmented reality system for exploring the urban environment," *Proc. Int'l Symp. on Wearable Computing*, pp.74–81, 1997.

[3] T. P. Caudell: "Introduction to augmented and virtual reality," *Proc. SPIE - Int. Soc. Opt. Eng.*, vol.2351, pp.272–281, 1994.

[4] T. P. Caudell and D. W. Mizell: "Augmented reality: An application of heads-up display technology to manual manufacturing processes," *Proc. 25th Hawaii Int'l Conf. on Sys. Sci.*, vol.2, pp.659–669, Jan. 1992.

[5] Y. Ban, K. Sato, and K. Chihara: "AR backpacker for manufacturing and inspection," *Proc. Int'l Conf. on Virtual System and Multimedia*, vol.1, pp.65–68, Sep. 1996.

[6] T. Starner, S. Mann, B. Rhodes, J. Levine, J. Healey, D. Kirsch, R. W. Picard, and A. Pentland: "Augmented reality through wearable computing," *Presence*, vol.6, no.4, pp.386–398, 1997.

[7] S. Mann: "Wearable computing: A first step toward personal imaging," *IEEE Computer*, vol.30, no.2, pp.25–32, Feb. 1997.

[8] J. Rekimoto and K. Nagao.: "The world through the computer: Computer augmented interaction with real world environments," *Proc. UIST*, pp.29–36, Nov. 1995.

Chapter 20

Wearing It Out: First Steps Toward Mobile Augmented Reality Systems

Steven Feiner
Blair MacIntyre
Tobias Höllerer

Columbia University, U.S.A.

20.1 Introduction

Over the past decade, there has been a ground swell of activity in two fields of user interface research: augmented reality and wearable computing. *Augmented reality* [1] refers to the creation of virtual environments that supplement, rather than replace, the real world with additional information. This is accomplished through the use of "see-through" displays that enrich the user's view of the world by overlaying visual, auditory, and even haptic, material on what she experiences. Visual augmented reality systems typically, but not exclusively, employ head-tracked, head-worn displays. These either use half-silvered mirror beam splitters to reflect small computer displays, optically combining them with a view of the real world, or use opaque displays fed by electronics that merge imagery captured by head-worn cameras with synthesized graphics. *Wearable computing* moves computers off the desktop and onto the user's body, made possible through the miniaturization of computers, peripherals, and networking technology. (While we prefer this general definition implied by the

363

term "wearable," some researchers have favored a more restrictive definition, requiring, for example, a capability for hands-free operation [2] or a physical appearance that makes the user and others consider the computer to be part of the user [3].)

We are interested in how augmented reality can be combined with wearable computing, with the ultimate goal of supporting ordinary users in their interactions with the world. To experiment with these ideas, we have been building the software infrastructure and application prototypes described in this chapter. We believe that the commercial success of future descendants of these systems will depend in large part on hardware issues, including the quality, size, comfort, and cost of wearable displays and computers. However, our research focus has been on developing experimental software infrastructure and user interfaces. To maintain this emphasis, we use commercially available hardware exclusively. Furthermore, rather than using a commercial belt-size wearable computer, in one project we rely on a much larger back-pack–sized computer. While this decision sacrifices current user comfort, in return we get access to the same processor, memory, operating system, and busses that we would expect on a desk-top system instead of the compromises imposed by the smaller form factor.

The applications that we address here provide users with information about their surroundings. One system presents a construction worker with instructions for assembling a modular building, while another creates a personal "touring machine" that assists the user in exploring our campus. There are two themes that we have stressed in this research:

- Presenting information about a real environment that is integrated into the 3D space of that environment.

- Combining multiple display and interaction technologies to take advantage of their complementary capabilities.

In Section 20.2 we discuss related work. Sections 20.3 and 20.4 describe our construction assistance and touring machine prototypes, respectively, including pictures generated by our testbed implementations. In Section 20.5, we describe our system-design approach and the hardware and software used. Finally, Section 20.6 presents our conclusions and future directions.

20.2   Related Work

Previous research in augmented reality has addressed a variety of application areas including aircraft cockpit control [4], assistance in surgery [5], viewing hidden building infrastructure [6], maintenance and repair [7], and parts assembly [8]. In contrast to these systems, which use see-through head-worn displays, Rekimoto [9] has used hand-held displays to overlay information on color-coded objects. Much effort has also been directed towards developing techniques for precise tracking using tethered trackers (e.g., [10]–[13]).

Work in wearable user interfaces has included several projects that allow users to explore large outdoor spaces. Loomis and his colleagues have developed an application that makes it possible for blind users to navigate a university campus by tracking their position with differential GPS and orientation with a magnetometer to present spatialized sonic location cues [14]. Petrie et al. have field-tested a GPS-based navigation aid for blind users that uses a speech synthesizer to describe city routes [15]. The CMU Wearable Computer Project has developed several generations of mobile user interfaces using a single hand-held or untracked head-worn display with GPS, including a campus tour [16]. Long et al. have used infrared tracking in conjunction with hand-held displays [17]. Mann [3] has developed a family of wearable systems with head-worn displays, the most recent of which uses optical flow to overlay textual information on automatically recognized objects. Starner [2] has explored the potential uses of wearable computers that can recognize what a user is doing to provide useful information to collaborating team mates.

Prior to the development of VRML, several researchers experimented with integrating hypertext and virtual environments [18]–[20]. All investigated the advantages of presenting hypertext on the same 3D display as all other material, be it head-worn or desktop. In contrast, some of the work described here exploits the different capabilities of our displays by presenting hypertext documents on a relatively high-resolution 2D hand-held display, which is itself embedded within the 3D space viewed through a lower-resolution head-worn display [21].

20.3   ARC: Augmented Reality for Construction

ARC (Augmented Reality for Construction) addresses the construction of space-frame buildings [22]. Spaceframes are made from a large number of modular components, typically cylindrical struts and spherical nodes. While the external appearance of many pieces may be identical for aesthetic reasons, the forces they will bear, and their inner diameters, vary depending on their position in the assembled structure. Thus, it is possible to assemble pieces in the wrong position, which, if undetected, could lead to structural failure. Furthermore, if too many or too few pieces are put together in a subassembly that is to be hoisted into its final position, it may break apart in mid-air. While workers who are trained on a particular spaceframe system are familiar with how to assemble individual components, the position and order in which components are assembled differ from one structure to another. ARC is an attempt to prototype a system that could guide construction workers to put the right piece in the right place at the right time. While the tracking system that we used (see Section 20.5) required a tethered implementation, and our demonstration venues were indoors, we designed our user interface for a mobile outdoor environment.

We have been working with a diamond-shaped, full-scale aluminum system, shown in Figure 20.1, manufactured by Starnet International. We taped to each physical component a set of labels that identify the component using both a human-readable number and a machine-readable barcode. The user wears a see-through head-worn display with integral earphones and a tool belt that holds a barcode reader. The spaceframe is assembled one component (strut or node) at a time. For each component, ARC performs a series of steps.

Figure 20.1   Prototype construction assistance system. The user wears a see-through head-worn display and a tool belt that serves as a holster for a hand-held barcode reader. Here, the user is following instructions to install a strut.

Figure 20.2   Prototype construction assistance system. User's view through the see-through head-worn display for the task shown in Figure 20.1. The user sees three real struts and a real node overlaid with a virtual strut (top) and textual instructions that instruct the user to install the strut in the location shown. A rotating arrow indicates the direction in which a fastener must be turned, and a wireframe sphere highlights the node to which the strut will be attached.

First, the worker is directed to a pile of parts and told which one to pick up. This is done by displaying textual instructions and playing a sound file of verbal instructions. Next, ARC confirms that the user has the right piece by having her scan a barcode on it. ARC then directs her to install the component, as shown in Figure 20.1. Overlaid graphics, such as those of Figure 20.2, indicate the correct location, and verbal instructions explain how to install it.[1] Finally, ARC verifies that the right piece is in the right place by asking the user to scan the component with the tracked barcode scanner (Figure 20.3).

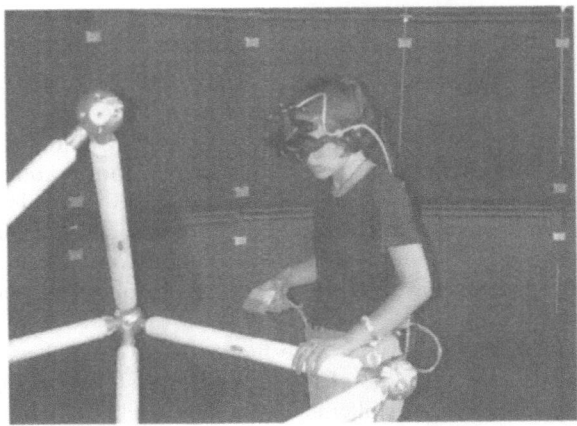

Figure 20.3   Prototype construction assistance system. The user scans one of the barcodes on a strut with a position-tracked barcode reader, to verify that the right strut has been installed in the right place.

ARC was first demonstrated at the *ASCE Third Congress on Computing in Civil Engineering* in June 1996, where dozens of attendees tried the system. Based on this experience, we modified the user interface in two ways for a demonstration given at the *ACM '97* conference in March 1997.

First, while construction workers are familiar with the convention of attaching a fastener nut by turning it clockwise, many of our demonstration participants were not. Some had difficulty following verbal directions describing the direction in which to turn a fastener or did not understand that a strut was to be fastened to a node. Therefore, we added to the graphics a rotating arrow and a representation of the node(s) to which a strut was to be attached, shown in Figure 20.2, effectively solving the problem.

Second, other attendees could see only the participant and, on a separate monitor, the participant's overlaid graphics. Since we used an optical see-through display, many attendees found the graphics uninteresting in isolation and had difficulty understanding that the participant was seeing the graphics overlaid on the real world.

---

[1] Figures 20.2, 20.5, and 20.6 were imaged with a dummy head whose right eyesocket contains a Toshiba IK-M41A 410,000 pixel miniature color CCD camera used to record through an eye of the see-through head-worn display.

We rejected the approach of displaying to the other attendees a synthesized composite of the real and virtual objects in the participant's view, on the grounds that it would be confusing—we were trying to emphasize that the participant's view was being augmented, not replaced. Instead, we created an additional, stand-alone display that provided a third-person remote supervisor's view of the task. Here, the participant is shown as a tracked mannequin, surrounded by the structure under construction and overlaid with information about the current task being performed. Attendees viewed this overview on a large projection display and could also see the participant's graphics on a separate display.

20.4 A Touring Machine

Our "touring machine" prototype assists a user exploring Columbia's campus, overlaying information about items of interest [23]. As a user moves about, she is tracked through a combination of satellite-based, differential GPS (Global Positioning System) position tracking and magnetometer/inclinometer orientation tracking. Information is presented and manipulated on a combination of a head-tracked, see-through, head-worn, 3D display, and an untracked, opaque, hand-held, 2D display with stylus and trackpad.

Consider the following scenario, whose figures were created using our system. The user is standing in the middle of our campus, as shown in Figure 20.4. His tracked see-through head-worn display is driven by a backpack computer, and he is holding a hand-held computer and stylus.

Figure 20.4 Prototype campus information system. The user wears a backpack and see-through head-worn display, and holds a hand-held display and stylus.

As the user looks around the campus, his see-through head-worn display overlays textual labels on campus buildings, as shown in Figures 20.5 and 20.6. (These images were shot through the head-worn display, and are somewhat difficult to read because of the low brightness of the display and limitations of the video recording technology.) Because we label buildings, and not specific building features, the relative inaccuracy of the trackers we are using is not a significant problem for this application.

Figure 20.5   View shot through the see-through head-worn display, showing campus buildings with overlaid names. Labels increase in brightness as they near the center of the display.

Figure 20.6   A view of the Philosophy Building with the "Departments" fly-down menu item highlighted. Since the "Departments" menu item has been selected, a copy was animated flying to the bottom of the display to indicate the presence of new information on the hand-held computer, and the department list for the Philosophy Building was displayed around the building's name.

At the top of the display is a menu of choices: "Columbia:", "Where am I?", "Depts?", "Buildings?", and "Blank". When selected, each of the first four choices sends a URL to a web browser running on the hand-held computer. The browser then presents information about the campus, the user's current location, a list of departments, and a list of buildings, respectively. The URL points to a custom HTTP server on the hand-held computer that generates a page on the fly containing the relevant information. The generated pages include links back to the server itself and to pages anywhere on the world wide web, to which we are connected through radio. The last menu item, "Blank", allows the head-worn display to be blanked when the user wants to view the unaugmented campus. Menu entries are selected using a trackpad mounted on the back of the hand-held computer. The trackpad's x coordinates are inverted to preserve intuitive control of the menus.

Labels seen through the head-worn display are grey, increasing in intensity as they approach the center of the display. The one label closest to the center is highlighted yellow. If it remains highlighted for more than a second, it changes to green, indicating that it has been selected, and a second menu bar is added below the first, containing the name of the selected building and entries for obtaining information about it. A selected building remains selected until the user dwells on another for more than a second as indicated by the color change. This approximation of gaze-directed selection can be disabled or enabled via a trackpad button.

When a building is selected, a conical green compass pointer appears at the bottom of the head-worn display, oriented in the building's direction. The pointer turns red if the building is more than 90 degrees away from the user's head orientation (i.e., behind the user). The pointer is especially useful for finding buildings selected from the hand-held computer. To do this, the user turns off gaze-directed selection, displays a list of all buildings via the "Buildings?" top-level menu entry, selects with a stylus the building he is looking for on the hand-held computer, and then follows the direction of the arrow pointer to locate that building. When the building's link is selected on the hand-held computer, the system immediately reflects the selection on the head-worn display. This is made possible by our custom HTTP server, which can interact with the backpack computer on URL selection.

A building's menu bar contains the name of the building plus additional items: "Architecture", "Departments", and "Miscellaneous". Selecting the building's name from the menu using the trackpad sends a relevant URL to the hand-held computer's browser. Selecting any of the remaining menu entries also sends a URL to the browser and creates a collection of items that are positioned near the building on the head-worn display. These items represent the information requested by the second-level menu entry selection and they stay in the same position relative to the building (and its label) until this menu level is left via a "Dismiss" entry.

When menu items that send URLs are selected, a copy of the menu item is translated down to and off the bottom of the head-worn display. This animated "fly-down" menu is intended to call the user's attention to the new material on the hand-held computer. For example, Figure 20.6 shows the Philosophy Building with the "Departments" menu item highlighted after its selection and animation has been completed. The building is annotated with the names of its departments and the hand-held computer's automatically-generated web page is shown in Figure 20.7(a).

Information about the selected building can be accessed in two ways. On the head-worn display, the user can select one of the surrounding items with the trackpad to present relevant information about it on the hand-held display. Alternatively, the user can select a corresponding item from the automatically-generated web page.

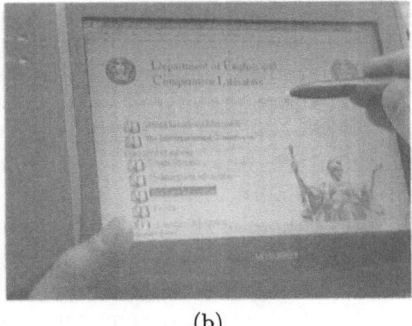

(a)   (b)

Figure 20.7   (a) Selecting the "Departments" menu item causes an automatically-generated URL to be sent to the web browser on the hand-held computer, containing the department list for the Philosophy Building. (b) Actual home page for the English and Comparative Literature department, as selected from either the generated browser page or the department list of Figure 20.6.

For example, Figure 20.7(a) shows the URL selection for the English Department in the Philosophy Building, resulting in the display of its regular web page in Figure 20.7(b).

Another way of accessing information about a specific department is through the global list of departments that is produced on the hand-held by selecting the top-level "Departments?" menu item on the head-worn display. In this case the associated building does not have to be selected beforehand.

20.5   System Design

The following subsections describe some of the hardware and software decisions we made in designing our prototypes.

20.5.1   Hardware

Computers.   Given the relatively small amount of overlaid graphics displayed in the ARC project, we were able to support the worker's view with a 90MHz Pentium PC, rendering entirely in software using Criterion RenderWare. In contrast, the supervisor's view is completely synthesized and includes a relatively complex view of a larger structure of which our demonstration's components are a subset; it required higher-end SGI and Sun workstations to achieve respectable performance.

Our touring machine's original backpack computer, obtained in early 1996, was a Fieldworks 7600 with a 133MHz Pentium processor and a cage that can hold 3 ISA and 3 PCI cards. While it was a big compromise in weight and size, it has significantly simplified our development effort, for example allowing us to use 3D graphics cards, such as our initial selection, the Omnicomp 3Demon OpenGL card, based on the Glint 500DTX chipset. We are currently replacing this unit with a smaller, more powerful, laptop. The hand-held computer is a Mitsubishi Amity SP, which has a 75MHz DX4, and 640×480 color display, and integral stylus. Control of the head-worn display menu is accomplished through a Cirque GlidePoint trackpad that we mounted on the back of the Amity.

Head-worn display. Both applications use the Virtual I/O i-glasses optical see-through head-worn display, a relatively inexpensive, but relatively low resolution (60,000 triads), color display. We have also experimented with a Virtual I/O 640×480 resolution greyscale display.

Trackers. Given the large amount of movable metal in the ARC demonstration, we decided on an optical, rather than electromagnetic, tracking technology: an Origin Instruments DynaSight optical radar tracker. The DynaSight tracks three LEDs on the head-worn display and a fourth LED on the hand-held barcode reader.

The touring machine uses a Trimble DSM GPS receiver to determine the position of its antenna, which is located on the backpack above the user's head. We subscribe to a differential correction service provided by Differential Corrections Inc., which allows us to achieve one-meter accuracy. We are currently replacing this GPS system with an Ashtech GG Surveyor GPS receiver, which provides real-time centimeter-level position fixes. Orientation tracking is accomplished using the head-worn display's built-in magnetometer, which senses the earth's magnetic field to determine head yaw, and a two-axis inclinometer, which uses gravity to detect head pitch and roll.

Power, Network, and Peripherals. With the exception of the computers, each of the touring machine's other hardware components has relatively modest power requirements of under 10 watts each. We run them all using an NRG Power-MAX NiCad rechargeable battery belt. To communicate with the rest of our infrastructure the touring machine uses Lucent WaveLan 2Mbit/sec wireless network cards in both the backpack and hand-held PCs, which operate with a campus network of base stations. In contrast, ARC's machines share a hardwired ethernet. Validation of part identity and position in ARC is accomplished with a PSC Inc. QuickScan barcode scanner that is position-tracked through a single attached DynaSight LED.

20.5.2   Software

Infrastructure. We use COTERIE [24], a system that provides language-level support for distributed virtual environments. COTERIE is based on the distributed data-object paradigm for distributed shared memory. Any data object in COTERIE can be declared to be a shared object that either exists in one process, and is accessed via remote-method invocation, or is replicated fully in any process that is interested in it. The replicated shared objects support asynchronous data propagation with atomic serializable updates, and asynchronous notification of updates. COTERIE runs on Windows NT/95, Solaris, and IRIX, and includes the standard services needed for building virtual environment applications, including support for assorted trackers. It is built on top of Modula-3 [25] and Repo [26], which is our extended variant of the lexically scoped interpreted language Obliq [27].

Graphics package. We use Repo-3D [28], a version of Obliq-3D [29], a scene-graph–based 3D graphics package, which we have modified to provide additional features needed for virtual environment applications and to achieve better performance.

Operating systems. We run Windows NT on the ARC worker's computer and the touring machine Fieldworks to benefit from its support for multitasking and assorted commercial peripherals. We run Windows 95 on the Amity because it does not support Windows NT.

Web browser. Information on the hand-held computer is currently presented entirely through a web browser. We selected Netscape because of its popularity within our university and the ease with which we can control it from another application. To obtain increased performance, we constructed a proxy server that caches pages locally across invocations. This has also been helpful during radio network downtime and for operation in areas without network coverage.

Application software. The original version of ARC was written in several hundred lines of Repo, and supported a single user: the worker. When we decided to add an additional overview display, implemented in a separate program, we needed it to share the task state with the worker's display. Therefore, we modified the ARC prototype to move its single state variable (representing the current task step) into a replicated object, and exported this variable to the network. We imported this variable into our overview display program, and allowed both programs to change the construction step. However, we noticed that this did not give us all the information the overview display needed, especially about when the worker performed incorrect actions. To distribute this information, we added routines to the replicated object that are called when various interesting conditions are noticed, such as the task being completed, the worker scanning the wrong part, and the worker scanning the right part in the wrong location. Note that none of this information is "typical" distributed virtual environment data that would be supported by a distributed VE toolkit, but Repo allows us to distribute the information and react to changes in the various programs in a few lines of code that took a few minutes to write.

The touring machine prototype comprises two applications, one running on each machine, implemented in approximately 3600 lines of commented Repo code. The tour application running on the backpack PC is responsible for generating the graphics and presenting it on the head-worn display. The application running on the hand-held PC is a custom HTTP server in charge of generating web pages on the fly and also accessing and caching external web pages by means of a proxy component.

One of the main reasons that we run our own HTTP server on the hand-held display is that it gives us the opportunity to react freely to user input from the web browser. For example, when a URL is selected on the hand-held display, the HTTP server can call a network object method that selects corresponding graphical items on the head-worn display. Thus data selection works in both directions: from the backpack PC to the hand-held PC (by launching relevant URLs from the head-worn display's menus) and vice versa (selecting buildings, departments, etc. on the head-worn display from a link on the hand-held's browser).

The HTTP server is initialized by the tour application running on the backpack PC. Each piece of information (buildings, departments, their whereabouts, and assorted URLs) in the tour data on the backpack PC is sent to the hand-held PC's HTTP server with an accompanying procedure closure. The closure executes a procedure on the backpack PC when the corresponding link is selected on the web browser. This makes it possible for the hand-held display to control the head-worn display, as described in Section 20.4.

The web browser on the hand-held PC is a totally separate process. It can be pointed at URLs from within the HTTP server, which we currently accomplish by forking off a separate URL pusher process. The web browser then issues a request back to the HTTP server to obtain either a locally generated, cached external, or uncached external HTML document.

The tour application continuously receives input from the GPS position tracker and the orientation tracker. It also takes user input from the trackpad that is physically attached to the back of the hand-held PC. Based on this input and a database of information about campus buildings, it generates the graphics that are overlaid on the real world by the head-worn display.

20.6   Conclusions and Future Work

We have described two prototype applications that explore approaches to the design of mobile augmented-reality systems. Many hardware issues must be resolved for commercial versions of such systems to become practical, including significant improvements in the quality of tracking and display technologies. We are working on several extensions to our work.

Overlaying virtual objects on the real world can be potentially confusing if they interfere with the user's view of the real world and of each other. For example, even the relatively sparse overlaid graphics of Figures 20.5 and 20.6 evidence problems caused by self-occlusion. We are currently incorporating the Snap-Together Math constraint-based toolkit [30] into our system to explore how automated satisfaction of geometric constraints among objects could help maintain display quality for a mobile user.

We are also working with colleagues in Columbia's Graduate School of Journalism to explore the touring machine's potential for developing and presenting situated 3D news stories that are embedded in the actual locations in which they occurred. Here, we are experimenting with the use of our system as a "mobile journalist's workstation" that presents images and audio on the head-worn display, coordinated with web pages and videos on the hand-held computer. Our first story is a documentary that describes Columbia's 1968 student revolt, as shown in Figure 20.8. This application raises many interesting issues about media coordination, storytelling, and multimedia authoring.

Figure 20.8 Mobile Journalist's Workstation. An image from a 3D news story about Columbia's 1968 student revolt. Protestors outside Hamilton Hall are shown in context of that building.

Acknowledgments

Tony Webster and his Building Technology Lab in Columbia's Graduate School of Architecture, helped develop the design for the ARC and Touring Machine projects. John Pavlik, and students at the New Media Center in Columbia's Graduate School of Journalism, helped develop the concept for the Mobile Journalist's Workstation. Xinshi Sha assisted in developing COTERIE, and Ruigang Yang created Windows 95 utilities. Christina Vernon and Alex Klevitsky helped create the campus databases. Rod Freeman, Jenny Wu, and Melvin Lew helped implement ARC. Jim Foley of MERL, a Mitsubishi Electric Research Laboratory, generously provided a Mitsubishi Amity and Reuven Koblock of Mitsubishi Electric ITA Horizon Systems Laboratory assisted us with the Amity; Marc Najork of DEC's System Research Center and Bill Kalsow of Critical Mass, Inc., assisted during the development of COTERIE.

This work was supported in part by the Office of Naval Research under Contracts N00014-94-1-0564 and N00014-97-1-0838, the Columbia Center for Telecommunications Research under NSF Grant ECD-88-11111, NSF Grant CDA-92-23009, NSF Gateway Engineering Coalition under NSF Grant EEC-9444246; a Columbia University Provost's Strategic Initiative Fund Award, and gifts from Microsoft, Mitsubishi, and Starnet International, Inc.

References

[1] R. Azuma: "A survey of augmented reality," *Presence*, vol.6, no.4, pp.355–385, Aug. 1997.

[2] T. Starner: "Wearable computing and context awareness," PhD thesis, Program in Media Arts and Sciences, MIT, Cambridge, MA, Feb. 1999.

[3] S. Mann: "Wearable computing: A first step toward personal imaging," *Computer*, vol.30, no.2, Feb. 1997.

[4] T. Furness: "The super cockpit and its human factors challenges," *Proc. Human Factors Society 30th Annual Meeting*, pp.48–52, Santa Monica, CA, 1986.

[5] A. State, M. Livingston, W. Garrett, G. Hirota, M. Whitton, E. Pisano, and H. Fuchs: "Technologies for augmented reality systems: Realizing ultrasound-guided needle biopsies," *Proc. SIGGRAPH '96*, pp.439–446, New Orleans, LA, Aug. 4–9, 1996.

[6] S. Feiner, A. Webster, T. Krueger, B. MacIntyre, and E. Keller: "Architectural anatomy," *Presence*, vol.4, no.3, pp.318–325, Summer, 1995.

[7] S. Feiner, B. MacIntyre, and D. Seligmann: "Knowledge-based augmented reality," *Comm. ACM*, vol.36, no.7, pp.52–62, Jul. 1993.

[8] T. Caudell and D. Mizell: "Augmented reality: An application of heads-up display technology to manual manufacturing processes," *Proc. Hawaii Int'l. Conf. on Sys. Sci*, pp.659–669, Hawaii, Jan. 1992.

[9] J. Rekimoto and K. Nagao: "The world through the computer: Computer augmented interaction with real world environments," *Proc. UIST '95*, pp.29–36, Nov. 14–17, 1995.

[10] M. Ward, R. Azuma, R. Bennett, S. Gottschalk, and H. Fuchs: "A demonstrated optical tracker with scalable work area for head-mounted display systems," *Computer Graphics (1992 Symp. on Interactive 3D Graphics)*, vol.25, pp.43–52, Mar. 1992.

[11] A. Janin, D. Mizell, and T. Caudell: "Calibration of head-mounted displays for augmented reality applications," *Proc. VRAIS '93*, pp.246–255, Seattle, WA, Sep.18–22, 1993.

[12] R. Azuma and G. Bishop: "Improving static and dynamic registration in an optical see-through HMD," *Proc. SIGGRAPH '94*, pp.197–204, Orlando, FL, Jul.24–29, 1994.

[13] A. State, G. Hirota, D. Chen, W. Garrett, and M. Livingston: "Superior augmented reality registration by integrating landmark tracking and magnetic tracking" *Proc. SIGGRAPH '96*, pp.429–438, New Orleans, LA, Aug.4–9, 1996.

[14] J. Loomis, R. Golledge, R. Klatzky, J. Speigle, and J. Tietz: "Personal guidance system for the visually impaired," *Proc. 1st Ann. Int'l. ACM/SIGCAPH Conf. on Assistive Technology*, pp.85–90, Marina del Rey, CA, Oct. 31–Nov. 1, 1994.

[15] H. Petrie, V. Johnson, T. Strothotte, A. Raab, S. Fritz, and R. Michel: "MoBIC: Designing a travel aid for blind and elderly people," *J. Navigation*, vol.49, no.1, pp.45–52, 1996.

[16] A. Smailagic and D. Siewiorek: "The CMU mobile computers: A new generation of computer systems," *Proc. COMPCON '94*, Feb. 1994.

[17] S. Long, D. Aust, G. Abowd, and C. Atkeson: "Cyberguide: Prototyping context-aware mobile applications," *CHI '96 Conf. Companion*, pp.293–294, Apr. 1996.

[18] P. Dykstra: "X11 in virtual environments," *Proc. IEEE 1993 Symp. on Research Frontiers in Virtual Reality*, pp.118–119, San Jose, CA, Oct. 25–26, 1993.

[19] S. Feiner, B. MacIntyre, M. Haupt, and E. Solomon: "Windows on the world: 2D windows for 3D augmented reality," *Proc. UIST '93 (ACM Symp. on User Interface Software and Technology)*, pp.145–155, Atlanta, GA, Nov. 3–5, 1993.

[20] I. Angus and H. Sowizral: "VRMosaic: WEB access from within a virtual environment," *Proc. IEEE Information Visualization '95*, pp.59–64, IEEE Computer Society Press, Oct. 30–31, 1995.

[21] S. Feiner and A. Shamash: "Hybrid user interfaces: Breeding virtually bigger interfaces for physically smaller computers," *Proc. UIST '91 (ACM Symp. on User Interface Software and Technology)*, pp.9–17, Hilton Head, SC, Nov. 11–13, 1991.

[22] A. Webster, S. Feiner, B. MacIntyre, W. Massie, and T. Krueger: "Augmented reality in architectural construction, inspection and renovation," *Proc. ASCE 3rd Congress on Computing in Civil Engineering*, pp.913–919, Anaheim, CA, Jun. 17–19, 1996.

[23] S. Feiner, B. MacIntyre, T. Höllerer, and A. Webster: "A touring machine: Prototyping 3D mobile augmented reality systems for exploring the urban environment," *Personal Technologies*, vol.1, no.4, pp.208–217, 1997.

[24] B. MacIntyre and S. Feiner: "Language-level support for exploratory programming of distributed virtual environments," *Proc. UIST '96*, pp.83–94, Seattle, WA, Nov. 6–8, 1996.

[25] S. Harbison: *Modula-3*, Prentice-Hall, 1992.

[26] B. MacIntyre: "Repo: Obiq with replicated objects. Programmer's guide and reference manual," Technical Report CUCS-023-97, Columbia University Dept. of CS, Aug. 1997.

[27] L. Cardelli: "A language with distributed scope," *Computing Systems*, vol.8, no.1, pp.27–59, Jan. 1995.

[28] B. MacIntyre and S. Feiner: "A distributed 3D graphics library," *Proc. SIGGRAPH '98*, pp.361–370, Orlando, FL, Jul. 19–24, 1998.

[29] M. Najork and M. Brown: "Obliq-3D: A high-level, fast-turnaround 3D animation system," *IEEE Trans. on Visualization and Computer Graphics*, vol.1 no.2, pp.175–145, Jun. 1995.

[30] M. Gleicher and A. Witkin: "Supporting numerical computations in interactive contexts," *Proc. Graphics Interface '93*, pp.138–146, Toronto, Ontario, Canada, Canadian Information Processing Society, May 1993.

Chapter 21

The Challenge of Making Augmented Reality Work Outdoors

Ronald T. Azuma
HRL Laboratories, U.S.A.

21.1 Background in Augmented Reality

Making Augmented Reality (AR) systems that work outdoors is a natural step in the development of AR toward the ultimate goal of AR displays that can operate anywhere, in any environment. To place this in context, this section gives a brief overview of the development of AR, my own contributions to this field, and our current research program at HRL Laboratories.

Despite its potential, Augmented Reality has not received nearly the amount of attention paid to its sibling, Virtual Environments (or Virtual Reality), despite the fact that both fields share a common ancestor. It is often forgotten that Sutherland's original HMD system was an optical see-through display. While the creators of that system did not explicitly attempt to register virtual 3-D objects with real-world objects, they did have an example of combining virtual and real. The motivation was to allow the user to issue commands. The problem was that the graphics engine did not have sufficient power to draw the menus and commands virtually at interactive rates. Therefore, they physically put large signs with the command names on a real wall, and allowed the user to virtually select one of the real signs by pointing at one with the hand controller. Despite this common root, most efforts following Sutherland's focused on Virtual Environments. It wasn't until the late 1980's and early 1990's that research in Augmented Reality began again in earnest.

While there are several problems in building Augmented Reality systems, many research efforts have focused on the registration problem (the proper alignment of virtual with real). This is a difficult problem because there are many sources of error, even small amounts of error are perceptible, and few commercially-available trackers provide the required performance. Without good registration, many AR applications will not be accepted. My own research history tackled this area. I contributed to a wide-area optical tracking system for HMDs [1] [2], and then built and analyzed an optical see-through HMD system that used head-motion prediction and other techniques to greatly reduce registration errors [3] [4]. I also wrote a survey paper [5] that summarizes the work done in this field.

What is the current state-of-the-art in registration today? A few prototype systems have demonstrated almost pixel-accurate displays for certain applications. These typically use a "closed loop" tracking approach [6], where the real-world images shown in a video see-through display are also used to aid the tracking, correcting remaining errors to ensure that the virtual is properly aligned with the real objects captured in the video images. A good example of this is [7]. Not every AR application requires pixel-accurate registration, and other effective demonstrations and prototype systems have been built for manufacturing and maintenance of aircraft parts [8] and even for entertainment [9]. Therefore, the current state-of-the-art is that effective prototype systems exist for a few applications, and at least one is getting close to actual deployment on a factory floor.

While there has been significant progress on the registration problem, it is still far from a solved problem. AR systems, especially those that rely on closed-loop tracking, required carefully controlled environments. The user cannot walk and look anywhere she pleases; the system only works for specific objects, as seen from a limited range of viewpoints. By constraining the problem in this manner, the registration problem is more tractable, but it also greatly restricts the flexibility of the system and makes it difficult to build an effective AR system without expert knowledge. A symptom of this is the unavailability of commercial systems. Today, one has a choice of several turnkey Virtual Environment systems, but nobody offers a turnkey Augmented Reality system. Contrast the situation today with the ideal situation: where one can put on an AR display and walk around anywhere, with no restrictions. Ideally, an AR display should work in all environments without the need to prepare rooms and objects ahead of time. There is a need to investigate AR systems that can work in unstructured, real-world environments, both indoors and outdoors. Some work along this direction has been done to reduce the calibration requirements [10] and the need to know the geometry of all observed objects at the start [11]. However, to continue making progress, we need to explore more difficult problem domains. We should explore AR in *outdoor* applications, rather than just the *indoor* applications that have dominated the field so far.

Exploring outdoor AR systems and applications is the current research direction in AR for personnel at HRL Laboratories. Funding is provided by a DARPA grant for a project called GRIDS: Geospatial Registration of Information for Dismounted Soldiers. The primary personnel involved with the project at HRL are myself, Bruce Hoff, Howard Neely, Ron Sarfaty, and Mike Daily. We are collaborating on this research with our partners at the University of North Carolina at Chapel Hill (Gary

Bishop, Vern Chi, and Greg Welch), the University of Southern California (Ulrich Neumann and Suya You), and Raytheon Systems Company (Rich Nichols and Jim Cannon). While the initial users this project is focusing on are infantrymen, this technology will also be suited for a variety of civilian users and applications. This chapter outlines the motivations and potential applications of AR in outdoor situations (Section 21.2), analyzes the problems we face in building outdoor AR systems, especially in the area of tracking (Section 21.3), and concludes by suggesting some general approaches (Section 21.4).

21.2   Motivation for Outdoor Augmented Reality

Augmented Reality systems that provide accurate registration outdoors are of interest because they would make possible new application areas and could provide a natural interface for wearable computers, an area of growing interest both in academia and industry. A user walking outdoors could see spatially located information directly displayed upon her view of the environment, helping her to navigate and identify features of interest. Today, a hiker in the woods needs to pull out a map, compass, and GPS receiver, convert the GPS and compass readings to her location and orientation, and then mentally align the information from the 2-D map onto what she sees in the 3-D environment around her. A personal, outdoor AR system could perform the same task automatically and display the trail path and landmark locations directly upon her view of the surrounding area, without the cognitive load. Soldiers could see the locations of enemies, friends, and dangerous areas like minefields that may not be readily apparent to the naked eye. Personal, outdoor AR systems would also be useful for groups of users working together. If the users are widely separated, it is difficult for them to establish common frames of reference to describe spatially located information. An instruction telling another team member to go to the "3rd white building to the left of the red building" may be useless if the listener sees the world from a different vantage point than the speaker. Personal AR displays provide an unambiguous method of sharing such information. Furthermore, personal outdoor AR displays may be a natural interface for wearable PC's. The standard WIMP interface does not map well onto wearable PCs, because the desktop metaphor is not appropriate for a user walking around outdoors who may not have her hands free or be able to allocate complete attention to the computer [12]. Augmented Reality may be a better approach for certain applications.

Another application area is the visualization of locations and events as they were in the past or as they will be after future changes are performed. Tourists that visit historical sites, such as a U.S. Civil War battlefield or the Acropolis in Athens, Greece, do not see these locations as they were in the past, due to changes over time. It is often difficult for a modern visitor to imagine what these sites really looked like in the past. To help, some historical sites stage "Living History" events where volunteers wear clothes of that time period and reenact historical events. A tourist equipped with an outdoor AR system could see a computer-generated version of Living History. The HMD could cover up modern buildings and monuments in the background and show, directly on the grounds at Gettysburg, where the Union and Confederate troops were at the fateful moment of Pickett's charge. The gutted interior of the modern Parthenon would be filled in by computer-generated representations of what it looked like in 430 BC, including the long-vanished gold statue of Athena in the middle. Tourists and students walking around the grounds with such AR displays would gain a much better understanding of these historical sites and the important events that took place there. Similarly, AR displays could show what proposed architectural changes would look like before they are carried out. An urban designer could show clients and politicians what a new building would look like as they walked around the adjoining neighborhood, to better understand how that skyscraper might affect nearby residents.

If truly effective outdoor AR displays existed, then some applications that we normally view as indoor applications may find new roles in outdoor situations. The maintenance and repair of vehicles in the field would be helpful for users stuck far away from traditional repair facilities. Aircraft technicians could use such displays when working on aircraft outdoors, at terminals. Paramedics might combine such displays with portable non-invasive 3-D scanners to provide a "3-D stethoscope" in the field. And the possibilities AR has for outdoor entertainment applications are unexplored.

While outdoor AR applications offer some interesting possibilities, few have attempted to build Augmented Reality systems that work outdoors. Steve Feiner's group at Columbia demonstrated the Touring Machine, which allows a user to view information linked to specific buildings on the Columbia campus as he walks around outside [13]. Some wearable computers, such as the CMU VuMan systems, have been used for vehicle maintenance applications in outdoor settings. However, none of these previous works focused on achieving accurate registration at a wide variety of outdoor locations, and to my knowledge, none offers the performance required for most outdoor applications.

21.3   Analysis of Problem Areas

What are the difficulties that prevent personal, outdoor AR systems from being deployed today? The ergonomic issues that face wearable PC systems apply to outdoor AR systems as well: size, weight, power, ruggedness, etc. Displays that have sufficient contrast to work in outdoor settings are required. But the biggest challenge lies in accurate tracking outdoors: determining the user's position and orientation with sufficient accuracy to avoid significant registration errors. This section analyzes these three problem areas.

21.3.1   Size, Weight and Power Issues

A major difference between outdoor and indoor AR systems is the amount of infrastructure and resources available. Indoor systems work in relatively luxurious conditions. Power, computation, and other resources are limited by the agency's budget rather than ergonomic constraints. An indoor application must worry about what the user wears on his head, but the user need not carry the computers or the power supplies. An extensive infrastructure can be available indoors, with a wider range of tracking technologies that may not be practical outdoors. The situation changes dramatically once you go outdoors. If the user is in a vehicle, then the computation and power budget may be relatively large. But if the user walks around outdoors on his own and must carry batteries, the computer, and all other resources, then size, weight and power issues become significant concerns.

The amount of gear that a user can be reasonably asked to carry depends on the type of person and the application. Helmets for infantrymen and military pilots typically weigh around four to five pounds. For civilian applications that support a wider user base than the military, the weight limit is probably far less. Minimizing weight on the head is more critical than minimizing weight in the backpack or belt, but overall weight is a concern as well. Developments in laptops, wearable PC's, PDA's and other upcoming portable devices may provide improvements that outdoor AR systems can leverage off of.

21.3.2   Displays

For indoor applications, we can control the lighting to match the display. But in outdoor applications, the AR display must work across a wide variety of lighting conditions, from bright sunlight to a moonless night. The contrast between these two conditions is huge, and most display devices cannot come close to the brightness required to match this range. This is no surprise to the laptop owner who tries to read his display in bright sunlight. For optical see-through displays this can be a big problem because the user sees the real world directly through the display. Certain upcoming displays, such as FED-based devices, may be bright enough to be visible in bright sunlight. But for the rest, it will be necessary to provide shading devices to cut down on the amount of light entering the display (much like a pair of sunglasses). Video see-through avoids some of this problem because one can set the exposure of the cameras to match the input. However, this comes at a price. The dynamic range of the real world must be compressed into the range of the output monitor, which is generally much smaller. Also, the resolution of the display monitor is much lower than the resolution of the human eye. These two combined will cause a loss of detail that may not be acceptable in some applications.

A primary motivation for using video see-through is that it allows closed-loop tracking approaches. Since the opportunity to do closed-loop tracking is reduced when we cannot modify the environment and place fiducial markers at known locations, the desirability of video see-through displays may be lower in outdoor applications. Video see-through also demands greater amounts of computation power and other resources, which are limited in outdoor applications.

21.3.3   Tracking

Accurate tracking indoors is hard enough; accurate tracking outdoors is even more daunting because of two main differences in the situation. First, we have less control over the environment. Second, we have fewer resources available – power, computation, sensors, etc. These differences mean that solutions for indoor AR may not directly apply to personal outdoor AR systems. For example, several indoor AR systems have achieved accurate tracking and registration by carefully measuring the objects in a highly constrained environment, putting colored dots over those objects, and tracking the dots with a video camera (see [5] for some references). This approach violates some of the constraints for an outdoor situation. We do not have control over the outdoor environment and cannot always rely on modifying it to fit the needs of the system. For example, in a military application it is not realistic to ask soldiers, friendly or enemy, to wear large, brightly colored dots to aid our tracking system. We should not expect to measure every object in the environment beforehand. Also, this approach may require more computational resources than is practical for a single outdoor user. Such systems have used an SGI Onyx or other high-end workstation with frame-grabbing capability and sufficient bandwidth to process captured frames at interactive rates.

Initial outdoor AR systems· may get by with less accurate registration because some outdoor applications may have less stringent requirements than their indoor counterparts. A doctor performing a needle biopsy requires that the virtual incision marker be accurate within a millimeter, but for a hiker walking around, perhaps even one degree of angular error is acceptable to cue the user to landmarks in the environment. However, the ultimate goal is still pixel-accurate registration under all circumstances.

We can analyze this problem by examining the various individual tracking technologies that are available outdoors. This will include a closer examination of some data taken from a sourceless outdoor orientation sensor.

- *GPS:* The Global Positioning System provides worldwide coverage and measures the user's 3-D position typically within 30 meters for regular GPS and about 3 meters for differential. It does not directly measure orientation. Differential accuracy is sufficient for viewing distant but not nearby objects; at 50 meters range a 3 meter position error results in 3.4 degrees of registration error. A typical update rate from a differential GPS receiver is 1 Hz, although I have seen some that claim 10 Hz. Again, this is sufficient for distant objects but not for nearby objects. For some vehicles, one can use several GPS receivers and establish sufficient baselines to recover vehicle orientation. But for a user walking around, that's not practical. The baseline required is larger than a user can be reasonably asked to support. New carrier-phase GPS receivers claim accuracy in centimeters. However, achieving such accuracies at interactive rates will be difficult due to multipath problems; avoiding those requires that the antenna be the highest object in the surrounding environment, which clearly is not practical under many circumstances [14] [15]. The startup delay of a GPS unit can vary from several seconds to several minutes, depending on whether or not the unit has a good initial guess of the current location and time. GPS requires direct line-of-sight to a sufficient number of satellites, so it works best in open areas. When inside urban areas, canyons, near hills and other terrain, GPS is often blocked. In military situations, GPS is easily jammed.

- *Inertial and dead reckoning:* Inertial sensors are sourceless and relatively immune to environmental disturbances. Their main problem is drift: the accumulation of error with time. Recovering orientation changes (through rate gyroscopes) is easier than position changes (through linear accelerometers) because the former requires only one integration step, while the latter requires two. Ships and aircraft have inertial measurement units that drift a fraction of a degree per hour in orientation and under one nautical mile per hour in position. However, these sensors are not appropriate for a single user walking around outdoors, due to cost and weight. Rate gyros of the form factor appropriate for a human exist and may be sufficient for shorter time intervals (minutes or seconds). However, no accelerometers exist that can provide the performance desired (millimeters of drift over long time periods). This requires accelerometers with bias accuracies several orders of magnitude better than what is commonly available today. Pedometers or other dead reckoning trackers may be better solutions for the individual user walking around.

- *Active sources:* Setting up active transmitters and receivers (using magnetic, optical, or ultrasonic technologies) is commonly done for indoor Virtual Environment systems. Modifying the environment in this manner outdoors may be feasible for a few applications, such as an outdoor maintenance application. However, this approach tethers the user a particular area outdoors (close to the active sources) and requires an infrastructure (power, supporting structures to hold sources in position, etc.) that may not be easy to construct outdoors. For many outdoor applications, such as hiking in the woods, it is not practical to modify the environment in this manner, especially given the user's range of operation.

- *Passive optical:* Using video sensors to track the user location based upon what is visible (the sun, the stars, and the view of the surrounding environment) provides the ability to do closed-loop tracking approaches. Video sensors are line-of-sight and will lose lock if the view is obscured (by buildings, vegetation, etc.) Computer vision techniques are not currently robust enough to provide a complete sole solution, and unless known landmarks are used, the tracking solutions are relative rather than absolute. Tracking based on real-world features and scenes is also more difficult and yields more brittle results than closed-loop tracking of fiducials that some indoor AR systems use. Video processing is also computationally intensive and is not feasible in real time on most wearable PC systems.

- *Electronic compass and tilt sensors:* One of the most common trackers in Virtual Environments is used by inexpensive HMD's: a sourceless orientation-only head tracker that consists of an electronic compass and two tilt sensors. These trackers can be small and inexpensive. The best units claim a yaw accuracy of ± 0.5 degrees, with smaller errors along roll and pitch. Such a sensor is attractive because it can potentially provide orientation measurements without the need of any supporting infrastructure. We have experimented with one such sensor, the Precision Navigation TCM2. The update rate is approximately 16 Hz. The main limitations we have encountered are the distortion, noise and delays in the sensor output.

The compass is highly sensitive to disturbances in the Earth's magnetic field. In hindsight, this is not surprising, because Earth's magnetic field is a relatively weak signal. Figure 21.1 shows the equipment we used to measure distortions

Figure 21.1 Turntable for measuring distortion in electronic compass (Precision Navigation TCM2). Turntable is made of Delrin to avoid creating an additional source of distortion.

in the field. The mechanical turntable is made of Delrin and other non-ferrous materials, to avoid placing sources of distortion near the compass. We measured the compass output with this device at many different locations and times. The compass was sampled at 5 degree intervals. Even when the compass was far from any apparent sources of distortion, the compass measurements would be in error by up to 3 degrees, and the distortion pattern varies significantly with time and location, as shown in Figure 21.2 . These are for measurements taken in "magnetically clean" environments, far from any known sources of distortion. However, in realistic AR systems, it is difficult to completely avoid all sources of magnetic distortion. In a more realistic situation, the peak-to-peak distortion errors were typically 20-30 degrees. The distortion makes it difficult to use such a sensor in an AR system without calibration or other compensation strategies. Because some of the curves in Figure 21.2 have radically different shapes than the others, developing adequate models that compensate for this distortion is a significant challenge. In other words, the magnetic distortion appears to be significantly different at different geographical locations, and even at different times at the same location.

The TCM2's output is noisy, both when sitting still and when the user moves around. When the sensor sits still, the output will typically have noise around 0.5 degrees, which is quite visible inside an AR display. Recall that the apparent width of the full moon is 0.5 degrees. When the TCM2 is moved around, the noise becomes unbearably large, on the order of several degrees. This is partially due to the "sloshing" of the inclinometers when in motion. Translation motion will change

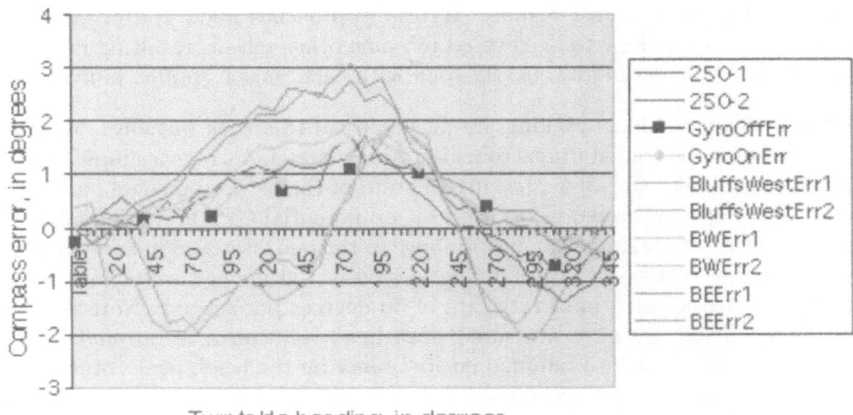

Figure 21.2 Distortion in TCM2. Data taken on turntable, shown in Figure 21.1 . X axis is yaw orientation of compass in degrees, Y axis is the relative error in degrees from the zero degree initial orientation (see color pages).

the measured orientation outputs, even if the true orientation has not changed. This is a basic limitation of using tilt sensors. Even the small motions imparted by a casual walk or by gusty winds are enough to render the output inaccurate. If one expects to use this sensor while walking around or riding in a vehicle, then the AR system must compensate for these noise sources.

Finally, the TCM2's output is delayed in time by ˜ 100 milliseconds. This is due to the settling time of the inclinometers (which affect the computed yaw reading from the compass) and the use of a serial line as the communication link, which is relatively slow. Sensor lags of this magnitude are serious in AR applications because system delay is the single largest contributor to registration error in most systems [16].

Some of these problems might be reduced by switching to a different compass and tilt orientation sensor, but the serious ones are endemic to any sensor using this technology. The magnetic distortion is different at different locations on the ground; more accurate sensors will not change that fact. And the distortion introduced when moving around will occur in any tilt-based sensor. While sourceless orientation sensors of this type are an interesting starting point for outdoor AR tracking, by themselves they will not be sufficient to achieve our ultimate goals.

21.4 Approaches and Conclusions

From this analysis of the tracking technologies available outdoors, we find that no single technology provides a complete solution. In the near term, no single technology is projected to provide the performance we require to meet our goals. Therefore, combining several tracking technologies, or hybrid tracking, is the only feasible approach for the near term. Hybrid approaches increase system complexity and cost but provide the most robust results. Hybrid approaches allow the weaknesses in each individual technology to be covered by some other sensor, resulting in an overall system that behaves more robustly than with each sensor applied individually.

What areas of outdoor tracking are most critical to attack initially? To answer that, we examine the contributions to registration error. Let's take a simple example, using a system similar to [13] as the current state-of-the-art for personal outdoor AR tracking. That system measures position by a differential GPS and orientation by a compass and tilt sensor similar in performance to the one discussed in Section 21.3.3 . Assume that the real-world object being viewed is 100 meters away, and that the user rotates his head at a moderate rate of 50 degrees per second. Note that even with a heavy HMD, we have measured peak head rotation rates around 80 - 120 degrees per second, and with minimal encumbrance on the head, peak rotation rates can exceed 300 degrees per second, so the assumption of 50 degrees per second is conservative. Assume the sensor itself has 100 milliseconds of delay internally, and the rest of the system adds another 100 milliseconds (for 200 ms total). The peak error of the differential GPS measurement is 3 meters, and the peak error of the compass is 3 degrees. Then the expected contributions to registration error are as shown in Table 21.1 :

Table 21.1   Quantifying contributions to registration error.

Contribution to registration error	Range of error
Position errors	0 - 1.7 degrees
Static orientation errors	0 - 3 degrees
Dynamic orientation errors	10 degrees

Clearly, the largest individual contributor is the dynamic orientation error, thus that is the most important one to tackle initially. This source can be attacked by reducing all sources of latency within the system and compensating for what remains via predictive means. However, the other two sources are also much larger than desired, so those must not be ignored either.

Although accurate tracking outdoors is a difficult problem, it is a worthwhile challenge because developing new tracking technologies that work in an outdoor realm will lead to improvements for AR and Virtual Environment systems overall. Research in this direction will help push AR systems away from the highly specialized, difficult-to-build prototype systems of today to more flexible, portable systems that can be deployed anywhere. I believe improvements in such areas must be made if AR systems are to appear commercially in wide numbers, and I hope that they will enable the applications described earlier in this paper, along with others we have yet to imagine.

Acknowledgments

Our work in this area is mostly funded by DARPA contract N00019-97-C-2013, from the DARPA ETO Warfighter Visualization program.

References

[1] R. Azuma: "Tracking requirements for augmented reality," *Comm. ACM,* vol.36, pp.50–51, July 1993.

[2] M. Ward, R. Azuma, R. Bennett, S. Gottschalk, and H. Fuchs: "A demonstrated optical tracker with scalable work area for head-mounted display systems," *Proc. Symp. on Interactive 3D Graphics,* pp.43–52, 29 March – 1 April 1992.

[3] R. Azuma and G. Bishop: "Improving static and dynamic registration in an optical see-through HMD," *Proc. SIGGRAPH '94,* pp.197–204, July 1994.

[4] R. Azuma and G. Bishop: "A frequency-domain analysis of head-motion prediction," *Proc. SIGGRAPH '95,* pp.401–408, August 1995.

[5] R. Azuma: "A survey of augmented reality," *Presence: Teleoperators and Virtual Environments,* vol.6, pp.355–385, August 1997.

[6] M. Bajura and U. Neumann: "Dynamic registration correction in video-based augmented reality systems," *Computer Graphics and Applications,* vol.15, pp.52–60, September 1995.

[7] A. State, M. A. Livingston, G. Hirota, W. F. Garrett, M. C. Whitton, H. Fuchs, and E. D. Pisano: "Techniques for augmented-reality systems: Realizing ultrasound-guided needle biopsies," *Proc. SIGGRAPH '96,* pp.439–446, August 1996.

[8] J. Nash: "Wiring the jet set," *Wired,* vol.5, pp.128–135, October 1997.

[9] T. Ohshima, K. Satoh, H. Yamamoto, and H. Tamura: "AR^2 Hockey: A case study of collaborative augmented reality," *Proc. VRAIS '98,* pp.268–275, March 1998.

[10] K. N. Kutukalos and J. R. Vallino: "Calibration-free augmented reality," *IEEE Trans. on Visualization and Computer Graphics,* vol.4, pp.1–20, 1990.

[11] S. L. Iu and K. W. Rogovin: "Registering perspective contours with 3-D objects without correspondence using orthogonal polynomials," *Proc. VRAIS '96,* pp.37–44, 30 March – 3 April 1996.

[12] B. Rhodes: "WIMP interface considered fatal," Position paper at IEEE VRAIS '98 Workshop on Interfaces for Wearable Computers, March 1998.

[13] S. Feiner, B. MacIntyre, and T. Hollerer: "A touring machine: Prototyping 3D mobile augmented reality systems for exploring the urban environment," *Proc. First Int'l Symp. on Wearable Computers*, pp.74–81, October 1997.

[14] G. Bishop: "Personal communication," January 1998.

[15] E. D. Kaplan: *Understanding GPS Principles and Applications*, Artech House Publishers, ISBN 0-89006-793-7, 1996.

[16] R. Holloway: "Registration Error Analysis for Augmented Reality," *Presence: Teleoperators and Virtual Environments*, vol.6, pp.413–432, August 1997.

Chapter 22

An Outdoor Augmented Reality System for GIS Applications

Byungtae Jang
Juwan Kim
Haedong Kim
Donghyun Kim

Electronics and Telecommunication
Research Institute, South Korea

22.1 Introduction

Augmented reality(AR) is a new technology in which a user's perception of the real environment is enhanced or augmented with additional information generated from computer system. The type of the additional information may be a text, a audio sound, or a graphics, etc. It is considered AR will be applied to a computer-aided surgery, a assembling or a disassembling of a complex machine, a architectural application,and the geographical information systems like a car navigation system or a personal navigation system. AR can be classified with a indoor system and a outdoor system according to the work bandwidth. Most of the AR system is limited to short range. The AR system can be consisted of a variety of computer graphics, computer vision, tracking, and video components. Especially the tracking system is very important factor to register the virtual information to the real scene. For an indoor AR system, many trackers were designed and produced. But in order to implement an outdoor AR system, GPS(global positioning system) may be a few tracker until now.

The purpose of this paper is to describe the outdoor and long rage AR system applied to the various GIS(geographical information system) applications using GPS. The term GIS is frequently applied to geographically oriented computer technology, integrated systems used in substantive applications and, most recently, there are so many kinds of GIS activities in many areas [1], But there is no the sense for the real because GIS provides a user with computer-oriented geographical information manipulated by computer in according to his requirements. When a user see computerized geographical information, it is hard to understand it at once. So as to enhance his perception of and interaction with a real world, we apply Augmented Reality(AR) technology to GIS because AR can allow the user to see the real world with virtual objects superimposed upon or composited with the real word, rather than completely replacing it [2] [3]. The virtual objects are geographical information that the user can't directly detect with his senses when he sees a real scene. For example, when a tourist want to sightsee the landscape in a tourist resort, to catch the information such as building name, road name and characteristics around, an atlas including the additional information may be required. Using it, he might still need much time to retrieve the information each time and may be inconvenient in case he wants to look at wide landscape at a glance. In addition to solve this problem effectively, there are many needs of long range AR systems to be applied to a vehicle's navigation application, unmanned airspace application, and so on.

In this paper, we propose a monitor-based augmented reality system in order to perceive remote geographical information. This system is composed of real scene acquisition system, tracking system for camera attitude, graphics rendering system and video mixer. For receiving live images in 3D space, we used the R/C helicopter fixed with a wireless CCD camera and the global positioning system (GPS) to gain a camera attitude data in real time. For real time rendering of virtual images and text information, a personal computer system (Pentium II)was used. Finally two types of images which were virtual information and real information were combined with the video mixer. As a result, we will discuss the integrated system which has total processing procedures. And we had the conclusion that GPS tracking system would be reasonable to apply outdoor AR system. The GPS's tracking error range was examined within 10 meters of position accuracy and within 0.3 degree of orientation accuracy. Conclusively we can enhance user's perception to remote geographical characteristics with AR system which has GPS tracker. In this paper, we introduce the outdoor AR system for GIS applications to perceive remote geographical characteristics. The rest of this paper is organized as follows. Section 22.2 Describes the related researches. Section 22.3 explains the design and implementation of this system and Section 22.4 describes the experiment. In Section 22.5, the conclusion of this research is presented.

22.2 The Related Works

22.2.1 Short Range AR

Augmented reality is a new technology but expanding area of research. We summarize the related research in this area. Feiner et al. has used augmented reality for laser printer maintenance task [4]. This system aids the user in the procedure required to open the printer and replace various parts. Wellner has developed an augmented reality system for office work in the form of a virtual desktop on a physical desk [5]. He can interact on this physical desk both with real and virtual documents. Bajura et al. has used augmented reality in medical applications in which the ultrasound imagery of a patient is superimposed on the patient's live image [6]. Besides these examples, many researches are carrying out in the fields of various applications [7].

22.2.2 Long Range AR

In AR, trackers to acquire the position and orientation to user's view are very important for more accurate registration of virtual images and real images. But few trackers are built for accuracy at long ranges, since many AR applications do not require long ranges. Most applications related with AR also tether the user to a limited working volume mostly [8].

To apply AR at long ranges, it is required that trackers draw out of the position and orientation of the user's view by the accuracy needed for visual registration. The commercial tracking devices that are useful in tracking a user at long ranges are motion capture system, scalable tracking systems for HMD and GPS (Global Positioning System) [9]. Motion capture systems are useful in tracking the position of a user's movement, but not the orientation. Scalable tracking systems can be expanded to cover any desired range, simply by adding more modular components to the system.

While scalable tracking systems can be effective, they are complex and, by their very nature have many components, making them relatively expensive to construct [10]. The GPS might be considered as tracking device for long range and outdoor AR systems. But the GPS is thought the accuracy is very low for AR application. We would like to solve these problems. Finally, we reached at a conclusion that some applications don't need very accurate image registration and then it is sufficient to use GPS as a long range tracker. In this paper, we will propose the design and implementation of AR system to be able to be used at outdoors.

22.3 Outdoor AR System for GIS Application

Figure 22.1 shows the conceptual diagram of this system. As it shows, this system consists of two parts: 1) real scene acquisition part, and 2) processing and visualization part.

The former acquires the real scenes of the remote region and sends them and additional information to the latter. It includes the wireless CCD camera, Tans Vector, RT-20, the industrial PC (Intel 80486) and the wireless data communication device.

The latter consisted of the wireless data communication device and PC, creates the augmented images using the real scenes and additional information received from the former. Figure 22.2 shows the data flow diagram of this system.

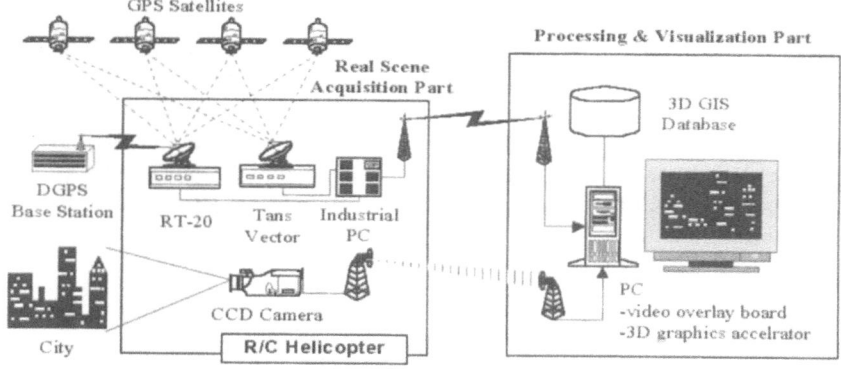

Figure 22.1   Conceptual diagram of outdoor AR system.

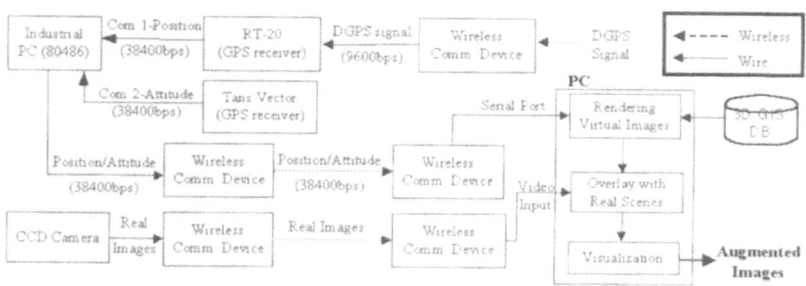

Figure 22.2   Data flow diagram.

22.3.1   Real Scenes Acquisition and CCD Camera Tracking

It is required to determine the attitude and position of the CCD camera attached with R/C helicopter in order to render the virtual images that represent geographical information, as they would appear if viewed from the wireless CCD camera. So two GPS receivers are applied to this system for it. The one is Tans Vector for attitude determination and the other is RT-20 for position determination. The another function of the wireless data communication device in the real scene acquisition part is that it receives the DGPS signal from DGPS reference station and sends to RT-20. As the position accuracy of GPS ranges from 10m to 100m, it must use DGPS (Differential GPS) correction mode using DGPS reference station to improve the position accuracy. The industrial PC (80486) calculates the position of it using raw data of RT-20 and filters off unnecessary data among the output of Tans Vector and RT-20. Its output, the position and attitude of it, is sent to the processing and visualization part by air. The real images from the wireless CCD camera are sent to the wireless communication device in the processing and visualization part.

22.3.2   Data Processing and Video Overlay

The wireless communication device in the processing and visualization part sends the received data to the PC through a serial port. And the real images from the wireless CCD camera are sent to the video overlay board in the PC. The received attitude and position data are different from those of the CCD camera because the GPS receivers measure the position and attitude based on their antennas. So it is required to adjust the inconsistency of their antennas with the CCD camera [11]. After changing the position and attitude of the viewport using the adjusted position, attitude data, the rendering program in the PC renders the virtual images that represent geographical information using 3D GIS database built in advance. The virtual images generated by the rendering program are blended with real images in the video overlay board and are displayed on a monitor screen.

22.4   Experiment

22.4.1   GPS Data Analysis

With DGPS mode, we had estimated GPS receiver applied our AR system. The experiment was performed during 1,000 second with resampling rate on 20Hz. In condition of a static status, the distribution of 2 dimensional RMS error and height RMS error is shown on Figure 22.3. The spatial RMS error's mean is about 0.88721 meter and the height RMS error's mean is about 1.1252 meter. Also, the orientation variation is shown on the Figure 22.4. When GPS receiver was fixed on a moving car, Figure 22.5 shows GPS receiver's error of the position and the orientation.

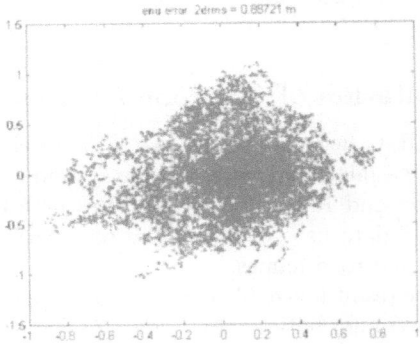

Figure 22.3   Long./lat. RMS error.

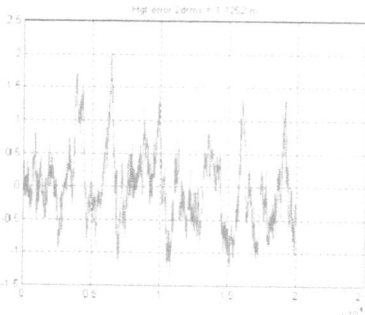

Figure 22.4   Height RMS error.

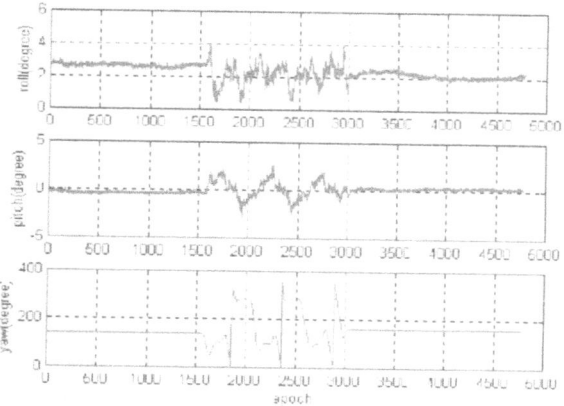

Figure 22.5   Orientation error.

22.4.2   Implementation of Outdoor AR System

Figure 22.6 shows the R/C helicopter with two GPS receivers, the wireless CCD camera and so forth. The pilot study area of this system is our research institute, ETRI(Korean Electronics and Telecommunication Research Institute) region in Korea. We make it 3D GIS data base to 3D spatial database and attribute database such as building names and road names.

Figure 22.7 shows the result image blended the real image from the wireless CCD camera with geographical information. As Figure 22.7 shows, a user can see geographical information synchronized with the real scenes. And besides, it is possible for him to select the geographical objects in augmented images and see the detailed information of them.

Figure 22.6   R/C helicopter integrated AR unit system.

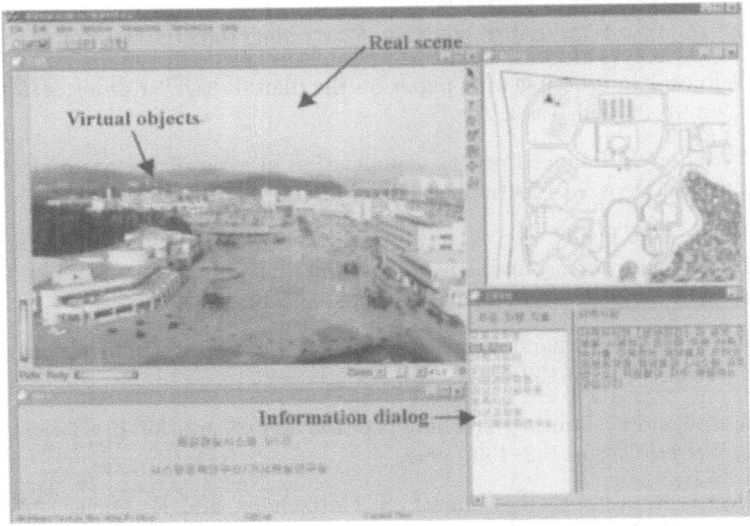

Figure 22.7   Augmented image combined the real image with geographical information.

22.5   Conclusions and Future Works

We have introduced the outdoor AR system for GIS application, which is a text based augmented information. Unlike most AR systems, the described system is working at outdoor and has wide activity area . As a result, we think that GPS is sufficient tracker system when was applied on long range AR system under the differential GPS mode. But to acquire more accurate registration, it has a few things to be improved. The most important one is the accurate registration problem between the real images and virtual images, because GPS receivers and 3D spatial database have error to themselves. And besides, each and every component such as the CCD camera, TANS Vector is not rigidly interconnected and attached to the R/C helicopter. In the future, it is necessary to develop the more effective methods for the accurate registration between real images and virtual images.

References

[1] D. J. Maguire: "Geographical information systems: Principles and applications," *Longman Scientific and Technical*, vol.1, pp.9–20, 1991.

[2] R. Azuma: "A survey of augmented reality," *Presence*, vol.6, no.4, pp.355–385, 1997.

[3] P. Milgram and F. Kishino: "A taxonomy of mixed reality virtual displays," *IEICE Trans. on Information and Systems*, vol.E77-D, pp.1321–1329, Sep. 1994.

[4] S. Feiner, B. MacIntyre, and D. Seligmann: "Knowledge based augmented reality," *Comm. ACM*, vol.30, no.7, pp.53–62, Jul. 1993.

[5] P. Wellner: "Interacting with paper on the digital desk," *Comm. ACM*, vol.36, no.7, pp.87–96, Jul. 1993.

[6] J. P. Mellor: "Enhanced reality visualization in a surgical environment," Master's thesis, Dept. of Electrical Engineering, MIT, 1995.

[7] M. Bajura, H. Fuchs, and R. Ohbuchi: "Merging virtual reality with the real world : Seeing ultrasound imagery within the patient," *Computer Graphics (Proc. SIGGRAPH 92)*, ACM Press, pp.203–210, 1992.

[8] S. Henray and J. Barners: "Tracking position and orientation in a large volume," *Proc. VRAIS'93*, pp.132–139, 1993.

[9] S. Henray and B. James: "Tracking position and orientation in a large volume," *Proc. VRAIS'93*, pp.132–139, 1993.

[10] M. Ward, R. Azuma, R. Bennett, S. Gottschalk, and H. Fuchs: "A demonstrated optical tracker with scalable work area for head-mounted display systems," *Proc. Symp. on Interactive 3D Graphics*, pp.43–52, 1992.

[11] J. Kim et al.: "Position and attitude calibration method of a CCD camera using the genetic algorithm," *SPIE Conf. on Acquistion, Tracking, and Pointing XII*, Orlando-USA, SPIE vol.3365, pp.329–336, 1998.

[12] T. P. Caudell and D. W. Mizell: "Augmented reality: An application of heads-up display technology to manual manufacturing processes," *Proc. IEEE Hawaii Int'l Conf. on Systems Science*, pp.659–669, 1992.

[13] P. Milgram, H. Takemura, A. Utsumi, and F. Kishino: "Augmented reality: A class of displays on the reality-virtuality continuum," *Proc. SPIE vol.2351: Telemanipulator and Telepresence Technologies*, pp.282–292, 1994.

[14] J. Rolland, R. Holloway, H. Fuchs, S. Henray, and J. Barners: "A comparison of optical and video see-through head-mounted displays," *Proc. SPIE vol.2351: Telemanipulator and Telepresence Technologies*, pp.293–307, 1994.

Index

2-view constraint, 95–98
3-view constraint, 95–98
3D Dome, 44
3D model, 39, 41, 43, 45, 49–51, 55, 131,
    148, 150, 152, 153, 155, 160,
    183, 192, 196, 218, 250, 266,
    286–289, 341, 348
3D Room, 44

A

abacus, 245
aberration, 73
absorption, 202, 203, 206
accelerometer, 101, 117, 385
    linear —, 117, 384
accommodation, 291–294, 309, 310, 312
accommodative
    — compensation, 291, 292, 294
    — convergence, 292, 311
    — cue, 311, 312
    — demand, 306, 307, 309–313
    — information, 314
    — relief, 308, 315
accuracy, 65, 68, 87, 102, 111, 115, 117–
    119, 123, 124, 132, 154, 273,
    305, 306, 309, 312, 372, 382,
    384, 385, 392–394
    dynamic —, 118
    static —, 118
acoustic
    — display, 206, 207, 209, 210
    — imagery, 206
    — signal, 206
    — tau, 203, 204
acuity, 33, 309, 314
affine
    — camera, 94–96
    — coordinate, 286, 289–291
    — epipolar geometry, 95, 97
    — image, 96
    — model, 94, 112, 114
    — motion, 96, 112, 114
    — parameter, 115
    — shape, 82
    — space, 94, 289
    — structure, 96
    — transform, 93, 289, 290
    — view, 96
alignment, 19, 111, 119, 125, 132, 150,
    307, 308, 313, 316, 341, 342,
    380
Alphatron, 209
altimeter, 101
ambient
    — display, 234, 241, 242, 244
    — fixture, 241–243
    — light, 33, 230, 342
    — media, 232–234, 241–245
    — sound, 208
    — space, 241
ambientROOM, 242, 243
anatomy, 288
anechoic environment, 203
annotation, 111, 113, 117, 118, 124, 267,
    279
ANOVA (analysis of variance), 310, 317,
    320
apparent distance, 312, 316
appearance model, 41, 42
AR^2 Hockey, 34, 65–68, 76, 215, 266, 275
ARTEMIS (Augmented Reality TEleMa-
    nipulation Interface System), 8,
    9
artificial intelligence (AI), 196, 348
aspect ratio, 96, 139, 337, 338
Aspen Movie Map, 83, 184
aspherical surface, 73
attitude, 16, 354, 356–358, 392, 394, 395
audible display, 230
audio communication, 264, 268, 269
auditory, 208
    — beacon, 208, 210

— distance, 200
— perception, 201, 202, 204, 205
— environment, 208
— information, 209, 269
— localization, 200
— masking, 208
— perception, 206
— space, 208
— stimuli, 76
— target, 204
augmentation, 60, 72, 232, 233, 267, 326, 327, 329, 330, 332, 334–336, 340–343
augmented reality (AR), 5–10, 12–14, 24–26, 31, 34, 37, 38, 60, 61, 63–67, 70, 74, 119, 215, 216, 225, 229–232, 253, 256, 294, 325–331, 333, 336, 340–343, 351, 360, 363, 364, 379–381, 391–393
— application, 32, 37–39, 105, 110, 327, 341, 380–382, 387, 393
— approach, 231
— assisted assembly, 330
— development, 24, 225, 379
— display, 19, 23, 24, 32, 38, 225, 379–383, 386
— environment, 19, 216, 225, 388
— experience, 326
— for construction (ARC), 365, 367, 371–373
— game, 76, 77
— image, 12, 110
— manufacturing, 110
— paradigm, 332
— research, 216, 225, 232, 245, 326
— set-up, 334
— system, 31, 32, 34, 37, 39, 63, 65, 67, 74, 76, 87, 105, 215, 221, 326, 327, 330, 333–339, 343, 363, 374, 379–388, 391–397
— technology, 32, 329, 360, 392
— telerobotics, 25
— tracking, 101, 118, 387, 388
— type amusement, 76
— user, 326, 343
— Viewer, 335
— work, 230, 379
collaborative —, 65, 66
HMD based —, 23
indoor —, 110, 382, 383, 385, 391
long range —, 392, 393, 397

monitor-based —, 334, 392
outdoor —, 112, 379–384, 387, 388, 391–394, 396, 397
real-time —, 333
short range —, 393
video see-through —, 65, 215
wearable —, 77
augmented virtuality (AV), 5, 7–10, 12–15, 26, 60, 61, 63, 69, 70, 72, 83, 232
aural axis, 202, 203
autonomy, 220, 221
azimuth, 119, 201–203

B

back-pack, 209, 302, 348, 349, 356–358, 360, 364, 368, 370–374 , 383
bandwidth, 60, 76, 233, 267, 270, 272, 274, 326, 384, 391
barcode, 365–367, 372
baseline, 47, 48, 88, 136, 137, 139, 140, 189, 190, 291, 316, 317, 384
Bauch & Lomb Orthorater stereo test, 309
beacon, 101, 102, 209, 210
auditory —, 208, 210
beam splitter, 265, 363
binary
— code, 337
— encoding, 337
— image, 337
binaural stimuli, 206
binocular
— condition, 306, 311, 312
— depth cue, 320
— disparity, 139, 291, 292
— display, 291, 306, 308, 321
— glasses, 167
— image, 308
— localization, 310
— parallax, 216
— presentation, 309
— rivalry, 306, 311
— stereo image, 145
— stimuli, 308
— viewer, 294
— viewing, 306, 309
— vision, 134, 257
biocular, see binocular
biopsy, 215, 262, 384
bore-sighting procedure, 321

C

calibration, 42, 43, 45, 63, 101, 102, 112, 117, 119, 122, 123, 149, 159, 225, 250, 253, 268, 274–278, 329, 334–336, 338–342, 354, 358, 359, 380, 386

camera
  — matrix, 87, 90, 277
  — model, 36, 47, 94, 120, 121
    perspective —, 275
    pinhole —, 121
  — offset, 35
  — parameter, 102, 132–134, 141, 142, 150, 178, 187, 334, 337
    external —, 338, 339
    internal —, 123, 276, 338, 339, 341
    affine —, 94–96

CAVE, 254, 256, 265

cell image, 153–155

center of gravity (COG), 135, 136, 295, 296, 355

centricity continuum, 17, 18, 23, 25, 26

chi-square, 318
  — analysis, 317
  — contingency, 318

chroma-keying, 34

chromatic separation, 75

CICC, 329, 344

ClearBoard, 233, 263, 264

cocktail-party effect, 269

COG, *see* center of gravity

collaboration, 35, 36, 39, 60, 65, 66, 71, 72, 76, 238, 261–270 , 272–275, 278–280, 326, 342, 365, 380

collaborative interface, 262–266, 268, 278–280

collaborative web interface, 268, 273

Collaborative Web Space, 267, 268

collimated light, 307

collision, 203, 205, 221

color matrix, 150, 151

color space
    HIS —, 139
    RGB —, 139

combinatorial optimization problem, 288

communication
  — device, 269, 393–395
  — link, 241, 387
  — protocol, 255
  — resource, 269

  — space, 165, 166, 262, 268, 269
  audio —, 264
  data —, 348, 357, 393, 394
  face-to-face —, 263, 264, 271, 272
  human —, 65, 285
  interpersonal —, 234, 238, 241
  multimedia —, 264, 285, 298
  visual —, 165–167, 180
  wearable —, 270
  wireless —, 269, 395

COMPAS, 38

completely real, 11, 12

completely virtual, 11, 12

composite image, 10, 11, 75, 132, 134, 139–143

compression, 44, 71, 148, 152–155, 157, 160, 165, 167, 172, 174, 204, 205, 383

computer graphics (CG), 2, 31, 37, 69, 83, 86, 101, 134, 137, 139, 142, 143, 168, 183, 196, 286, 292, 294, 295, 305, 342, 351, 358, 391

computer vision (CV), 2, 43, 62, 65, 69, 82, 85–87, 98, 99, 132, 143, 262, 274, 277, 286, 287, 339, 341, 349, 360, 385, 391

computer-aided
  — design (CAD), 14, 51, 265, 291, 295, 328, 348, 357
  — engineering (CAE), 33
  — instruction (CAI), 261
  — maintenance, 348
  — manufacturing (CAM), 357
  — surgery, 391

computer-generated image (CGI), 8, 32, 61, 72, 179

congruence, 18, 19, 21, 24–26

constraint
  — based toolkit, 374
  — geometric, 374
  2-view —, 95–98
  3-view —, 95–98
  epipolar —, 136
  ergonomic —, 382
  functional —, 262
  matching —, 95
  real-time —, 325
  space —, 37, 109
  spatial —, 262
  temporal —, 262

control order, 19
control-display
    — compatibility effect, 16, 20
    — congruence continuum, 19, 23
    — issue, 22, 23
    — mapping, 5, 18, 20, 21
convergence, 292, 308, 310, 311, 313–319, 321
    — distance, 308
    accommodative —, 292, 311
    ocular —, 316
    static —, 315, 316, 319, 321
cooperation, 87, 267
coordinate
    — conversion, 355
    affine —, 286, 289–291
    calibration frame —, 275–278
    camera-centered —, 121
    camera —, 105, 120, 122, 134, 136–138, 274–277, 355
    Euclidean —, 45
    image —, 87, 95, 96, 112, 121, 153, 158, 159, 169, 170, 172, 354
    inertial-centered —, 121
    linear —, 95
    local —, 50
    marker —, 277, 278
    object-centered —, 354
    texture —, 50
    vertex —, 50
    world —, 45, 120, 121, 137, 138, 159, 274, 354, 355, 358
coordinated
    physical —, 278
    real-world —, 8, 134, 354
    robot joint —, 8
    screen —, 135, 136, 274–278
corrector, 111
correlation, 47, 49, 115, 123, 154, 155, 291, 318, 321, 338, 341
correspondence, 43, 47, 55, 92, 96, 111, 112, 126, 291, 293, 341, 353
    — detection, 47
    — search, 47, 87, 88, 90, 120
    correct —, 47, 89
    false —, 87, 89
    true —, 90
coupling of bits and atoms, 233
cross correlation, 341
crosstalk, 75

CRT (cathode ray tube), 32, 37, 38, 220, 307, 348
CSCW (computer supported collaborative work), 261–266, 268, 280
cue
    accommodative —, 311, 312
    depth —, 36, 308, 310, 320
    spatial —, 261, 267, 268, 278
    visual —, 166–169, 174, 180, 204, 268
curvature, 150
CyberMirage, 70, 71, 83, 179, 180
cyberspace, 59, 60, 62, 65, 71, 200, 229, 233, 242, 244, 250
cyclopean position, 306

D
DataGlove, 225
DC motor, 218, 219, 225, 239, 243
DCT (Discrete Cosine Transform), 172, 287–289
dead reckoning tracker, 385
degree of freedom (DOF), 102, 105, 118, 120, 122, 217–220, 225, 255, 270
Delaunay triangulation, 158
delay, 14, 36, 65, 72, 102, 186, 240, 263, 384, 385, 387, 388
depth
    — buffer, 50
    — cue, 36, 308, 310, 320
    — discontinuity, 48, 360
    — effect, 321
    — estimation, 47, 48, 134, 138, 142
    — information, 87, 98, 192
    — map, 47–50, 87, 143
    — perception, 292
    — recovery, 47
    — sensation, 141
    apparent —, 308, 310
    judged —, 316–319
    perceived —, 311
    stereo —, 310
depth of field, 35, 36
desktop
    — display, 218, 225, 360
    — environment, 352
    — metaphor, 230, 381
    physical —, 230
    virtual —, 393
dexterity, 251, 330
diffraction, 202, 203, 206

digital mixing, 208
DigitalDesk, 230
diopter, 308, 309, 311–313, 315
DIP switch, 349, 355, 356
directional localization, 201–203
directness, 19
DirectSound, 272
DirectX, 271
discomfort, 37, 309
disorientation, 33, 269
disparity, 47, 48, 88–90, 172, 178, 306,
    308, 311, 314, 319
    — map, 88–91, 178
    binocular —, 139, 291, 292
    exofixation —, 315, 321
    fixation —, 315, 318
    stereo —, 321
display
    3-D —, 62, 63, 72, 75, 76, 167, 180,
    265, 286, 291, 298, 365, 368
    acoustic —, 206, 207, 209, 210
    ambient —, 234, 241, 242, 244
    audible —, 230
    billboard —, 14
    binocular —, 306, 308, 321
    flat —, 306
    focused-beam array —, 167
    force —, 200, 215–223, 225, 297
    hand-held —, 231, 364, 365, 368,
    370, 373, 374
    haptic —, 200, 230
    head-mounted —, 2, 6, 23–25, 31–
    36, 39, 59–62, 64, 66–68, 72–
    74, 82, 98, 132–134, 140, 143,
    167, 220, 221, 225, 245, 254,
    256, 258, 266–268, 270, 272–
    278, 280, 286, 291, 294, 298,
    306, 332–334, 339, 340, 342, 348,
    356–358, 360, 363, 365, 369–
    374, 379–381, 385, 388, 393
    closed view —, 72, 75, 132, 133
    see-through —, 32–36, 60, 65, 67,
    68, 72, 73, 132, 133, 141, 215,
    216, 231, 269, 275, 305, 307,
    308, 321, 339, 357, 364–369, 372
    head-up —, 6, 31, 308, 330
    holographic —, 167, 168
    lenticular —, 167
    lookup —, 348, 349
    monocular —, 306, 308, 321
    MR —, 5, 16, 23, 26, 61, 72, 280

opaque —, 265, 363, 368
    parallax barrier —, 167
    stereoscopic —, 3, 59, 63, 75, 76,
    133, 140, 141, 285, 291, 302,
    306, 308, 309, 321, 358
    volumetric —, 265
dissimilarity, 88, 89
distance
    — cue, 203, 310
    — effect, 310
    — localization, 203, 312
    — perception, 201, 202, 204
    accommodation —, 292, 293
    apparent —, 312, 316
    auditory —, 200–202, 204, 205
    convergence —, 308
    egocentric —, 203, 206
    focal —, 36, 293
    inter-fiducial —, 109
    interpupillary —, 35, 307, 309
    perceived —, 204–206
    transformation —, 295
    vergence —, 292, 293
distortion, 16, 73, 74, 114, 119, 121, 122,
    188–190, 206, 308, 354, 385–
    387
distortion-free image, 190
DIVE, 265
divergence, 316–318
DIVO, 140
DOF, *see* degree of freedom
dominant eye, 308
drift, 117–120, 123–125, 187, 340, 384,
    385
DSP, 112

E
echoic environment, 203
edge-collapse decimation algorithm, 49
ego-referenced, 17, 18, 21, 22
egocentric distance, 203, 206
eigen texture, 147, 148, 152, 155, 157, 160
eigenspace, 152–155
eigenstructure decomposition, 154
eigenvalue, 154
eigenvector, 93, 154, 155
elasticity, 221
electric parts, 348, 358
elevation, 14, 201–203
Emmert's Law, 311
endscopic surgery, 25

environment
    mixed —, 15, 16, 62, 131, 132, 134,
        201, 208, 261, 266
    real —, 6–8, 10, 12, 13, 38, 131, 132,
        134, 137, 143, 206, 208, 253,
        266, 348, 357, 364, 391
    remote —, 251, 254, 258, 259, 266
    virtual — (VE), 2, 6, 7, 11–13, 16,
        17, 26, 36, 60, 63, 69, 72, 131,
        132, 183, 195, 200, 215, 220,
        221, 232, 252–254, 256–258, 265–
        268, 286, 295, 348, 357, 363,
        365, 372, 373, 379, 380, 385,
        388
EPI, see epipolar plane image
epipolar
    — constraint, 136
    — geometry, 95, 97
    — line, 47
    — plane image (EPI), 69, 170
Euclidean
    — coordinate, 45
    — motion, 94, 96
    — parameter, 96, 97
    — solution, 97
    — structure, 94, 96
exocentric, 17–20, 22–26
exofixation
    — disparity, 315, 321
exophoria, 321
exoskeleton, 216, 218, 219, 225, 251
extent of world knowledge (EWK) con-
        tinuum, 6, 7
eye
    — contact, 262–264, 268
    — position, 167, 339

F
facial expression, 266, 267, 286–289
facial image, 93
false target, 88
Fastrak, 220, 272
fatigue, 37, 286, 291, 292, 347
fattening, 48, 49
FED, 383
Feel-through, 215–226
fidelity, 306, 315
fiducial, 82
fiducial identification, 102, 105, 277
field of view (FOV), 15, 35–37, 109, 119,

        187, 220, 256, 268, 270, 272–
        274, 307, 309, 337
field sequential, 75
fixation disparity, 315, 318
flight simulator, 38
flythrough, 22, 53, 115
focal
    — distance, 36, 293
    — length, 47, 105, 107, 118, 119,
        121, 123, 137, 139, 170, 338
    — point, 275
focus, 6, 23, 35, 36, 42, 118, 132, 167,
        175, 218, 232, 234, 242, 257,
        262, 263, 265, 270, 295, 306,
        315, 321, 347, 350, 364, 379,
        380, 382
force
    — display, 200, 215–223, 225, 297
    — feedback, 193, 200, 215–217, 221–
        225, 239, 286, 297
    — sensation, 215, 220
    gravitational —, 225
    synthetic —, 216
FOV, see field of view
fractal, 172, 174
frame
    — buffer, 139
    — grabber, 34, 384
    — rate, 15, 51, 238, 339, 340, 342
    confidence evaluation —, 114, 115
    coordinate —, 121, 122, 275, 277,
        278
    ego-reference —, 21
    inter —, 341
    video —, 307
    wire —, 64, 286, 287, 308, 311, 312,
        316, 366
friction, 326
fusion, 49, 61, 65, 120, 126, 306, 350, 360

G
gaze, 140, 272, 273, 370
    — awareness, 263, 264, 273
    — detector, 293
genetic algorithm (GA), 288, 289
geographical information, 386, 392, 394–
        397
geometric model, 7, 43, 44, 69, 70, 77,
        165, 167, 168, 172, 179, 183,
        342, 348

geometry, 36, 37, 42, 43, 50, 62, 63, 65, 67, 69, 70, 76, 77, 85, 87, 95, 97–99, 114, 131, 132, 134, 136, 153, 157, 158, 183, 186, 237, 238, 255, 289, 290, 306, 337, 342, 348, 374, 380

gesture, 265–268, 272, 274, 279, 280, 286, 295, 298, 332

GIS (Geographical Information System), 209, 391–393, 395–397

glasses, 37, 38, 61, 69, 72, 75, 167, 221, 265, 315, 354, 357, 383

glove, 19, 217, 225, 295, 296, 348

GPS (Global Positioning System), 101, 102, 184, 186, 187, 209, 365, 372, 374, 381, 384, 392–398

　carrier-phase —, 384

　differential — (DGPS), 209, 365, 368, 384, 388, 394, 395, 397

　Kinematic —, 187

graspable, 232–234, 238, 244, 245

gravity, 117, 118, 135, 136, 221, 295, 296, 326, 355, 372

GreenSpace, 265

GUI (graphical user interface), 229, 230, 232, 242, 245, 255

gyroscope, 65, 67, 101, 117–119, 124–126, 184, 187, 357, 358

　rate —, 117, 118, 384, 385

H

hand silhouette image, 295

haploscope, 307, 309

haptic

　— device, 298

　— display, 200, 230

　— icon, 221

　— information, 42

　— interaction, 229

　— interface, 200, 286

　— interpersonal communication, 234, 238, 241

　— model, 221

　— renderer, 220

　— sensation, 297

　— stimuli, 76

HapticMaster, 217, 225

HapticScreen, 218, 219

Hardiman, 251

HCI, 230, 232, 234, 245

head-related transfer function (HRTF), 202, 203, 207

head-stabilized image, 268

HMD (head-mounted display), *see* display

HMP (head-mounted projector), 257–259

hologram, 165, 168, 236

holographic, 168

　— film, 238

　— layout, 234, 237

　— method, 167

　— optical system, 265

　— recording, 237, 238

holography, 236

Holomedia, 69

HRP, 254, 256, 260

human

　— error, 348, 349, 353, 360

　— interface, 14, 60, 86, 252, 261

　— motion, 286

　— perception, 233, 234, 294

human-computer interface, 261

hybrid, 69, 101, 117, 120, 121, 124, 125, 132, 174, 336, 387, 388

hypertext, 255, 365

I

i-glasses!, 220, 270, 308, 333, 372

I/O Bulb, 235, 236, 238

IBR, *see* image-based rendering

IC (integrated circuit), 351, 355

IIRO (Intelligent Interactive Remote Operations), 13–15, 20, 23, 25, 26

Illuminating Light, 234, 236–238, 244

illumination, 37, 42, 75, 148–153, 155, 157–159, 234, 236–238 , 244, 308, 309, 315, 317, 337, 338

illusion

　stereo —, 37

image

　— acquisition, 37, 65, 70, 149, 157, 158, 287–289, 354, 355

　— analysis, 147, 277, 298, 334, 354

　— annotation, 117

　— capturing, 16, 34–36, 54, 65, 66, 69, 71, 72, 87, 90, 104, 119, 132, 134, 140, 157, 165, 168, 171, 184, 187, 188, 190, 192, 196, 274, 363

　— database, 61, 177, 184, 187, 193

　— feature motion, 122

— plane, 43, 97, 98, 119, 121, 123, 148, 290

— position, 123, 124

— processing, 69, 109, 121, 134, 167, 196, 277, 338

— quality, 49, 50, 63, 64, 67, 68, 87, 176, 193

— registration, 274, 280, 393

— resolution, 118, 119

— sequence, 51, 112, 115, 117, 135, 136, 153, 155, 186, 335, 341, 354

— velocity, 117, 123

— warping, 93, 114, 153

image-based

— rendering (IBR), 2, 69, 90, 148, 153, 168, 183, 196

— stereo, 41, 45

ImersaDesk, 38

immersive, 36, 40, 44, 101, 239, 265–267, 274, 280, 298, 326, 340, 350, 360

Immersive Video, 44

in-Touch, 234, 238–241, 244

inclinometer, 270, 368, 372, 386, 387

indoors, 77, 365, 380, 382, 383

inertia, 101, 117–126, 274, 296, 307, 336, 348, 349, 357–359, 384, 385

infrared, 7, 68, 288, 289, 354, 355, 365

inspection, 96, 347–360

integral photography, 165

intensity matrix, 150, 151

interaction

human-computer —, 326, 332

user —, 332, 334

interactive surface, 233

interaural intensity difference (IID), 202, 206, 207

interaural time difference (ITD), 202, 206, 207

interface

— design, 18, 234, 280

— metaphor, 261, 280

— object, 274

— paradigm, 325

Internet, 60, 221, 240, 242, 254, 259, 270, 272

interpolated image, 22, 69, 148, 188, 189

interpolation, 14, 43, 148, 151, 155, 166, 172–179, 187–190, 196, 239

—area, 192

bilinear —, 123

fractal —, 174

image —, 22, 69, 188, 189

light ray —, 174, 177, 178, 180

linear —, 174, 189, 190

view —, 43

interpupillary distances (IPD), 35, 307, 309

InterSense, 118

intrinsic parameter, 96, 121, 122

iRX, 242, 243

iso-surface, 45, 49

J

Jacobian, 111

joystick, 19, 20, 193, 195, 217, 225, 316

K

Kaiser Electro Optics, 34, 36

Kalman filter, 111, 339, 340

Extended (EKF) —, 111

keyhole effect, 15, 16

keystone correction, 308

kinetic depth, 321

Koenderink and Van Doorn's rotation representation, 97

L

lag

— reduction, 340

temporal —, 35

time —, 63–65, 67, 72

LAN (local area network), 240, 244, 270

landmark, 23, 67–69, 76, 208, 210, 381, 384, 385

LCD (liquid crystal display), 32, 59, 64, 73–75, 265, 266, 294

least square fitting, 150

LED (light emitting diode), 98, 308, 312, 316, 337, 354, 355, 372

lens set, 307

LHX, 220, 221

light

— reflection, 149, 326, 341–343

ambient —, 33, 230, 342

collimated —, 307

environment —, 36

light field rendering, 69, 83

likelihood, 22, 36, 319

linear combination, 91–93, 155, 156, 289

localization, 201, 209, 305, 306, 308, 310–315, 318, 321

— accuracy, 306, 312
— error, 318
— performance, 306
— task, 305
— technique, 309
auditory —, 200
binocular —, 310
directional —, 201–203
distance —, 203
extracranial —, 206
intracranial —, 207
operator —, 305
spatial —, 321
surface —, 49
long range, 254, 392, 393, 397
look-and-feel, 347
lookup table (LUT), 354, 355
Lumigraph, 69, 83
luminance, 62, 308, 312, 315
Luminous Room, 235, 236

M

Macintosh, 230
Maddox Rod test, 321
magnetic distortion, 386, 387
magnetometer, 270, 365, 368, 372
maintenance, 24, 110, 269, 302, 348, 364,
    380, 382, 385, 393
manipulation, 8, 16–19, 22, 25, 42, 44,
    97, 204, 206, 207, 215, 218, 229,
    232, 234, 236, 238, 239, 254,
    255, 261, 265, 274, 295–297, 325,
    327, 328, 330, 341, 348, 352,
    368, 392
    collaborative —, 238
    remote —, 252
    simultaneous —, 239
    visual-manual —, 305
manual, 64, 93, 124, 286, 305, 347–350,
    353, 355, 356, 359, 360
marching cubes algorithm, 49
marker, 68, 132–136, 138, 141–143, 274,
    277–279, 327, 328, 336–339, 349–
    351, 354, 355, 383, 384
master slave, 252, 254
matching window, 88–90
matrix
    camera calibration —, 277
    camera —, 87, 90
    coefficient —, 95
    color —, 150, 151

intensity —, 150, 151
intrinsic parameter —, 96
Jacobian —, 111
model-view —, 134, 137–139, 142,
    143, 278
observation —, 150, 151
perspective transformation —, 278
projection —, 95, 96, 133, 134, 278
rotation —, 96–98, 121, 138
transformation —, 92, 122, 137, 275–
    277
viewing —, 66
mean error, 222
media, 232–234, 241–245
medium technology, 85–87
membership function, 102–104
    fuzzy —, 102
mid-sagittal plane, 202
mirror
    — reflection, 132, 342
    half-silvered —, 32, 37, 363
mixed reality (MR), 5, 10, 13–16, 18–20,
    23, 25, 26, 31, 59–65, 70, 72,
    74, 76, 77, 86, 87, 90, 98, 99,
    131–134 , 140, 142, 143, 147,
    148, 157, 161, 165, 166, 179,
    180, 196, 229, 232, 245, 253,
    259, 261, 262, 265, 266, 268,
    269, 274, 275, 277–280, 347–
    358, 360
    — aided engineering, 347
    — centricity, 13
    — control issue, 13
    — control-display compatibility, 16
    — definition, 5, 12, 13, 231
    — design issue, 15, 20
    — display, 5, 16, 23, 26, 61, 72, 280
    — environment, 15, 16, 131, 132,
        134, 261, 266
    — excavator, 13–15, 25
    — integrated instruction, 349
    — interface, 261, 262, 265–269, 272,
        274, 278–280
mobile, 184, 187, 216, 225, 326, 337, 343,
    356, 363, 365, 374, 375
model
    appearance —, 41, 42
    camera —, 36, 47, 94, 120, 121, 275
    geometric —, 7, 43, 44, 69, 70, 77,
        165, 167, 168, 172, 179, 183,
        342, 348

projection —, 290
reflectance —, 148, 152, 159, 160
volumetric —, 44, 49, 52, 53
model-based, 70, 72, 110, 147–149, 151–
153, 155, 157, 160, 253
model-view matrix, 134, 137–139, 142,
143, 278
modelled
completely —, 6, 8, 9, 12
partially —, 7
moment, 8, 16, 41, 44, 307, 381
monocular
— afterimage, 311
— condition, 306, 310, 311, 313
— display, 306, 308, 321
— field, 307
— helmet, 306
— image, 294, 306
— overlap, 307
— superimposition, 309
— viewing, 310, 311
— vision, 133
morphing, 43, 90, 184, 187–192, 195, 196
motion
— analysis function, 112
— based calibration, 122
— capture, 393
— correction, 115
— direction, 22, 340
— estimation, 114, 115, 117, 119,
286
— flow, 115
— mask, 44
— model, 114, 115, 339, 340
— parallax, 69, 203, 321
— prediction, 112, 124, 326, 338–
340, 380
— range, 187, 255
— sequence, 52, 54
— track, 112, 220, 326, 336
— vector, 115, 339
body —, 269, 272
erratic —, 326, 338, 343
Euclidean —, 94, 96
head shaking —, 339
head —, 220, 269, 296, 326, 338–
340, 343, 357, 359, 380
linear —, 125
lip —, 94
mouse —, 340

Newton-Euler dynamic — equation,
42
optical flow —, 114
physical —, 244
spiral —, 53
structure-from — theory, 90
trackball —, 271
translation —, 115, 386
mouseless interface, 340
MR Living Room, 67, 76
multi-baseline stereo, 47, 48, 88, 291
multi-view image, 165, 168, 177, 178
multimedia, 264, 285, 298, 374

N

natural language, 349
navigation system, 208, 210, 348, 391
nearest neighbor, 172, 173
neurosurgery, 37, 38
nominal viewpoint, 16–18, 25
NTSC (national television standards committee), 50, 220, 294, 307

O

observation matrix, 150, 151
obstacle avoidance, 23, 208
occluding boundary, 87, 89
occlusion, 13, 36, 77, 87, 89, 90, 93, 117,
125, 133, 134, 138, 143, 159,
177, 193, 208, 223, 257, 295,
297, 319–321, 326, 337, 342, 374
— detectable stereo, 87, 99
— mask, 89, 90
ocular
— convergence, 316
oculomotor
— bias, 321
— cue, 319
— effect, 319
— explanation, 320
— response, 319
off-axial, 73
off-line digitization, 44
Office of the Future, 38, 39
omni-directional image, 148, 157–159
on-line, 61, 241, 326, 331, 338, 343, 351
open-loop, 204
OpenGL, 220, 221, 278, 335, 371
optical
— axe, 87, 136, 140, 175, 189, 190
— flow, 112, 114, 115, 189, 192

— mixing, 208

— position encoder, 239

— see-through, 2, 23, 31–37, 61, 64–66, 68, 72–74, 122, 132, 140, 220, 274, 275, 367, 372, 379, 380, 383

— system, 32, 33, 36, 64–66, 73, 122, 236, 257, 265, 274, 337, 338, 380

— tracking, 336, 338, 343, 372, 380

optimization, 35, 112, 172, 174, 288, 306, 335

outdoors, 64, 77, 101, 102, 112, 115, 205, 206, 302, 327, 329, 331, 348, 356, 357, 364, 365, 379–385, 387, 388, 391–394, 396, 397

P

panorama, 69, 184, 187–189, 195, 196

panoramic image, 184, 187–189, 195

parallax

— barrier method, 75, 167

binocular —, 216

motion —, 69, 203, 321

PCB (printed circuit board), 348, 349, 352–355, 358, 360

PDA (personal digital assistant), 245, 383

PDDM (Palmtop Display for Dextrous Manipulation), 297

pedometer, 101, 385

Peloton, 10, 16

penetration, 326

perceived distance, 204–206

perception

auditory —, 201, 202, 204, 206

depth —, 292

distance —, 201, 202, 204

human —, 233, 234, 294

peripheral

— image, 88, 89

— vision, 36

periphery, 233, 234, 241

Personal Guidance System, 208, 209

phenomenology, 202

photoreality, 83, 160, 183, 184, 193, 196

physical cursor, 308, 311, 312

Pinwheels, 234, 241–244

pitch, 118, 119, 187, 307, 372, 385

plenoptic function, 168

Polhemus Sensor, 66, 67, 220, 225, 272, 359

polinocular stereo, 87

polycarbonate

— mirror, 307

polygon connectivity, 50

polygonal approximation, 45

position, 14, 17–19, 26, 32, 36, 37, 42, 43, 47, 50, 54, 63, 65, 67, 68, 87, 88, 90, 98, 99, 102, 104, 109–113, 117, 119, 121, 123, 124, 131–134, 136–140, 143, 155, 158, 159, 167, 168, 170, 172, 173, 175, 184–189, 192, 195, 196, 206, 208, 216, 223, 236, 238–240, 255, 256, 267, 271, 274, 275, 277–279, 287, 288, 291, 294–297, 305–310, 312, 314, 315, 318, 320, 321, 326, 329, 333, 336–342, 354, 356–359, 365, 367, 370, 372, 374, 382, 384, 385, 388, 392–395

potentiometer, 219, 225

prediction

linear —, 340

motion —, 112, 124, 326, 338–340, 380

target —, 112, 296

tracking —, 123

predictor, 111, 340

presence, 206, 243, 258, 264, 269, 312, 316, 317, 369

principal component analysis, 92

prism, 32, 33, 73, 74, 294, 307

projection

— matrix, 95, 96, 133, 134, 278

— model, 290

orthographic —, 92

para-perspective —, 290

perspective —, 121, 122, 290

pin-hole —, 47, 290

weak perspective —, 110

— screen, 265, 289

back —, 38, 41, 44, 119, 124

front —, 38

stereo —, 136, 265

Q

QuickTime VR, 69, 148, 184

R

R-Cube, 254, 255, 259, 260

R/C helicopter, 392, 394, 396–398

RAC (Recursive Average of Covariances), 111
radiance, 158, 159
   — distribution, 148, 157, 159, 161
   — map, 157–160
   illumination —, 159
RAID, 189, 193
range
   — based image, 14
   — camera, 14
   — estimation, 45, 49, 208
   — finder, 7, 77, 149
   — image, 12, 14, 41, 43, 45, 48–50, 149–153, 155, 157, 160
   — intensity, 7
   — map, 41
   — scanning hardware, 43
   dynamic —, 383
   far —, 72
   mid —, 72
   short —, 72, 391, 393
ray-based, 70–72, 165, 167, 168, 174, 180, 238
Ray-Space, 69, 70, 168
real
   — environment, 6–8, 10–13, 26, 38, 131, 132, 134, 137, 142, 143, 201, 205, 206, 208, 230, 232, 253, 266, 326, 348, 357, 364, 380, 391
   — space, 237, 245, 267, 289
   — time, 6, 31, 42, 44, 59, 61–64, 67, 71, 77, 86, 94, 99, 104, 112, 119, 132, 143, 172, 174–176, 180, 187, 189, 196, 220, 221, 254, 259, 286, 291, 298, 325–327, 330, 333, 336, 339–343, 349, 351, 353, 372, 385, 392
   — recording, 44
   — world, 5, 6, 8, 13, 15, 16, 20, 23, 24, 33, 34, 36, 42, 59–65, 69, 71, 76, 77, 85, 87, 90, 99, 131, 132, 134, 138, 140, 147, 183, 196, 215, 219, 230, 231, 238, 245, 253, 261–263, 266–272, 274, 277, 278, 280, 325, 326, 328, 332, 340, 342, 343, 349, 350, 352, 354, 358, 363, 367, 374, 379, 380, 383, 385, 388, 392
reality

   — model, 46, 52, 341
augmented — (AR), 5–10, 12–14, 19, 23–26, 31, 32, 34, 37–39, 60, 61, 63–67, 70, 74, 76, 77, 87, 101, 105, 110, 112, 118, 119, 215, 216, 221, 225, 229–232, 245, 253, 256, 294, 325–343, 351, 360, 363–365, 367, 371–374, 379–388, 391–397
enhanced —, 66, 219, 262, 264
mixed — (MR), 5, 10, 12–16, 18–20, 23, 25, 26, 31, 59–65, 70, 72, 74, 76, 77, 86, 87, 90, 98, 99, 131–134 , 140, 142, 143, 147, 148, 157, 161, 165, 166, 179, 180, 196, 229, 231, 232, 245, 253, 259, 261, 262, 265–269, 272, 274, 275, 277–280, 347–358, 360
photo —, 160, 183, 184, 193, 196
synthesized —, 151, 160, 253
transmitted —, 253
virtual — (VR), 7, 33, 34, 36, 41–45, 50, 51, 53, 54, 59, 60, 65, 75, 118, 147, 148, 229–232, 256, 259, 266, 268, 280, 285, 295–298, 327, 333, 348, 356, 357, 379
virtualized —, 41, 42, 44, 45, 50, 51, 53, 54
reality-virtuality (RV) continuum, 2, 3, 6, 7, 10, 11, 13, 23–26, 60
receiver, 86, 172, 209, 220, 372, 381, 384, 385, 394–396, 398
reflectance
   — analysis, 148
   — characteristics, 148
   — model, 148, 152, 159, 160
   — parameter, 147–152, 160
   — property, 149
reflection, 26, 32, 37, 62, 73, 149–151, 157, 202, 206, 236, 243, 256–259, 297, 309, 311, 326, 341–343, 363, 370
   — model, 148, 149, 151, 167
   — surface, 73
   diffuse —, 149–151
   mirror —, 132, 342
   specular —, 149–151
   Torrance-Sparrow — model, 151, 159
registration
   — algorithm, 66

— error, 68, 131, 280, 380, 382, 384,
    387, 388
— method, 65, 274, 277, 278
— problem, 380
— requirement, 267, 274
dynamic —, 67, 123
geometric —, 65, 82, 85
geospatial —, 380
hybrid —, 69
photometric —, 63, 77, 82
pixel-accurate —, 380, 384
real-time —, 62, 63
static —, 67, 77, 123
vision-based —, 132
visual —, 393
world-stabilized image —, 274
relocalization, 316
RenderWare, 371
resistance, 225, 355, 356
Responsive Workbench, 265, 328
retro-reflective, 256–259
reverberation, 203, 206, 207
RF, 101
rivalry problem, 306
RMS, 117, 118, 395
robot, 6, 8, 9, 14, 25, 42, 85, 86, 98, 149,
    218, 252–255, 258–260
roll, 118, 187, 372, 385
rotation matrix, 96–98, 121, 138
RS232C, 220, 334

S

SAD (Shape Approximation Device), 257
sampling
    — rate, 184, 187, 220
scene modeling, 42, 43, 339, 341, 342
scene-centered description, 42
SEA (Stereo by Eye Array), 87–91, 99
seamless, 3, 14, 59–63, 69, 71, 72, 74, 82,
    131, 147, 165, 229, 233, 262,
    263, 265, 266, 274, 280, 341
— interface, 264
searching window, 136
see-through
    — application, 339
    — AR, 6, 19, 24, 65, 67, 215
    — display, 24, 31–36, 122, 200, 269,
        305, 307, 321, 363, 367, 379,
        380, 383
    — feeling, 74
    — function, 75, 140

— HMD, 2, 3, 32–36, 60, 65, 67, 68,
    72, 73, 132, 133, 141, 215, 216,
    231, 266, 269, 275, 302, 308,
    339, 357, 364–369, 372, 380
— optics, 347
— view, 6
optical —, 2, 23, 31–37, 61, 64–66,
    68, 72–74, 122, 132, 140, 220,
    274, 275, 367, 372, 379, 380,
    383
video —, 2, 23, 31, 32, 34–37, 61,
    64, 65, 68, 72, 82, 131–134, 140,
    142, 143, 215, 380, 383
See-through Hand, 296–298
sensation, 131, 133, 141, 167, 215, 220,
    251, 258, 297
sensor
    — fusion, 65
    acceleration —, 187, 357, 358
    altimeter —, 101
    compass —, 385, 387, 388
    electro-magnetic —, 272
    gyroscope —, 65, 67, 119, 187, 357,
        358
    hybrid —, 388
    inertial —, 101, 117–120, 384
    magnetic —, 65, 76, 132, 184, 186,
        187, 296, 356, 359
    magnetometer —, 372
    mechanical —, 132
    pedometer —, 101
    rate gyroscope —, 117
    tilt —, 385, 387, 388
    ultrasonic —, 65, 67, 208, 356
SGI (Silicon Graphics Inc.), 65, 67, 109,
    140, 180, 221, 238, 289, 307,
    333–335, 371, 384
shading, 63, 71, 149, 153, 157, 383
shadow
    cast —, 64, 148, 159, 341, 342
    hindrance —, 257
    soft —, 158, 160
    water ripple —, 241
shaft-encoder, 308
shape-from-silhouette, 44
shared, 66, 193, 221, 239, 262–264, 266,
    268, 272, 273, 372
shutter glasses, 37, 38, 69, 75, 221, 265
SIGGRAPH, 10, 34, 66, 67, 219, 289
smoothing, 90, 143, 188
solenoid, 242, 243

sound
    — attenuating hearing protector, 207
    — direction, 203
    — frequency, 202
    — level, 203, 204
    — source, 201, 204, 206
    3D —, 76
    ambient —, 208
    audio —, 391
    earphone-delivered —, 206, 208
    spatialized —, 210
    virtual —, 206, 208–210
Spaceframe, 325, 365
spatialized sound, 210
speech synthesizer, 208–210, 365
standard deviation, 115, 150, 222
Star, 230
stereo, 35, 37, 38, 42, 45, 47–50, 75, 87,
        88, 99, 131, 133–137, 139, 140,
        142, 143, 145, 157–159, 165, 207,
        221, 265, 275, 291, 292, 308–
        311, 315, 330, 340, 357
    — acuity, 309
    — disparity, 321
    — illusion, 37
    — image, 45, 134, 139, 142, 275
    — matching, 45, 48, 50, 54, 134,
        136, 139, 143, 291, 295
    — vision, 131, 134, 143, 257, 312
    image-based —, 41, 45
stereophonic
    — earphones, 209
stereoscopic
    — camera, 61, 134, 137, 140, 143
    — cue, 310, 314
    — display, 3, 59, 63, 75, 76, 133,
        140, 291, 302, 306, 308, 309,
        321, 358
    — HMD, 59, 272, 358
    — horopter, 315
    — image, 45, 69, 74, 134, 141, 142,
        220, 265, 275, 307, 357
    — localization, 310
    — presentation, 309
    — projection, 136, 265
    — system, 135, 141
    — video see-through HMD, 133, 141
    — view, 75, 266, 274
    — view condition, 312
    — viewing, 75, 140, 306, 315
    — visual effect, 165

    — VRML viewer, 76
stiffness, 224
structure from motion, 90
sum of squared differences, 47
superimpose, 6–10, 20, 24, 32, 61, 64, 77,
        148, 208, 215, 226, 230, 266,
        271, 274, 311, 341, 348, 349,
        352–354, 357, 360, 392, 393
surface normal, 150, 151
surgical simulator, 37
Sutherland, 32, 33, 379
sync signal, 44
synchronization, 35, 36, 41, 44, 61, 132,
        142, 238, 239, 396
Synchronized Distributed Physical Objects,
        238, 239
synthesized image, 90, 93, 94, 148, 149,
        151, 152, 155–157, 160, 171, 187,
        188, 192, 193, 285, 298
synthetic image, 32, 37, 115, 216

T
Table KINGYO, 98, 99
tangibility, 245
tangible user interface (TUI), 229, 232–
        234, 238, 245
TCL, 242
TCP/IP, 221
Team WorkStation, 263
telecommunication, 250, 396
teleconference, 268, 295
teleoperation, 14, 23, 218, 254
telephony, 272
telepresence, 39, 42, 61, 75
telerobotics, 25, 250
telesurgery, 37
telexistence, 250–254, 258, 259, 357
tessellation, 49
tether metaphor, 18
texture
    — blending, 91, 93, 94
    — coordinate, 50
    — map (mapping), 2, 9, 10, 12, 43,
        50, 72, 91, 93, 286, 342
    color —, 286
    eigen —, 147, 148, 152, 155, 157, 160
    surface —, 167
    video —, 175, 176
    view dependent —, 41, 43, 50
TFT, 74, 75, 294
thermal image, 288, 289

thinning, 48, 49

time code, 44, 184, 186

Torrance-Sparrow model, 150, 151, 159

tracking, 32, 35, 36, 67, 69, 98, 101, 102, 105, 107, 109–112, 114, 115, 117–120, 122–126, 132, 134–136, 143, 216, 220, 225, 235, 236, 254, 266, 274, 279, 280, 326, 327, 329, 333–341, 343, 353–355, 365, 367–369, 372, 374, 380–385, 387, 388, 391–394, 397

    closed-loop —, 112, 126, 380, 383, 385

    eye position —, 167, 315

    head —, 37, 61, 69, 206, 220, 266–269, 271, 272, 274, 339, 363, 364, 368, 385

    hybrid —, 117, 120, 336, 387

    optical —, 336, 338, 343, 372, 380

    orientation —, 32, 118, 122, 267–270, 272, 274, 368, 372, 374, 393

    position —, 32, 36, 37, 111, 267, 274, 341, 365, 367, 368, 372, 374, 393

    scalable —, 105, 393

Tracking Vision, 98

trackpad, 368, 370, 372, 374

tradeoff, 22, 102, 193, 256, 326, 327, 337, 343, 357

transform

    affine —, 93, 289, 290

    discrete cosine — (DCT), 172, 287–289

    distance —, 295

    frequency domain —, 287

    mental —, 19, 23

    perspective —, 278

    rotation —, 122

    signal —, 307

    spatial —, 53

    view —, 43

transmitter, 356, 385

transparency, 74

transparent, 33, 139, 232, 306, 308, 332, 342, 343, 357

triangle mesh, 49, 50

triangular mesh, 158

triangular patch, 93, 153, 286

triangulation, 45, 149, 158, 204

TTT (things that think), 245

ubiquitous, 229, 232

UDP (User Datagram Protocol), 240

ultrasound image, 262

update rate, 15, 118, 142, 221, 307, 384, 385

user interface (UI), 210, 221, 261, 280, 285, 286, 295, 298, 325, 341, 350, 363–365, 367

vector quantization, 172

vergence, 292, 293, 310, 311, 315, 316, 318, 320, 321

    perspective —, 323

    proximal —, 319, 321

VERO (Virtual Environmented for Remote Operations), 13

vertex coordinate, 50

VGA, 74

video

    — conference, 250, 264, 268, 269, 278, 279, 333

    — see-through, 2, 23, 31, 32, 34–37, 61, 64, 65, 68, 72, 82, 131–134, 140, 142, 143, 215, 380, 383

video-rate

    — stereo machine, 42

view

    — dependent texture, 41, 43, 50

    — generation, 43, 90, 286, 289, 291, 292, 298

    — point, 5, 13, 15–18, 20, 22, 23, 25, 26, 41, 43, 44, 49, 50, 54, 61, 63, 65, 69, 102, 111, 132, 168, 169, 171, 180, 184, 186, 189, 190, 192, 193, 196, 265, 267, 268, 271, 272, 286, 290, 293, 349, 380

    — synthesis, 90, 91, 94, 168, 169, 171, 172, 176, 178, 179

    — transform, 43

    — warping, 176, 178

    inside-out —, 18

    outside-in —, 18

View Interpolation, 43

View Morphing, 43, 189

view-based, 42, 44

viewing

    — angle, 47, 68, 73–75, 186

    — area, 190, 360

— axis, 35
— channel, 307
— condition, 151, 153, 306, 308–315
— direction, 122, 134, 150, 151, 186, 190, 290
— geometry, 306
— lens, 310
— location, 204
— matrix, 66
— mode, 140
— orientation, 131, 132
— perspective, 17
— point, 132
— position, 17, 43, 131, 132, 354
— situation, 311
— transformation, 66
— vector, 306
cross —, 140
egocentric —, 17
exocentric —, 17
optimal —, 35
optimal —, 34
parallel —, 134, 140, 306
stereo —, 75, 140, 306, 315
virtual
— 3D coordinate system, 277
— acoustic display, 206, 207, 209, 210
— acoustics, 202, 206
— annotation, 267, 279
— auditory beacon, 208, 210
— avatar, 267, 269, 279
— beacon, 209, 210
— beam, 216, 223–226
— camera, 18, 53, 171, 328, 341
— communication, 285, 286
— connection, 240
— desktop, 393
— document, 393
— end effector, 8
— environment (VE), 2, 6, 7, 11–13, 16, 17, 26, 36, 60, 63, 69, 72, 131, 132, 183, 195, 200, 215, 220, 221, 232, 252–254, 256–258, 265–268, 286, 295, 348, 357, 363, 365, 372, 373, 379, 380, 385, 388
distributed —, 372, 373
— equipment, 76
— excavator, 14, 15
— graphical display, 360

— image, 8, 12, 16, 20, 43, 50, 54, 61, 65, 73, 147, 148, 152, 157, 160, 208, 265–267, 271, 272, 274, 277, 278, 280, 293, 306–309, 312, 313, 315, 317, 318, 359, 392–395, 398
— incision marker, 384
— information, 208, 272, 274, 342, 391, 392
— instrument, 349, 358
— landmark, 208
— metamorphosis system, 285–287, 298
— model, 11, 14, 22, 53, 54, 266, 328, 341, 342
— object, 10, 12, 20, 36, 38, 39, 42, 44, 63–66, 77, 87, 90, 98, 132–134, 137–143, 147, 148, 155, 157–160 , 166, 215, 216, 218–222, 224–226, 230, 266, 274, 275, 277, 278, 291, 295–297, 305–322, 326, 327, 329, 330, 334, 336, 340–343, 359, 368, 374, 392
— reality (VR), 7, 33, 34, 36, 41–45, 50, 51, 53, 54, 59, 60, 65, 75, 118, 147, 148, 229–232, 256, 259, 266, 268, 280, 285, 295–298, 327, 333, 348, 356, 357, 379
— robot, 8, 9, 254
— scene, 179, 285, 286, 356
— sound, 206, 208–210
— space, 59–61, 63, 65, 66, 69, 180, 192, 193, 195, 216, 220, 245, 271, 296, 298
— telexistence, 253
— view, 54, 168, 169, 171, 172, 175, 176, 179
— viewpoint, 50, 54, 171
— web, 272, 274
— window, 39
— world, 5, 8, 9, 12, 16, 23, 60, 62–65, 72, 76, 132, 134, 167, 172, 179, 180, 183, 184, 187, 196, 217, 221, 225, 231, 233, 250, 253, 261, 266, 267, 280, 295, 342, 350, 353, 355, 358
Virtual Dome, 184, 188
Virtual i-O, 270, 272, 273, 308, 333
Virtual Kabuki, 286, 289
Virtual Workbench, 38

virtualized reality, 41, 42, 44, 45, 50, 51, 53, 54
viscosity, 220, 221
visibility, 42, 253, 273
vision
    binocular —, 134, 257
    computer — (CV), 43, 62, 65, 69, 85–87, 98, 99, 132, 143, 262, 274, 277, 286, 287, 339, 341, 349, 360, 385, 391
    machine —, 236
    monocular —, 133
    peripheral —, 36
    stereo —, 131, 134, 143, 257, 312
    x-ray —, 262, 331
visual
    — acuity, 314
    — angle, 308
    — augmentation, 232
    — communication, 165–167, 180, 264
    — connection, 278
    — cue, 166–169, 174, 180, 204, 268
    — data, 166, 167, 172, 180, 331
    — display, 17, 21, 220, 221, 256, 297
    — effect, 64, 165, 177
    — evidence, 319
    — fatigue, 291
    — feature, 119
    — feedback, 221–224, 238
    — fidelity, 315
    — field, 25, 33, 238, 315
    — image, 215, 216, 220, 221, 318
    — information, 76, 85, 86, 99, 264, 269, 321
    — markers, 349
    — overlay, 231
    — perception, 204
    — presentation, 353
    — property, 230
    — range, 75
    — reference point, 124
    — registration, 393
    — representation, 243, 269, 272
    — sensation, 167
    — sensing, 112
    — simulation, 67
    — suppression, 306
    — target, 204, 310
    — zone, 168, 169, 171
visualization, 31, 60, 261, 343, 348, 381, 393, 395

VITC (Vertical Interval Time Code), 44
voice-driven interface, 334
volumetric
    — display, 265
    — integration, 43, 45, 48–50
    — merging algorithm, 49
    — model, 44, 49, 52, 53
voxel, 49, 221
    — coloring, 55
    — representation, 41, 49, 55
    — resolution, 49
    — space, 49
VPL, 59
VR sickness, 65
VRML (Virtual Reality Modeling Language), 70, 71, 76, 167, 180, 327, 328, 365
VuMan, 382

W

walk-through, 70, 183, 184, 188, 190, 193, 195
warp, 93, 114, 153, 158, 175, 176, 178
Water Lamp, 234, 241–244
wearable
    — AR, 77
    — communication, 270
    — computer, 77, 232, 269, 270, 275, 278–280, 360, 363–365, 381, 382
    — device, 225
    — display, 364
    — force display, 216, 225
    — hardware, 271
    — information space, 274
    — interface, 272, 280
    — MR, 77, 360
    — PC, 381–383, 385
    — system, 365, 382, 385
    — user, 269, 271, 274, 364
    — user interface, 364
Wearable Master, 225, 226
WearCom, 267–270, 272, 274, 278
web (WWW), 8, 26, 76, 244, 255, 267, 268, 272–274, 369–371, 373, 374
wide area, 72, 193, 380
WIMP interface, 280, 381
Windows, 47, 88, 220, 221, 229, 230, 232, 233, 266, 270, 287, 334, 343, 372, 373
wire frame, 64, 286, 287, 308, 311, 312, 316, 366

wired, 229, 274, 372
world referenced, 17, 18, 20, 22, 23
world-stabilized image, 274, 278
WWW, *see* web (WWW)
WYSIWYG (what you see is what you
       get), 230

X
X'tal vision, 256–258
x-ray, 7, 343
     — vision, 262, 331

Z
z-buffer, 139, 140, 342